AN INTRODUCTION TO ELECTROMAGNETIC THEORY

AN INTRODUCTION TO ELECTROMAGNETIC THEORY

P. C. CLEMMOW

Lecturer in Theoretical Electrodynamics and Radio
University of Cambridge

CAMBRIDGE
AT THE UNIVERSITY PRESS
1973

Published by the Syndics of the Cambridge University Press
Bentley House, 200 Euston Road, London NW1 2DB
American Branch: 32 East 57th Street, New York, N.Y.10022

Library of Congress Catalogue Card Number: 73-77174

ISBNs
0 521 20239 6 hard covers
0 521 09815 7 paperback

Printed in Great Britain
at the University Printing House, Cambridge
(Brooke Crutchley, University Printer)

CONTENTS

5 Electromagnetic waves

PREFACE

Electromagnetism is central to so many aspects of physics, engineering and applied mathematics that it must form a significant part of the curriculum of each of these disciplines. On the other hand the continual broadening of the scope of these disciplines as knowledge advances means that progressively less time can be devoted to the study of their more traditional parts: and whilst electromagnetism is very much alive, and itself contributes to the demand for the admittance of new topics into educational courses, there is a need to put across its fundamentals more cripsly than has been the fashion in the past.

This book aims to give reasonable coverage of first and, depending on the syllabus, some second and possibly third-year university work in an introductory text that is comparatively short but fully explanatory. Some sacrifice was required, and it may be helpful to indicate where the axe has fallen.

The main omissions arise in two ways. First, the account is confined to theory, and that mostly field theory; it does not include to any significant extent descriptions of experimental phenomena or technical applications. Secondly, the treatment assumes a rather specific mathematical competence on the part of the reader; namely, that he is conversant with the standard results of vector algebra and vector calculus summarized in the Appendix.

Which said, it must be emphasized that the treatment is in no sense at a sophisticated theoretical level aloof from practicalities. On the contrary, it aims to foster an awareness of orders of magnitude in practical units; and whilst omitting any discussion of the mathematics of vector fields *per se*, which so often takes up space in texts on electricity and magnetism, it also omits material that requires other significant mathematical expertise. Thus Legendre polynomials, Bessel functions and the like are kept out, in the belief that for illustrative purposes there are enough problems soluble in terms of elementary functions: the only concession is the brief appearance of elliptic integrals in the discussion of inductance. Furthermore, the commitment to standard vector analysis is total. The temptation to introduce cartesian tensors was strong, and resisting it played some part in prescribing the coverage; in particular, by effectively curtailing discussions of electromagnetic momentum and of anisotropic media, which are best conducted in tensor terms.

The topics covered are by and large conventional, as a glance at the Contents shows. The foundations of electromagnetic theory were laid over a century ago, and must be mastered before applications, whether old or new, can be appreciated: even Feynman, with reference to the second year of his *Lectures on Physics*, wrote 'In the first part of the course, dealing with electricity and magnetism, I couldn't think of any really unique or different way of doing it – any way that would be particularly more exciting than the usual way of presenting it.' But this is not to say that yet another introductory text cannot offer something worthwhile. As Bob Hope observed, when asked what Dorothy Lamour had that others hadn't, 'Nothing, but she groups it better'.

In the present book a comparatively unusual feature of the way the topics are grouped is the early development of Maxwell's equations. The traditional route that leads to them is taken in Chapter 2, without the prior *detailed* treatments of electrostatics, magnetostatics and induction that so easily eat up the pages, and whose many and varied individual aspects may divert attention from the fundamental unity of the theory. Time independent, and slowly varying situations are considered in Chapters 3 and 4, of which perhaps all that need be noted here is the introduction of dielectric and magnetic media on a purely macroscopic, phenomenological basis. Chapter 5 deals with electromagnetic waves, covering important aspects that do not need to invoke mathematics beyond the limits already indicated. and again considering media in the simplest phenomenological way.

Finally, Chapter 6 introduces media in essentially classical microscopic terms. The dipole moment densities **P** and **M** make their first appearance, and a number of elementary models are investigated. The theory is necessarily less clear cut than that in previous chapters, partly because of the inherent difficulty of attempting to describe classically what may well be truly quantum effects. There is also the complication of defining an 'average' field at points inside a molecular medium. This is one of those questions that can be a perennial thorn in the flesh of an honest taecher. Varied treatments in the literature seem insufficiently incisive, or at least too complicated, to include in an elementary book: yet to gloss over the matter is likely to lead to confusion. Opportunely, however, a comparatively straightforward procedure has recently been proposed,† and the gist of this is incorporated in §6.1.3.

Some problem solving is an essential part of learning, and a set of problems has been provided which is quite closely related to the text; hopefully, their solution demands an understanding of the material rather than, save here and there, mathematical dexterity or perseverance.

It is rare to feel of a text book 'that's how I would have done it', and it would be rash to suppose that any of my colleagues will so react to this one. Nevertheless, I recognize gratefully that many of them, both at Cambridge and elsewhere, have influenced the outcome. To them my thanks is col-

† A derivation of the macroscopic Maxwell equations, G. Russakoff (1970), *American Journal of Physics* **38** (10), 1188–95.

lectively and sincerely offered. Specifically I am indebted to Professor Colin Hines for the quoted reference and forceful advocacy, of its viewpoint which he had arrived at independently. Thanks are also due to the University Press for their expertise in dealing with the typescript and great patience in awaiting its arrival. Responsibility for the delay was entirely mine, and it could have been longer but for the efficient typing of Miss Hilary Holden and Miss Susan Whitelegg.

Cambridge
May 1973

P.C.C.

1

BASIC CONCEPTS

1.1 Charged particles, force and field

1.1.1 *The force between charged particles*

At a certain level of sophistication much of physics can be conceived in terms of elementary particles, such as the electron, proton and neutron, between which there are forces of interaction. These forces are classified as gravitational, electromagnetic and nuclear; and broad areas of physics can be distinguished by the type of force with which they are chiefly concerned. In this sense the immensely diverse phenomena of electromagnetism, ranging from the orientation of a compass needle to the reception of radiation from the remotest regions of the universe, are basically manifestations of the electromagnetic force.

The electromagnetic force of interaction that exists between certain elementary particles is regarded as a force between electric charges; electric charge is thus intrinsic to these particles. There is some analogy with the gravitational force between masses, and in particular the magnitude of the force between two charged particles is proportional to the product of the charges. However the characters of the two types of force are in general quite distinct. One obvious difference is in the reversibility of the direction of the electromagnetic force, and this is accounted for by attributing alternative signs to charge. The electron and proton have charges equal in magnitude but opposite in sign, and by convention the electron charge is negative: the neutron is uncharged. It may also be noted that the repulsive electromagnetic force between two protons at rest exceeds the attractive gravitational force between them by a factor of the order of 10^{36}; and the proton mass is 1836 times the electron mass.

In laboratory scale phenomena the electron, proton and neutron are usually the only fundamental particles that need be considered, because together they constitute overwhelmingly the greater part of charge and matter. Moreover, they are effectively immutable, so that charge, of each sign, and matter are conserved. Thus, for example, the amount of positive charge in some specified region of space is determined by the number of protons the region contains; and any increase in the amount, synonymous with an increase in the number of protons, can only arise from a corresponding influx of protons across the boundary of the region. The net charge in a region is, of course, represented by the excess of protons over electrons; and an increase in net charge can arise from an efflux of electrons as well as from an influx of protons.

1.1.2 *The electromagnetic field*

Although in principle the law of force between elementary charged particles can be taken as the ultimate physical basis of electromagnetic theory, it goes without saying that development of the theory to bring it into close contact with applications is required. This is the more important because the general force law is extremely complicated; the familiar Coulomb law, that the force between two particles is inversely proportional to the square of the distance between them and acts along the line joining them, only applies to the special case when the particles are at rest.

A major step in forwarding the development of the theory is the introduction of the concept of the electromagnetic field. Any specific charged particle interacts with other charged particles, and the force on it due to these other particles is the vector sum of the individual forces of interaction. It is convenient to ascribe this force to the fact that the 'test' particle, as it may be called, is in an electromagnetic *field* that arises from the presence of the other charged particles. This field at position $\mathbf{r}$ at time t is defined in terms of the force per unit charge on a test particle, located at $\mathbf{r}$ at time t, in the following way.

Consider, first, a test particle of charge e held at rest; then the electromagnetic force on it is written

$$e\mathbf{E}, \tag{1.1}$$

and this defines the electric field vector $\mathbf{E}$. Suppose, next, that the particle is moved with uniform velocity $\mathbf{v}$. Then it is a matter of experience that the electromagnetic force can be written

$$e(\mathbf{E}+\mathbf{v}\wedge\mathbf{B}), \tag{1.2}$$

where $\mathbf{B}$ is independent of $\mathbf{v}$; and the expression (1.2), known as the *Lorentz force*, defines the magnetic field vector $\mathbf{B}$, often called the magnetic induction.

There need be no hesitation in claiming that a charged particle at rest or in uniform rectilinear motion exerts no electromagnetic force on itself. The quantities $\mathbf{E}$ and $\mathbf{B}$, in general functions of position and time, can therefore be granted an existence independent of the test charge. Together they represent the electromagnetic field due to all the charges except the test charge.

If the velocity $\mathbf{v}$ of the test particle is not constant the expression (1.2) still represents the force on it to an adequate degree of approximation for nearly all purposes. More precisely, though, a contribution should be included from the field of the accelerated charged particle reacting on the particle itself; and this is no simple matter to investigate, since it involves the question of the 'structure' of the particle. However the contribution is usually very small, and the cases for which it is significant are outside the scope of this book, so there is no occasion to refer to it again.

Another point concerning particle structure is that in some sense the charge of the electron, say, should be regarded as 'spinning', so that the charge would be in circulatory motion even when the electron as a whole is at rest. Again, however, this feature, though fundamentally important, has a quite negligible effect in most circumstances. It is therefore convenient to maintain the concept of a charged particle free of spin, and this is understood throughout the book unless there is an explicit statement to the contrary.

The second term of (1.2) indicates that the force associated with the magnetic field is proportional to the speed and at right angles to the motion of the particle. These distinctive features are peculiar to electromagnetism, and play a major part in many characteristic phenomena. The predictions of (1.2) have been tested countless times in experiments on charged particle dynamics, some simple but important examples of which are indicated in problems 1.1 to 1.9.

That the force on a charged particle is the vector sum of the forces of interaction with other charged particles implies, of course, that the electromagnetic field **E**, **B** is the vector superposition of the fields of individual charges. Consideration of the field of a single charged particle now follows. This is in effect a discussion of the law of force between a pair of particles based on a knowledge of the dependence of the force on the motion of one of the pair as expressed in (1.2).

1.1.3 *The field of a charged particle*

The inverse square law of force between charged particles at rest was established by direct methods (Coulomb 1785), and contemporaneously by indirect methods (Cavendish) which were subsequently developed to great accuracy; from an experiment in 1936 it was concluded that the index in a power law could not differ from -2 by more than one part in 10^9. In terms of the field concept the law can be stated as follows: the field of a particle of charge e at rest in an unbounded vacuum is purely electric, and the electric vector **E** is proportional to

$$e\mathbf{r}/r^3, \tag{1.3}$$

where **r** is the position vector of the field point from the particle.

There is no reason to doubt that, in the context of classical physics, this statement is in effect exact. The experiments just referred to involve familiar 'laboratory scale' measurements, but results from atomic and nuclear physics indicate that the law continues to hold for microscopic values of r, perhaps even down to 10^{-15} m.

For a charged particle in *motion* no such exact statement of comparable simplicity can be made about the field, which is both electric and magnetic. However in certain circumstances, the background to which will be clarified shortly, it can be said that, for a particle of charge e with velocity **v**, it is approximately true that the additional field due to its motion is magnetic, and the magnetic field **B** is proportional to

$$e\mathbf{v}\wedge\mathbf{r}/r^3. \tag{1.4}$$

This vector product containing the velocity of the particle producing the field may be compared with that in (1.2) containing the velocity of the particle on which the field acts.

The result (1.4) cannot claim as direct an experimental basis as Coulomb's law, chiefly because, as will be seen, the magnetic force between charged particles is likely to be negligible compared with the electric force. It is introduced here mainly to give an immediate insight into the fundamental connection between electric and magnetic fields. The historical development of the theory of the magnetic field of moving charges followed experiments by Ampère, Biot and Savart (1820); these are taken up in due course in Chapter 2, where the reasoning that links them to the present statement will be found.

1.1.4 *Units*

The constant of proportionality by which the expressions (1.3) and (1.4) must be multiplied to give, respectively, the electric and magnetic field of a charged particle depend on the units adopted. In one conventional notation, used in this book, the formulae are written

$$\mathbf{E} = \frac{e}{4\pi\epsilon_0}\frac{\mathbf{r}}{r^3}, \tag{1.5}$$

$$\mathbf{B} = \frac{\mu_0 e}{4\pi}\frac{\mathbf{v}\wedge\mathbf{r}}{r^3}. \tag{1.6}$$

There is no incentive to debate whether or not to include the factors $1/(4\pi)$ explicitly; if they are omitted here, 4π factors appear elsewhere, and it is hardly more than a matter of taste where one prefers to see them. If they are retained in (1.5) and (1.6) the units are said to be *rationalized*.

On the other hand, what values and dimensions are to be ascribed to ϵ_0 and μ_0 does offer scope for argument, and has been widely debated.

In discussing this point it should first be noted that, since the ratio of the electric to the magnetic force on a particle is a dimensionless number, an implication of the combination of (1.2), (1.5) and (1.6) is that $(\epsilon_0\mu_0)^{-\frac{1}{2}}$ is a velocity (to be pedantic, a speed). There is, then, a fundamental velocity

$$c = (\epsilon_0\mu_0)^{-\frac{1}{2}} \tag{1.7}$$

built into electromagnetic theory. Development of the theory in fact shows that c is the velocity of propagation, in vacuum, of electromagnetic effects; briefly, it is the speed of light. The specification of one of ϵ_0, μ_0 therefore fixes the other through the experimental determination of c.

Of the available systems of units only two are currently in wide use; the mks (rationalized), and the gaussian (unrationalized). The basic distinction between them concerns the unit and dimensions of charge.

In the unrationalized gaussian system, which omits the $1/(4\pi)$ factors in (1.5) and (1.6), the unit of charge is fixed by specifying $\epsilon_0 = 1$; and since ϵ_0

is also assumed dimensionless, the force law saddles charge with dimensions $(\text{mass})^{\frac{1}{2}}$ $(\text{length})^{\frac{3}{2}}$ $(\text{time})^{-1}$. With $\epsilon_0 = 1$ the implication of (1.7) is $\mu_0 = 1/c^2$. However it happens that in the gaussian system the second term of Lorentz force (1.2) is conventionally written $e\mathbf{v} \wedge \mathbf{B}/c$, giving $\mathbf{B}$ the same dimensions as $\mathbf{E}$; hence (1.5) and (1.6) appear as $\mathbf{E} = e\mathbf{r}/\mathbf{r}^3$, $\mathbf{B} = e\mathbf{v} \wedge \mathbf{r}/(cr^3)$.

On the other hand, in the mks system charge is assigned dimensions independent of those of mass, length and time, so that the force law ascribes dimensions to both ϵ_0 and μ_0, those of μ_0 being $(\text{charge})^{-2}$ mass length. Moreover the numerical value of μ_0 is determined by the choice of the amp for the unit of current, as described in §1.3.2.

The rationalized mks system is adopted in this book, and it is worth emphasizing the simplicity with which any relation can be converted into the unrationalized gaussian system if required. From what has just been said, the basis of the conversion is merely the replacement of ϵ_0 by $1/(4\pi)$, of μ_0 by $4\pi/c^2$, and of $\mathbf{B}$ by $\mathbf{B}/c$. A few other consequential changes must be included, but only to cater explicitly for quantities not yet defined; the complete list is given in §A.3 of the Appendix.

The mechanical units in the gaussian system are the centimetre, gram and second (cgs). The unit of charge is therefore $\text{g}^{\frac{1}{2}}\,\text{cm}^{\frac{3}{2}}\,\text{sec}^{-1}$, called the electrostatic unit (esu). The main reason for introducing the mks system was to bring the units for electromagnetic quantities into line with those commonly used in applied calculations, the so-called practical units; specifically, the amp for current, the coulomb (amp sec) for charge, the volt for potential difference, and so on. It was noticed that this aim would be largely achieved if the mechanical units were chosen to be the metre, kilogram and second; hence the appellation mks.

The charge of an electron (or proton, or positron) is the smallest quantity of charge known to be recognizable as an individual element. The electrostatic unit of charge is a very much greater quantity, and the practical unit, the coulomb, much greater again, reflecting the fact that commonplace 'laboratory scale' electromagnetic phenomena involve enormous numbers of electrons and protons. In such a situation it is, of course, out of the question to consider keeping track of each particle; some 'averaging' or 'smoothing out' process must be part of the description, and this requires the concept of charge density, which will now be introduced.

1.2 Charge and current density

1.2.1 *Charge density*

The concept of charge density is entirely analogous to that of mass density. The charge density ρ at any point is the charge per unit volume at that point; that is to say, if $\rho'\,\delta\tau$ is the charge in an element of volume $\delta\tau$ that includes a fixed point P, then ρ at P is the limit of ρ' as $\delta\tau \to 0$ through the element shrinking to the point P.

What emerges if this mathematical definition is applied to the 'actual' physical situation? The charge density is zero everywhere except in the regions occupied by charged particles. Since the linear dimensions of the electron and proton, when described in classical terms, are estimated to be of the order of 10^{-15} m, these regions are tiny and comparatively isolated; moreover, their locations are continually changing.

To adopt the actual charge density is, of course, tantamount to accepting the impossible task of keeping track of each particle. A reasonable alternative procedure is to work in terms of a macroscopic density ρ, from which spatial variations on the scale of interparticle distances are removed by taking an average of the actual density over regions of space that are small on the macroscopic view, but whose linear dimensions are large compared with the particle spacing. This means that $\rho\,\delta\tau$ continues to give, with negligible error, the charge in a small volume element $\delta\tau$, provided the element contains a large number of charged particles.

It should be mentioned that there is a difficulty in calculating the average charge density that does not arise when considering mass density. The reason for the difference lies in the existence of charge of either sign. Commonly, of course, positive charge (protons) and negative charge (electrons) are closely knit; and if they are in the form of atoms whose net charge is zero it may well be that volume elements containing a large number of charged particles will simply contain a large number of neutral atoms. It would then appear that the average charge density is zero. However this is not necessarily the case, basically because the individual amounts of positive and negative charge involved are so enormous (see §1.3.3) that their separation by distances small even on the scale of an atom gives a macroscopic effect. An adequate method of averaging has therefore to be quite sophisticated; it needs explicit consideration in Chapter 6, but not before.

It is because the difference between the fields associated with the actual and the macroscopic charge densities would not ordinarily be detectable that comparatively simple deductions from electromagnetic theory can claim to given an adequate account of electromagnetic phenomena. For example, the mathematical theory of electrostatics can safely proceed on the basis of a charge density assumed constant, even though the actual charge density must vary because of the thermal agitation of electrons and protons.

The advantage of a description of electromagnetic theory in terms of charge density rather than charged particles is thus clear. Fundamentally both descriptions carry the same information, but the former enables problems to be treated macroscopically by permitting the charge density to be idealized in a manner consistent with the other idealizations inherent in any macroscopic treatment. The field calculated from a macroscopic charge density is, of course, correspondingly an 'average' field; it is what would be found by a laboratory scale measurement.

1.2.2 *Current density*

The special cases in which, on average, the charge may be considered at rest are of comparatively limited interest. In general, charge is in motion, and it is the associated current that becomes the physical quantity of prime significance. This is the reason why the practical units are determined by a definition of the amp (see § 1.3.2) rather than the coulomb.

For the 'actual' physical situation, described in terms of elementary charged particles, each element of charge has a specific velocity, and the current density is defined as the product of charge density and velocity; in symbols $\mathbf{J} = \rho\mathbf{v}$.

Current on the atomic scale has no measurable effect unless there is some macroscopic ordering of the motion; in electrostatic phenomena, for example, no measurable magnetic field is produced, despite the thermal motion of individual electrons and protons. The macroscopic current density is given by averaging the actual microscopic current density, just as for charge.

For particles of a single species the contribution to the average current density is the product of the average charge density and the average velocity, the latter being simply the vector sum per particle of the individual velocities of the particles in a volume element that is macroscopically small but contains many particles. If, then, both electrons and protons, say, are involved, contributing macroscopic densities ρ_e and ρ_p, and having average velocities $\mathbf{v}_e$ and $\mathbf{v}_p$, respectively, the corresponding net current density is

$$\mathbf{J} = \rho_e \mathbf{v}_e + \rho_p \mathbf{v}_p. \tag{1.8}$$

It should perhaps be emphasized that in general $\mathbf{J}$ is not, of course, the product of $\rho = \rho_e + \rho_p$ with some average velocity. In particular it may well be the case that $\rho = 0$ but $\mathbf{J} \neq 0$; for example, current might be constituted solely from the mean motion of electrons ($\mathbf{v}_e \neq 0$, $\mathbf{v}_p = 0$) with the electron charge density annulled by protons ($\rho_p = -\rho_e$).

It is evident that, at any point P, $\mathbf{J} \,.\, \mathrm{d}\mathbf{S}$ is the rate at which charge crosses an infinitesimal plane surface element represented by the vector $\mathrm{d}\mathbf{S}$. This statement refers equally to the microscopic and macroscopic situations, the net rate of flow of charge being understood in the latter case.

Once the relation of the mathematical formalism to the physics has been grasped there is for the most part no need in the development of the theory to distinguish explicitly between actual and average densities. Their role in the theory will be based on the assertions that $\rho\,\mathrm{d}\tau$ gives the charge in an infinitesimal element of volume $\mathrm{d}\tau$, and $\mathbf{J} \,.\, \mathrm{d}\mathbf{S}$ gives the rate at which charge crosses an infinitesimal element of area $\mathrm{d}\mathbf{S}$. On this basis mathematical expression is now given to the fact, to which attention was drawn in § 1.1.1, that charge is conserved.

1.2.3 *The charge conservation relation*

Conservation of charge is synonymous with the fact that the rate of increase of charge within an arbitrary closed surface is equal to the net rate at which charge flows across the surface. That is,

$$\frac{\mathrm{d}}{\mathrm{d}t}\int_V \rho\,\mathrm{d}\tau = -\int_S \mathbf{J}.\mathrm{d}\mathbf{S}, \tag{1.9}$$

where the volume V is bounded by the closed surface S, and $\mathrm{d}\mathbf{S}$ is along the outward normal to the surface.

An important alternative form of the relation is obtained in the following way. Transform the right hand side to a volume integral by the divergence theorem (A. 18),

$$\int_S \mathbf{J}.\mathrm{d}\mathbf{S} = \int_V \operatorname{div}\mathbf{J}\,\mathrm{d}\tau,$$

and combine it with the left hand side after the time derivative has been taken underneath the integral sign. Thus

$$\int_V \left(\frac{\partial\rho}{\partial t}+\operatorname{div}\mathbf{J}\right)\mathrm{d}\tau = 0,$$

where the partial time derivative of course signifies the rate of increase of ρ at a fixed point. Since the integral vanishes when taken over any region it follows that

$$\operatorname{div}\mathbf{J}+\dot{\rho} = 0, \tag{1.10}$$

where the dot stands for $\partial/\partial t$; for if the left hand side of (1.10) were not everywhere zero there would be a region throughout which it was of one sign, and its integral over this region would not vanish.

Equation (1.10) is the differential relation between ρ and $\mathbf{J}$ equivalent to the integral relation (1.9), and gives a concise mathematical expression of the fact that charge is conserved.

Of course the wider statement that electrons and protons separately are conserved contains (1.10) by implication. In the notation of (1.8), conservation of electrons means

$$\operatorname{div}(\rho_\mathrm{e}\mathbf{v}_\mathrm{e})+\dot{\rho}_\mathrm{e} = 0; \tag{1.11}$$

the corresponding equation for protons has suffix p replacing e, and the addition of the two equations gives (1.10).

1.3 Practical units and magnitudes

1.3.1 *The field of a steady rectilinear current*

Current flows readily in metals, therefore said to be good *conductors*, because the atomic structure is such that some of the electrons, though confined to the metal, are not bound to parent atoms and can move comparatively freely through the structure. Under the influence of a electric field these so-called *conduction* electrons acquire an average velocity in the direction of the field, and thereby constitute a macroscopic current density.

Many man-made electrical devices depend on the flow of current in wires. Often the value of the cross-sectional area of a current carrying wire is immaterial, and it is permissible to adopt the idealization of a *line* current. The concept of a line current implies an infinite current density, and its magnitude must of course be specified simply by the total current that flows along the line at each point on it.

The experimental convenience and importance of current carried by wires hardly needs stressing. The study of the magnetic fields of such currents, through the measurement of forces of interaction, played a vital part in the development of the theory and practice of electromagnetism. In particular, it happens that the practical electrical units stem from a definition of the unit of current based on the force between line currents. The simple facts necessary for an understanding of the definition are now given.

Consider an infinite straight line along which flows a constant, uniform current I. The line is supposed uncharged, so that the associated field is purely magnetostatic.

The field can be calculated from (1.6) by treating the current as a uniform line distribution of charged particles all moving with the same velocity. If N is the number of particles per unit length of line, v their velocity, and e the charge on each particle, the current is

$$I = Nev. \tag{1.12}$$

Now the application of (1.6) shows that, at any point P, the contribution to $\mathbf{B}$ from a charged particle at Q is at right angles to the plane containing P and the line current, and of magnitude

$$\frac{\mu_0 ev \sin\theta}{4\pi R^2}, \tag{1.13}$$

where R is the distance QP, and θ is the angle between QP and the line current. Since the direction of the contribution is independent of the position of Q on the line it follows at once by superposition, using (1.12), that the magnitude of $\mathbf{B}$ is

$$B = \frac{\mu_0 I}{4\pi}\int_{-\infty}^{\infty}\frac{\sin\theta}{R^2}\,\mathrm{d}\xi, \tag{1.14}$$

where ξ is the distance of Q from the foot of the perpendicular from P to the line (see figure 1.1). From figure 1.1 it is evident that $R\,\mathrm{d}\theta/\mathrm{d}\xi = r/R$; hence

$$B = \frac{\mu_0 I}{4\pi r}\int_0^{\pi}\sin\theta\,\mathrm{d}\theta = \frac{\mu_0 I}{2\pi r}. \tag{1.15}$$

The calculation therefore shows that at a point P distance r from a rectilinear line current I (of infinite extent) the magnetic field is inversely proportion to r, and is perpendicular to the plane containing P and the line

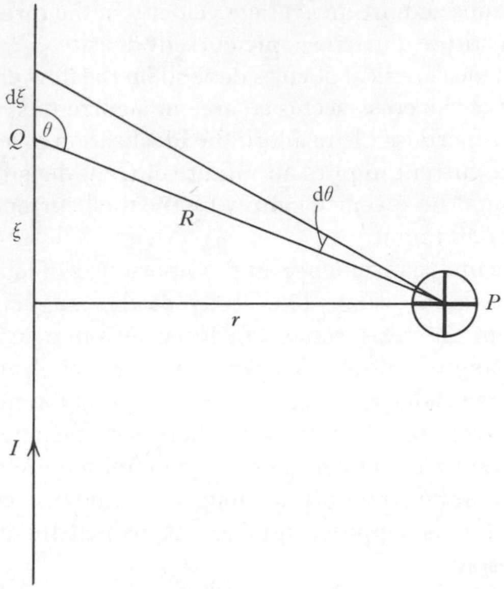

Figure 1.1

current in the sense of a right-handed screw about the current (into the paper, in figure 1.1, as indicated by the arrow tail ⊕ at P).

It should be emphasized that this is an exact result, despite the derivation being based on the approximate equation (1.6). In fact, equation (1.6) is inexact in so far as it implies instantaneity and thus fails to take account of the finite velocity c of electromagnetic effects. But for steady current flow the situation is time independent, and the error is immaterial; it cancels out in the superposition process. The application of (1.6) to steady currents is essentially equivalent to the so-called Biot–Savart law (see §2.2.4, and problem 2.17).

1.3.2 *The units of current and charge*

Suppose two rectilinear currents I and I' are parallel and distance r apart. The magnetic field of one acts on the moving charged particles that constitute the other, and by applying the force formula (1.2) in conjunction with (1.15) it is apparent that the force between the two currents is

$$\mu_0 II'/(2\pi r) \tag{1.16}$$

per unit length, being an attraction if the currents are in the same sense and a repulsion if they are in opposite senses.

The mks unit of current is the practical unit, namely the amp. It is specified in terms of this force of interaction in the following way.

Units based on fundamental formulae are called absolute, and the absolute unit of current, the abamp, was defined by the statement that parallel line currents, each of 1 abamp, when 1 cm apart experience a force per centimetre length of 2 dynes; which prescription came from the force formula $2II'/r$, the form (1.16) takes without the factor $\mu_0/(4\pi)$, in conjunction with cgs units. To have a more convenient magnitude it was then decided that the amp should be one tenth of the abamp. Translated directly into mks units the definition of an amp therefore states that parallel line currents, each of 10 amp, when 10^{-2} m apart experience a force per 10^{-2} metre length of 2×10^{-5} newtons (a newton being 1 kg m sec^{-2}, which is 10^5 dynes); or expressed more aptly, parallel line currents, each of 1 amp, when 1 m apart experience a force per metre length of 2×10^{-7} newton. That (1.16) conform to this specification requires

$$\mu_0 = 4\pi \times 10^{-7}\,\text{newton amp}^{-2}, \tag{1.17}$$

which is thus an exact value by definition.

The unit of charge corresponding to amp is coulomb, defined simply as amp sec: a current of 1 amp in a wire carries charge along the wire at the rate of 1 coulomb sec^{-1}.

The unit of electric field is newton coulomb^{-1}.

The relation (1.7) has already been noted, from which, with μ_0 defined in (1.17), ϵ_0 is given through the experimental determination of c. The value of c has been found by many methods, from measuring the velocity of light or radio waves on one hand, to the comparison of electric and magnetic forces on the other: the result can be accepted to at least four significant figures, and is

$$c = 2.998 \times 10^8\,\text{m sec}^{-1}. \tag{1.18}$$

Hence

$$\epsilon_0 = 0.8854 \times 10^{-11}\,\text{newton}^{-1}\,\text{coulomb}^2\,\text{m}^{-2}. \tag{1.19}$$

It is often adequate to use the convenient approximations $c = 3\times 10^8$, $\epsilon_0 = 10^{-9}/(36\pi)$.

1.3.3 *Numbers and magnitudes*

In the discussion so far it has been taken for granted that the electron and proton are particles whose size, mass and charge are all extremely small compared with respective macroscopically measurable quantities. Some figures are now given to make this point more precisely, and to indicate the sort of magnitudes involved in electromagnetic phenomena.

As has been seen, the practical units of electricity come from experience with currents, and a feeling for the quantity of current represented by an amp can be got from familiarity with commonplace electrical devices: for example, the current in an electric heater is of the order of 10 amp, in a light bulb 0.5 amp, and in a radio valve 10^{-3} amp.

What rate of flow of electrons constitutes one amp of current? There are various ways of measuring the charge of an electron. Early (*c.* 1915) comprehensive experiments were those of Millikan, who observed individual charged oil droplets held against gravity by an electric field, and the effect on their motion of the acquisition or loss of an electron. The present best estimate of the magnitude of the electron charge is

$$1.602 \times 10^{-19}\,\text{coulomb}. \qquad (1.20)$$

A coulomb therefore comprises about 6.3×10^{18} electrons, and a current of 1 amp in a wire represents, on average, the passage at any point of the wire of this number of electrons each second.

Suppose the cross-section of the wire is $A\,\text{m}^2$, the number of conduction electrons per unit volume is $N\,\text{m}^{-3}$ and their average velocity is $v\,\text{m}\,\text{sec}^{-1}$; then for a 1 amp current

$$NAv = 6.3 \times 10^{18}\,\text{sec}^{-1}. \qquad (1.21)$$

For a cross-section of a square millimetre this gives

$$Nv = 6.3 \times 10^{24}\,\text{m}^{-2}\,\text{sec}^{-1}. \qquad (1.22)$$

Perhaps surprisingly, (1.22) implies a low average velocity v. This is because N is so large. For example, the best conductors, silver, copper and gold, have N roughly $10^{29}\,\text{m}^{-3}$, so that v is about $0{\cdot}6 \times 10^{-4}\,\text{m}\,\text{sec}^{-1}$. This average velocity is, of course, very much less than the thermal velocities at ordinary temperatures: the mean thermal speed is given by

$$(3KT/m)^{\frac{1}{2}}, \qquad (1.23)$$

where T is the absolute temperature,

$$K = 1.38 \times 10^{-23}\,\text{joule}\,\text{deg}^{-1} \qquad (1.24)$$

is Boltzmann's constant, and

$$m = 9.108 \times 10^{-31}\,\text{kg} \qquad (1.25)$$

is the electron mass; and at $T = 300\,^\circ\text{K}$ (1.23) is about $10^5\,\text{m}\,\text{sec}^{-1}$. The mean thermal speed is therefore of the order of 10^9 times the average velocity v; remarkable enough, since it is v alone that accounts for the macroscopically detectable feature, namely the current.

A coulomb is evidently a quite orthodox amount of charge when it flows as a current. In current flow, however, charge of one sign is completely, or almost completely, neutralized by charge of the opposite sign. Were a coulomb of charge to be isolated from charge of the opposite sign it would represent an enormous quantity. For consider two small bodies each carrying a total charge of 1 coulomb. If they were 1 m apart, and stationary, the force of interaction would be, from Coulomb's law, $1/(4\pi\epsilon_0)$. This is 9×10^9 newtons, being about a million tons weight.

Put another way, the degree of charge separation achieved in ordinary electrostatic phenomena is minute compared with the individual positive and negative charge densities actually present. As already stated, the number of conduction electrons in a good conductor is roughly 10^{29} per cubic metre, and this alone represents about 1.6×10^4 coulomb of electric charge in each cubic centimetre, a quantity that is for practical purposes effectively unlimited.

That the amounts of charge giving rise to measurable electrostatic effects are very small compared with the amounts that flow to produce measurable magnetic effects is, of course, a manifestation of the fact that the electric force between two charged particles greatly exceeds the magnetic force if the speeds of the particles are much less than c. The formulae (1.2), (1.5) and (1.6) predict that the forces differ by a factor of the order of vv'/c^2, where v and v' are the particle speeds. Evidently, for collections of charged particles of average velocity much less than c, a high degree of charge neutralization must occur before the total magnetic force is comparable to the total electric force.

1.4 Development of the theory

1.4.1 *Complexity of the force law*

It has already been remarked that expressions (1.5) and (1.6) for the field of a charged particle are approximations of limited applicability, and that generally valid expressions must take account of the finite speed of propagation of electromagnetic effects. It would be expected that the finite speed of propagation implies that the field at any point P at time t is given by the position at Q, and the motion, of the charged particle at time $t - PQ/c$. This is indeed the case, so that (1.5) and (1.6) can be good approximations only if $\mathbf{r}$ and $\mathbf{v}$ do not alter significantly in the time r/c; that is, only if v/c and $r|\dot{\mathbf{v}}|/(cv)$ are negligible compared with unity. The complication of the expressions for $\mathbf{E}$ and $\mathbf{B}$ when these quantities are not negligible is such that, had they not been derived by indirect means, it is unlikely that they would have been arrived at either by direct experiment or by inspired guesswork. Hence there is little justification for taking an exact statement of the field of a moving charged particle as the starting point from which the development of the theory proceeds by deduction: nor, indeed, would such a presentation be the most expeditious or illuminating. However, before describing the preferred alternative it is worth noting two further points in connection with the force law.

Consider the interaction between two charged particles e and e', with respective velocities $\mathbf{v}$ and $\mathbf{v}'$. According to (1.2) and (1.6) the magnetic force on e is

$$\frac{\mu_0}{4\pi} ee' \frac{\mathbf{v} \wedge (\mathbf{r} \wedge \mathbf{v}')}{r^3}, \tag{1.26}$$

where $\mathbf{r}$ is the vector from e to e'; whereas the magnetic force on e' is

$$-\frac{\mu_0}{4\pi} ee' \frac{\mathbf{v}' \wedge (\mathbf{r} \wedge \mathbf{v})}{r^3}. \tag{1.27}$$

Now (1.27) is the negative of (1.26) only in the special case when $\mathbf{v}$ is parallel to $\mathbf{v}'$. The magnetic force of interaction, in contrast to the Coulomb electric force, thus violates Newton's third law. So, therefore, does the total force; and this remains true even when the exact expressions for the electric and magnetic forces are used. There are no particular grounds for supposing that Newton's third law must apply to electromagnetic forces. On the other hand it does turn out, as a comparatively advanced piece of theory, that the law of conservation of momentum of a closed system can be retained by ascribing momentum to the electromagnetic field.

The second point concerns the division of the field into an electric part and a magnetic part. It is evident from the appearance of $\mathbf{v}$ in the formula (1.6) that $\mathbf{B}$ is subjective to the extent that it is dependent on the velocity of the observer; for an observer O_1 with respect to whom a charge is stationary measures no magnetic field, whereas an observer O_2 who moves relative to O_1, and relative to whom, therefore, the charge is in motion, does measure a magnetic field. This *relative* nature of $\mathbf{B}$ (and also of $\mathbf{E}$ in its strict expression) is at first sight a peculiarity in electromagnetism. In fact, though, it is a manifestation of relativity principles that are now accepted as governing all physics, about which a little more is said in § 2.5.4. Essentially the same point arises from the dependence on $\mathbf{v}$ of the Lorentz force (1.2); observers in relative motion can agree on the force only by differing in their measurement of the field.

1.4.2 *A route to Maxwell's equations*

The main route in the historical development of electromagnetic theory may be traced, in the broadest terms, as follows:

(*a*) experiments with stationary charge distributions, and with magnets and steady currents, led to an understanding of the force laws, and to the introduction of the concepts of electric and magnetic fields, in time independent situations;

(*b*) the consequences of the force laws in electrostatics and magnetostatics were expressed mathematically as partial differential equations relating the fields and sources;

(*c*) these partial differential equations were generalized, partly on the basis of experiment and partly by theoretical reasoning, to be applicable to arbitrary time varying situations;

(*d*) the generalized equations, proposed by Maxwell in 1864, were then subject to intensive scrutiny. Their mathematical structure was explored, and the nature of their physical predictions exposed. The cumulative evidence leaves no doubt that the equations accord with experience and constitute a definitive statement of the fundamental laws of electromagnetism.

The feature of the sequence (*a*) to (*d*) that should be emphasized is that while the theory is initiated by the acceptance of comparatively simple force laws established experimentally, the general form of the theory is reached in terms of differential relations (or equivalent integral relations) between field quantities. The exact force law for moving charged particles is then accepted as just one of the many deductions that can be made from these relations.

Most introductory accounts of electromagnetic theory, at the relevant level, give a treatment somewhat on these lines, and the present book is no exception. The development of electrical science takes place within the disciplines of mathematics, physics and engineering, and it does seem that the most convenient description for general purposes is obtained by giving pride of place to Maxwell's equations. The equations could, of course, be baldly stated as a 'law', but this would deprive the student of the intellectual satisfaction of witnessing their emergence, and of the opportunity to gain an understanding of the concepts while doing so.

There is naturally considerable scope for variations of treatment within the broad pattern. In this text deductions from postulated laws, such as the Coulomb force law, are made with reference to simple models; and it should be recognized in what sense the results of such deductions are accepted as 'proved' for the physical situation that is modelled. Whereas a simple model will not adequately represent all aspects of a physical situation, it can often safely be presumed adequate in some respects, and some deductions from it can therefore be expected to agree with experience (see the remarks in the last two paragraphs of § 1.2.1). Strictly, each deduction should be supplemented by reference to its experimental standing; but for a basic, well established result this may be taken for granted. The expression (1.15) for the magnetic field of a rectilinear current may be cited as an example. The model used in the derivation is that in which charged particles are uniformly distributed and move with a common velocity. This obviously misrepresents actual current flow in many ways, but not in ways that would be expected to affect (1.15), which is indeed an experimental result of long standing.

1.4.3 *The role of quantum theory*

That electrons and protons are classical particles with a definite force law is a fruitful concept for the development of electromagnetic theory, and is unquestionably an adequate concept in the right context. On the other hand there are various electromagnetic phenomena of which no proper explanation can be given without a quantum theory treatment. It is beyond the aims of this book to introduce material of this kind, but it is desirable at the outset to have some idea of which parts of the subject are affected.

The most obvious area in which a quantum treatment is necessary is the detailed explanation of the electromagnetic properties of media, particularly where atoms are comparatively closely packed. Thus, for example, only rudimentary explanations of the conductivity and magnetic properties of solids can be given without quantum ideas, and a proper discussion of these topics must be reserved for a course in solid state theory. Many technical devices of recent introduction, such as transistors and lasers, also depend for their operation on quantum effects, and their discussion would form part of the newly recognized discipline of quantum electronics.

Another general point may be made. All types of radiation, such as radio waves, heat, light, X-rays, are electromagnetic fields which differ in character only by virtue of their different frequencies. For example, medium wave broadcasting has a frequency band from 2×10^5 to 5×10^5 cycles sec^{-1}, whereas that of visible light extends from about 5×10^{14} to 7×10^{14} cycles sec^{-1}. Now the origins of quantum theory arose in connection with a problem of electromagnetic radiation, from which it emerged that such radiation transfers energy in quanta of amount

$$h\nu, \tag{1.28}$$

where ν is the frequency and h is Planck's constant. If the quantum (1.28) is so small that its discreteness is of no consequence the classical theory of radiation is valid; and since

$$h = 6.625 \times 10^{-34}\,\text{joule sec}, \tag{1.29}$$

a joule being the mks (and practical) unit of energy (10^7 erg), it is not surprising that this is often the case. Broadly speaking, the failure of classical theory is more likely the higher the frequency.

Problems 1

(The symbols e and m stand for the magnitude of the charge and for the mass of an electron.)

1.1 An electron moves in a constant, uniform electric field $\mathbf{E}$, and its position vector $\mathbf{r}$ and velocity $\mathbf{v}$ have initial values $\mathbf{a}$ and $\mathbf{u}$ respectively. Show that at time t

$$\mathbf{r} = -\frac{1}{2}\frac{e}{m}\mathbf{E}t^2 + \mathbf{u}t + \mathbf{a}.$$

Show also that

$$\tfrac{1}{2}m(v^2 - u^2) + e\mathbf{E}.(\mathbf{r} - \mathbf{a}) = 0.$$

1.2 An electron travelling along the negative x axis with velocity $(u, 0, 0)$, where $u > 0$, trasverses the electric field

$$\mathbf{E} = \begin{cases} (0, E, 0) & \text{for} \quad |x| < \tfrac{1}{2}a, \\ 0 & \text{for} \quad |x| > \tfrac{1}{2}a, \end{cases}$$

where E is constant and uniform. Show that the path in $x > \tfrac{1}{2}a$ is the line

$$y = -\frac{eEa}{mu^2}x, \quad z = 0.$$

1.3 Show that an electron moving in a constant magnetic field $\mathbf{B}$ has constant speed.

Consider the case in which $\mathbf{B}$ is also uniform. Show that the electron can remain in a plane perpendicular to $\mathbf{B}$, and that its path is then a circle described with angular velocity eB/m; deduce that in general its path is a helix.

1.4 An electron travelling along the negative x axis with velocity $(u, 0, 0)$, where $u > 0$, traverses the magnetic field

$$\mathbf{B} = \begin{cases} (0, 0, B) & \text{for} \quad |x| < \tfrac{1}{2}a, \\ 0 & \text{for} \quad |x| > \tfrac{1}{2}a, \end{cases}$$

where B is constant and uniform, and $a < mu/(eB)$. Show that the path in $x > \tfrac{1}{2}a$ is the line

$$y = \frac{\gamma}{(1-\gamma^2)^{\frac{1}{2}}}x, \quad z = 0$$

where $\gamma = aeB/(mu)$.

1.5 Electrons (with negligible interaction) emerge with a common speed from a small aperture into a constant, uniform magnetic field **B**. Show that those electrons whose velocities are nearly parallel to **B** are brought to a focus at distance $2\pi mv/(eB)$ from the aperture.

1.6 Electrons (with negligible interaction) travelling along the negative x axis with speeds close to $(u, 0, 0)$, where $u > 0$, suffer successive *small* deflections on traversing first the electric field

$$\mathbf{E} = \begin{cases} (0, E, 0) & \text{for} \quad |x| < \tfrac{1}{2}a, \\ 0 & \text{for} \quad |x| > \tfrac{1}{2}a, \end{cases}$$

and then the magnetic field

$$\mathbf{B} = \begin{cases} (0, 0, B) & \text{for} \quad |x-l| < \tfrac{1}{2}a, \\ 0 & \text{for} \quad |x-l| > \tfrac{1}{2}a, \end{cases}$$

where $l > a$, and E and B are constant and uniform. Show that, provided $2E/(uB)$ lies between $1/(1+2l/a)$ and 1, the electrons are brought to a focus at

$$x = l\bigg/\left(1-\frac{2E}{uB}\right), \quad y = \frac{eEa}{mu^2}x, \quad z = 0.$$

1.7 An electron, initially at the origin with velocity $(0, 0, u)$, moves in superposed uniform, constant fields $\mathbf{E} = (E, 0, 0)$, $\mathbf{B} = (B, 0, 0)$. Show that at time t the electron is at

$$\left\{-\frac{1}{2}\frac{e}{m}Et^2, \quad -\frac{u}{\Omega}[1-\cos(\Omega t)], \quad \frac{u}{\Omega}\sin(\Omega t)\right\},$$

where $\Omega = eB/m$.

Show that, for different initial speeds u, the locus of the point of impact of the electron on a fixed plane $z = l$, where $l \ll u/\Omega$, is the parabola

$$y^2 = -\frac{e}{m}\frac{l^2B^2}{2E}x.$$

1.8 An electron moves with velocity $\mathbf{v}$ in the field of a fixed proton, from which its vector distance is $\mathbf{r}$. Show that $\tfrac{1}{2}mv^2 - e^2/(4\pi\epsilon_0 r)$ and $\mathbf{r} \wedge \mathbf{v}$ are constant.

Deduce that
$$\mathbf{K} \equiv m\,\mathbf{v} \wedge (\mathbf{r} \wedge \mathbf{v}) - \frac{e^2\mathbf{r}}{4\pi\epsilon_0 r}$$
is also constant.

1.9 Suppose that in problem 1.8 the electron at $r = \infty$ is fired with velocity u towards the proton along a line l from which the distance of the proton is b (the *impact parameter*). By equating the values of **K** at either end of the path show that when the electron again reaches $r = \infty$ it is travelling in a direction making angle θ (the *angle of deflection*) with l, where

$$\tan\left(\tfrac{1}{2}\theta\right) = \frac{e^2}{4\pi\epsilon_0 mbu^2}.$$

1.10 In an accelerating machine electrons travel in the same circular orbit, of radius 10 m, with a speed very close to the speed of light. Evaluate the current if the total number of electrons is 10^{10} and they are distributed evenly round the orbit.

1.11 An infinitely long straight line carries a charge q per unit length. Use the inverse square law to obtain the electric field.

Now suppose the charge travels along the line with constant speed u. Obtain the magnetic field, and compare its magnitude with that of the electric field.

1.12 Each of a pair of infinitely long, parallel straight lines carries a charge q per unit length which travels along the line with speed u. Show that the ratio of the magnetic to the electric force per unit length on either line is u^2/c^2.

1.13 In an isolated homogeneous conductor the charge density at each interior point decreases according to the law

$$\rho(\mathbf{r}, t) = \rho_0(\mathbf{r})\, e^{-\gamma t},$$

where γ is a positive constant characteristic of the conductor, and $\rho_0(\mathbf{r})$ is the initial density: ultimately, therefore, any net charge initially within the conductor appears solely on the surface. Obtain the associated current density in the case when the conductor is a sphere carrying a net quantity of charge q which is initially concentrated at its centre. How does a quantity of charge q arrive at the surface of the sphere without any charge appearing between the centre and the surface?

1.14 Show that for charge density ρ and current density $\mathbf{J}$,

$$\int_V J_x \mathrm{d}\tau = \frac{\mathrm{d}}{\mathrm{d}t}\int_V x\rho\, \mathrm{d}\tau + \int_S x\mathbf{J}\,.\,\mathrm{d}\mathbf{S},$$

where S is the surface bounding the region V. Deduce that, if $\mathbf{J}$ is spatially bounded,

$$\int \mathbf{J}\,\mathrm{d}\tau = \frac{\mathrm{d}}{\mathrm{d}t}\int \mathbf{r}\rho\,\mathrm{d}\tau,$$

where the integration is over all space.

1.15 Electrons having a common time-dependent velocity $\mathbf{v}(t)$ are initially distributed with a space-dependent number $N_0(\mathbf{r})$ per unit volume. Write down the associated macroscopic charge density ρ and current density $\mathbf{J}$, and confirm explicitly that they satisfy the conservation relation $\operatorname{div}\mathbf{J} + \partial\rho/\partial t = 0$.

1.16 From the approximation (1.6) for the magnetic field of a moving point charge show that, if $\mathbf{F}_1$ is the magnetic force on a particle of charge e_1 with velocity $\mathbf{v}_1$ due to a particle of charge e_2 with velocity $\mathbf{v}_2$, and $\mathbf{F}_2$ the corresponding force on e_2 due to e_1, then

$$\mathbf{F}_1 + \mathbf{F}_2 = \frac{\mu_0 e_1 e_2}{4\pi r^3}\mathbf{r} \wedge (\mathbf{v}_1 \wedge \mathbf{v}_2),$$

where $\mathbf{r}$ is the position vector of e_2 relative to e_1.

A rigid framework OAB, where $OA = OB = a$ and $A\hat{O}B = \frac{1}{2}\pi$, carries equal charges q fixed at A and B, and rotates in its own plane about O with angular velocity ω. Find the contribution to the total force on the framework due to the interaction between the charges.

1.17 Calculate the magnitude $\mathscr{E}$ of the potential energy $-e^2/(4\pi\epsilon_0 r)$ of an electron in the field of a fixed proton when $r = 10^{-10}$ m (the order of an atomic radius). For what values of the temperature T and frequency ν are the mean energy of thermal motion of an electron, $\frac{3}{2}KT$, and the quantum of radiation energy, $h\nu$, equal to $\mathscr{E}$?

2

THE VACUUM FIELD OF CHARGE AND CURRENT: MAXWELL'S EQUATIONS

2.1 The field of static charge

2.1.1 *The field*

This chapter is concerned with setting out the mathematical development of electromagnetic theory on the lines indicated in § 1.4.2, and ends with Maxwell's equations relating the fields **E**, **B** to the charge and current densities ρ, **J**. For the purposes of basic theory the charge and current densities are presumed to exist in what is otherwise a vacuum; in principle this is not a restriction, since any material medium has electromagnetic properties only by virtue of the charged particles it contains. The aim is to present a chain of reasoning leading to Maxwell's equations, and it is left to later chapters to exemplify and extend the mathematics, to show the application to particular physical problems, and to introduce further concepts.

In the present section the simplest case is treated, in which ρ is independent of time and **J** is everywhere zero. The theory, electrostatics, can then be based solely on Coulomb's law. Each infinitesimal volume element $d\tau$ contains charge $\rho\, d\tau$, so the application of (1.5) gives the field in the form of a volume integral over the region occupied by the charge, namely

$$\mathbf{E} = \frac{1}{4\pi\epsilon_0}\int \rho \frac{\mathbf{R}}{R^3}\, d\tau, \qquad (2.1)$$

where **R**, of magnitude R, is the vector from the element $d\tau$ to the field point P at which **E** is evaluated (see figure 2.1).

In the general investigation of the mathematical consequences of (2.1) it is assumed that ρ is a sufficiently well behaved function of position to meet any condition that may be required to justify the analysis. In particular, it is supposed that ρ vanishes everywhere beyond some large enough distance from an arbitrary origin. Such restrictions on ρ are certainly acceptable physically; sometimes, admittedly, it is useful to discuss idealized models in which they are violated, but these can be regarded as limiting cases.

It is therefore presumed that there is a bounded region of space in which ρ is non-zero, except possibly at particular points, and outside which ρ is everywhere zero. For any field point P outside this region the integrand in (2.1) is continuous and infinitely differentiable, and so likewise is **E**. But for a field point P inside the region the integral is improper, because of the singularity of the integrand at $R = 0$, and requires closer examination. However, that (2.1) does converge within the region is easily established by

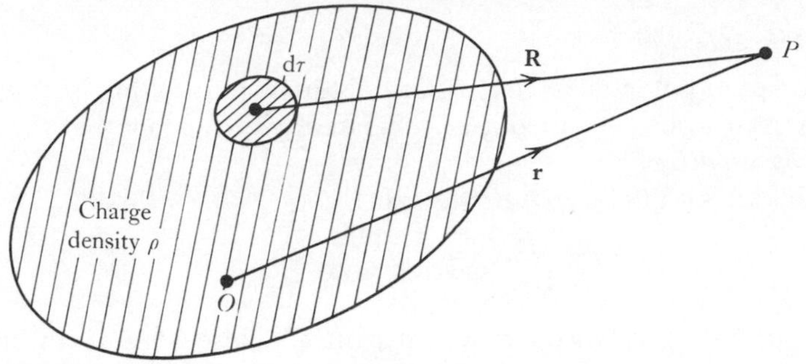

Figure 2.1

writing $d\tau = R^2\,dR\,d\Omega$ where $d\Omega$ is an element of solid angle subtended at P; for then the R^2 factor in $d\tau$ is seen to remove the R^{-2} singularity in the integrand.

An important corollary to these remarks should be noted. If it is desired to form the partial derivative of $\mathbf{E}$ with respect to a distance coordinate it is legitimate to take the differentiation under the integral sign in (2.1) only for points outside the region in which ρ is non-zero. This is because the operation of differentiation introduces an R^{-3} singularity in the integrand, so that the integral does then diverge at interior points.

At first sight it might seem that the theory of electrostatics could begin and end with (2.1), apart from the technical details of evaluating volume integrals. That this is far from the case is chiefly due to the fact that only in very special circumstances can ρ be taken as given: in general the charge distribution is not known initially, since it is itself affected by the electric field. If comparison is made with the Newtonian theory of gravitation, also governed by the inverse square law, it is seen that an electrostatic problem with a given charge density is equivalent to the familiar gravitational problem with a given mass distribution. The more realistic problem, in which the charge density has to be determined, might be compared with an equilibrium gravitational problem involving liquid or gaseous masses; but the nature of the other forces that must come into play to maintain equilibrium are likely to be quite different. In electrostatics the usual case is that in which the electric force is balanced by binding forces that constrain the charge to remain within fixed conductors, whereas in gravitational theory the balance might well be maintained by pressure forces. In electrostatics the inverse square law therefore operates in distinctive physical contexts.

Implicit in (2.1) are two general characteristics of the electric field that are of the utmost importance in facilitating its study. These are now described in turn.

2.1.2 *The potential*

The concept of potential in electrostatics is analogous to that in gravitation. In a way now to be explained it is associated with the fact that the electric field is *conservative*.

The field is said to be conservative if

$$\oint \mathbf{E}.\mathrm{d}\mathbf{s} = 0 \tag{2.2}$$

when the line integral is taken round an arbitrary closed circuit of which $\mathrm{d}\mathbf{s}$ is a vector element of arc; or, stated in physical terms, if the total work done by the electric force on a point charge is zero when the charge is carried round a circuit and ultimately returned to its original position.

Transformation of the left hand side of (2.2) by Stokes' theorem (A 21),

$$\oint \mathbf{E}.\mathrm{d}\mathbf{s} = \int (\mathrm{curl}\,\mathbf{E}).\mathrm{d}\mathbf{S}, \tag{2.3}$$

evidently yields the equivalent statement that the field is conservative if

$$\mathrm{curl}\,\mathbf{E} = 0 \tag{2.4}$$

everywhere.

Another synonymous criterion is that the field can be expressed as the gradient of a single-valued scalar function of position, the *potential* function ϕ. The relationship is conventionally written

$$\mathbf{E} = -\mathrm{grad}\,\phi, \tag{2.5}$$

and the following argument shows that (2.2) and (2.5) are different ways of expressing the same fact.

Corresponding to an infinitesimal displacement $\mathrm{d}\mathbf{s}$ of the field point the increment in any function of position ϕ is

$$\mathrm{d}\phi = (\mathrm{grad}\,\phi).\mathrm{d}\mathbf{s}. \tag{2.6}$$

Hence, if (2.5) holds,

$$\oint \mathbf{E}.\mathrm{d}\mathbf{s} = -\oint \mathrm{d}\phi.$$

But this is simply the total decrease in ϕ as the field-point moves round a circuit and returns to its original position, which total decrease must be zero if ϕ is single-valued. Alternatively put, if (2.5) holds, (2.4) follows and consequently (2.2) also, by virtue of the fact that the curl of the gradient of any function is identically zero.

Conversely, if (2.2) holds, then, for an arbitrary fixed point A,

$$\phi = -\int_A^P \mathbf{E}.\mathrm{d}\mathbf{s} \tag{2.7}$$

defines a function that is single-valued at any field point P, because the line integral (2.7) is independent of the path from A to P by virtue of (2.2). With this definition the increment in ϕ corresponding to an infinitesimal displacement $\mathrm{d}\mathbf{s}$ of the field point is

$$\mathrm{d}\phi = -\mathbf{E}.\mathrm{d}\mathbf{s};$$

and since the direction of $\mathrm{d}\mathbf{s}$ is arbitrary comparison with (2.6) implies (2.5).

To see that the field given by (2.1) is indeed conservative, and to find the potential function, observe that, if $\mathbf{r}$ is the position vector from some origin, then

$$\operatorname{grad} 1/r = -\mathbf{r}/r^3. \tag{2.8}$$

The field of a point charge e at the origin therefore has potential

$$\phi = \frac{e}{4\pi\epsilon_0}\frac{1}{r}. \tag{2.9}$$

Likewise the field of a charge density ρ has potential

$$\phi = \frac{1}{4\pi\epsilon_0}\int\frac{\rho}{R}\mathrm{d}\tau; \tag{2.10}$$

for, as the remarks made in § 2.1.1 indicate, (2.10) converges at all points, and *single* partial differentiation with respect to any distance coordinate is permissibly taken under the integral sign to give immediate confirmation that (2.5) agrees with (2.1).

For a given field $\mathbf{E}$ the specification of the potential function by (2.5) does, of course, leave it arbitrary to the extent of an additive constant. In particular, the initial point A in (2.7) may be located anywhere. However, unless there is some special reason for choosing otherwise, it is usual to take A at infinity, so that the potential at any point P is the work done against the electric force in bringing a unit point charge from infinity to P. This makes the potential zero at infinity, as in (2.9) and (2.10); in fact, if the distance r of the field point from some origin tends to infinity, $1/R = 1/r + O(1/r^2)$ (see figure 2.1), and (2.10) therefore gives $\phi \sim e/(4\pi\epsilon_0 r)$, where e is the total charge.

The mks and practical unit of potential is volt. One volt is the difference in potential at two points 1 metre apart in a uniform electric field of 1 newton coulomb^{-1}. The practical definition relates to circuitry, and states that 1 volt is the potential difference that maintains a current of 1 amp at a rate of working of 1 watt, the units newton coulomb^{-1} m and watt amp^{-1} being identical.

Since the volt figures so largely in practical measurements the magnitude of an electric field is commonly stated in volt m^{-1}. Fields of the order of 10^6 volt m^{-1} over distances of some metres are achieved in the laboratory.

2.1.3 *Gauss' theorem*

The second important consequence of (2.1) concerns the *flux* of $\mathbf{E}$; the inverse square law, in contrast to any other law, can be identified with a simple general property of this quantity.

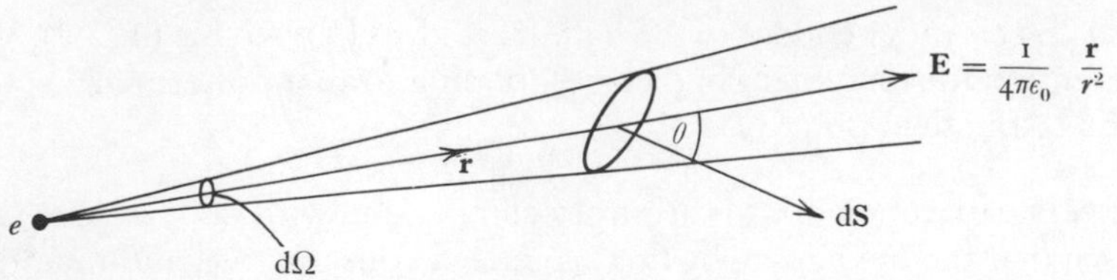

Figure 2.2

By the flux of **E** across a vector element of area d**S** is meant the quantity **E**.d**S**. For a single point charge e at the origin O it is evident (see figure 2.2) that

$$\mathbf{E}.\mathrm{d}\mathbf{S} = \frac{e}{4\pi\epsilon_0}\frac{\mathrm{d}S\cos\theta}{r^2} = \frac{e}{4\pi\epsilon_0}\mathrm{d}\Omega, \tag{2.11}$$

where $\mathrm{d}\Omega$ is the solid angle subtended by d**S** at O, being positive if the angle θ between **r** and d**S** is acute, and negative if θ is obtuse. This is merely an expression of the fact that the area of the base of a cone of infinitesimal vertical solid angle is proportional to the square of the height of the cone. The implication is that the total flux **E** out of a closed surface **S**,

$$\int_S \mathbf{E}.\mathrm{d}\mathbf{S}, \tag{2.12}$$

is zero or e/ϵ_0 according as the point charge is outside or inside the surface (see figure 2.3).

If now, instead of a single point charge, a distribution of charge with density ρ is considered, the corresponding statement is that the total flux of $\epsilon_0\mathbf{E}$ out of a closed surface is equal to the total charge inside the surface. That is

$$\epsilon_0\int \mathbf{E}.\mathrm{d}\mathbf{S} = \int \rho\,\mathrm{d}\tau, \tag{2.13}$$

the right hand side being the integral of ρ over the region bounded by the closed surface S.

Equation (2.13) is known as Gauss' theorem. It should perhaps be mentioned that the generalization from the case of a single point charge, by 'superposition', does take for granted that when **E** in (2.13) is written in the form (2.1) then the order of the integrals may be reversed. This is, in fact, permissible, even when the surface S intersects the region in which ρ is non-zero so that both the exterior and the interior charge abut against S as in figure 2.4.

It is noteworthy that the integral relation (2.13) is reminiscent of the charge conservation relation (1.9). By precisely the mathematical argument used in § 1.2.3, (2.13) is seen to be equivalent to the differential relation

$$\operatorname{div}\mathbf{E} = \rho/\epsilon_0. \tag{2.14}$$

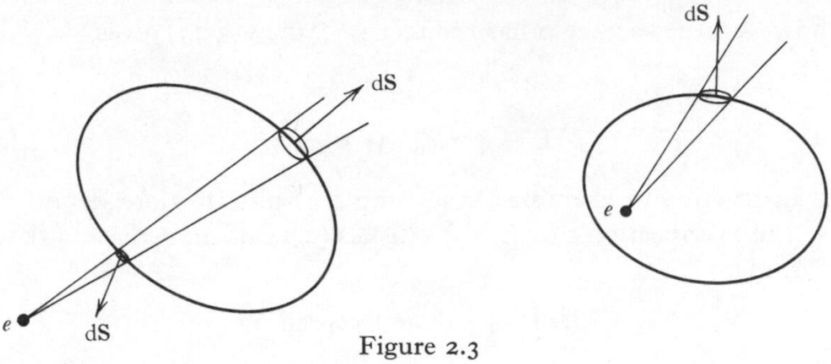

Figure 2.3

In particular, at all points where the charge density is zero,

$$\operatorname{div} \mathbf{E} = 0. \tag{2.15}$$

Two examples are now discussed which illustrate the application of the relations so far derived, and to which reference is made in the further development of the theory.

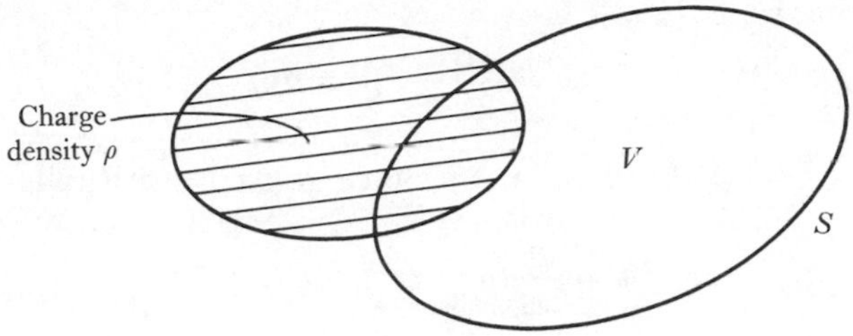

Figure 2.4

2.1.4 *Spherically distributed charge*

One of the simplest charge distributions imaginable is that in which ρ is uniform throughout a sphere and zero elsewhere. Let the sphere have centre O and radius a, and take $\rho > 0$. Then it is evident by symmetry that the inverse square law gives a field $\mathbf{E}$ that, at any point P, is directed along OP and whose magnitude depends only on the distance OP.

In these very special circumstances the easiest way of calculating the field is to use the integral form (2.13) of Gauss' theorem. For if the surface S is taken to be spherical, centre O and radius $r > a$, then (2.13) gives

$$4\pi r^2 E = \tfrac{4}{3}\pi a^3 \rho/\epsilon_0, \tag{2.16}$$

since $4\pi a^3\rho/3$ is the total charge. Thus, for points outside the charge distribution

$$\mathbf{E} = \frac{\rho a^3}{3\epsilon_0}\frac{\mathbf{r}}{r^3} \quad (r > a), \tag{2.17}$$

being the same as if the total charge were concentrated at O.

If, however, the surface S has radius $r < a$, then (2.13) gives

$$4\pi r^2 E = \tfrac{4}{3}\pi r^3 \rho/\epsilon_0, \tag{2.18}$$

so that

$$\mathbf{E} = \tfrac{1}{3}\rho\mathbf{r}/\epsilon_0 \quad (r < a). \tag{2.19}$$

It is instructive to examine these simple results in more detail. Since $\operatorname{div}\mathbf{r} = 3$ the verification that (2.19) satisfies (2.14) is immediate. Likewise, using (A. 13),

$$\begin{aligned} \operatorname{div}\frac{\mathbf{r}}{r^3} &= \frac{1}{r^3}\operatorname{div}\mathbf{r} + \mathbf{r}\,.\operatorname{grad}\frac{1}{r^3} \\ &= \frac{3}{r^3} - \mathbf{r}\,.\frac{3\mathbf{r}}{r^5} \\ &= 0, \end{aligned} \tag{2.20}$$

which confirms that the field outside the charge distribution, namely (2.17), satisfies (2.15).

What is the potential ϕ? When $r > a$ it is, of course, the point charge potential; that is (cf. (2.9))

$$\phi = \frac{\rho a^3}{3\epsilon_0}\frac{1}{r} \quad (r > a). \tag{2.21}$$

For $r < a$, $-\operatorname{grad}\phi$ has to be (2.19). Since $\operatorname{grad} r^2 = 2\mathbf{r}$ it follows that

$$\phi = \frac{\rho}{2\epsilon_0}(a^2 - \tfrac{1}{3}r^2) \quad (r < a), \tag{2.22}$$

where the constant part is determined by the condition that (2.22) be equal to (2.21) at $r = a$. That both ϕ and $\mathbf{E}$ are continuous across $r = a$, despite the jump in charge density, is obviously guaranteed by the integral representations (2.10) and (2.1).

It is natural to ask how easily (2.21) and (2.22) can be derived from (2.10). The integration is reasonably straightforward but the results are not obtained as quickly as in the method just given. However, it should be stressed that Gauss' theorem is but a single scalar equation, and is sufficient to yield the vector field $\mathbf{E}$ in the present case only because the symmetry of the problem leaves the magnitude of $\mathbf{E}$ as the sole unknown.

Gauss' theorem could evidently be applied equally effectively to the case when the charge distribution is spherically symmetric, but not necessarily uniform; that is, ρ is some function of r. The field inside the charge distribution would then depend on the specific form of ρ, having magnitude

$$\frac{1}{\epsilon_0 r^2}\int_0^r \rho(r')\,r'^2\,\mathrm{d}r';$$

but that outside would still be as though the whole charge were concentrated at O. In this sense the inverse square law holds for arbitrary spherically symmetric charge distributions.

These results have, of course, exact analogues in the theory of gravitational attraction.

2.1.5 *The dipole*

A single point charge is sometimes called a *pole*. Two point charges, equal in magnitude but opposite in sign, constitute a *dipole*.

If the point charges are e and $-e$, and if $\mathbf{l}$ is the position vector of e relative to $-e$, then

$$\mathbf{p} = e\mathbf{l} \tag{2.23}$$

is the *vector moment* of the dipole.

The concept of a dipole plays an important part in electromagnetic theory, particularly in the description of neutral atoms or molecules, which contain equal quantities of positive and negative charge. If the charge distribution is not spherically symmetrical (which can happen, for example, when an atom is in an externally applied electric field, so that the positive and negative charges are pulled apart) it gives rise to an electric field that approximates to a dipole field. Since such *polarization* takes place on the atomic scale the usual case of interest is $l/r \ll 1$, where r is the distance of the field point from the centre of the dipole. With the assumption of this inequality the dipole field is relatively simple. Its calculation proceeds by combining the contributions of the two point charges, and affords a good example of the use of the potential, since it is somewhat more straightforward to add two scalars and then differentiate than to combine two vectors.

The potential is

$$\phi = \frac{e}{4\pi\epsilon_0}\left(\frac{1}{r_1} - \frac{1}{r_2}\right), \tag{2.24}$$

where r_1 and r_2 are the respective distances of the field point P from the charges e and $-e$ (see figure 2.5). But

$$r_1^2 = r^2 - lr\cos\theta + \tfrac{1}{4}l^2,$$

where r, θ are the polar coordinates of P referred to the centre of the dipole as origin and the line of the dipole as axis. Hence

$$\begin{aligned}\frac{1}{r_1} &= \frac{1}{r}\left(1 - \frac{l}{r}\cos\theta + \frac{l^2}{4r^2}\right)^{-\frac{1}{2}}\\ &= \frac{1}{r}\left[1 + \frac{l}{2r}\cos\theta + O\left(\frac{l^2}{r^2}\right)\right].\end{aligned}$$

With $1/r_2$ obtained by replacing θ by $\pi - \theta$, it is seen that (2.24) gives

$$\phi = \frac{e}{4\pi\epsilon_0}\frac{l\cos\theta}{r^2}\left[1 + O\left(\frac{l}{r}\right)\right]. \tag{2.25}$$

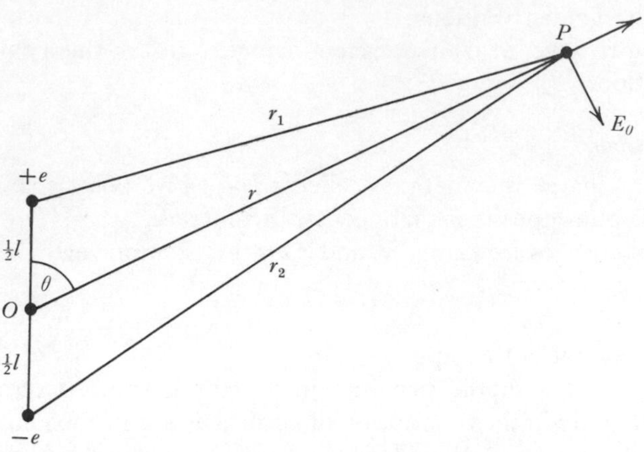

Figure 2.5

In terms of the dipole moment $p = el$, and with l/r neglected in comparison with unity, this is

$$\phi = \frac{p}{4\pi\epsilon_0}\frac{\cos\theta}{r^2}. \tag{2.26}$$

The potential therefore falls off as the inverse square of the distance.

There are two useful alternative ways of writing (2.26). First, it can obviously be presented in terms of the vector moment (2.23) as

$$\phi = \frac{1}{4\pi\epsilon_0}\frac{\mathbf{p}.\mathbf{r}}{r^3}. \tag{2.27}$$

Secondly, it is evident from (2.8) that (2.27) can in turn be expressed as

$$\phi = -\frac{1}{4\pi\epsilon_0}\mathbf{p}.\operatorname{grad}\frac{1}{r}. \tag{2.28}$$

The expression (2.28) could, in fact, almost be written down by inspection. For the potential of the charge e is $e/(4\pi\epsilon_0 r)$ at the point whose position vector is $\frac{1}{2}\mathbf{l}$ relative to P; and therefore it is, to the first order in l/r,

$$\frac{e}{4\pi\epsilon_0}\left[\frac{1}{r} - \tfrac{1}{2}\mathbf{l}.\operatorname{grad}\frac{1}{r}\right]$$

at P itself. The combination of this with the corresponding contribution from the charge $-e$ gives (2.28).

To find the field it need only be noted that, from (2.26), the non-zero spherical polar components of $-\operatorname{grad}\phi$ give

$$E_r = -\frac{\partial\phi}{\partial r} = \frac{p}{4\pi\epsilon_0}\frac{2\cos\theta}{r^3}, \tag{2.29}$$

$$E_\theta = -\frac{1}{r}\frac{\partial\phi}{\partial\theta} = \frac{p}{4\pi\epsilon_0}\frac{\sin\theta}{r^3}. \tag{2.30}$$

The field is therefore symmetric about the dipole axis, lies in the plane through the axis, and falls off as the inverse cube of the distance.

An alternative form for **E** which avoids specific components can be obtained from (2.27). Thus

$$E = -\frac{1}{4\pi\epsilon_0}\operatorname{grad}\left(\frac{\mathbf{p}.\mathbf{r}}{r^3}\right)$$

$$= -\frac{1}{4\pi\epsilon_0}\left[\frac{1}{r^3}\operatorname{grad}(\mathbf{p}.\mathbf{r}) - (\mathbf{p}.\mathbf{r})\frac{3\mathbf{r}}{r^5}\right];$$

and since $\operatorname{grad}(\mathbf{p}.\mathbf{r}) = \mathbf{p}$ this gives

$$\mathbf{E} = \frac{1}{4\pi\epsilon_0 r^3}\left[\frac{3(\mathbf{p}.\mathbf{r})}{r^2}\mathbf{r} - \mathbf{p}\right], \tag{2.31}$$

which is the sum of vectors parallel to **r** and to **p** respectively. It is easy to check the equivalence of (2.31) and (2.29), (2.30).

In (2.26), and subsequent expressions, higher order terms in l/r have been neglected. It is often convenient for theoretical purposes to introduce the concept of an *infinitesimal* or *point* dipole, defined by letting $l \to 0$, but assuming also that $e \to \infty$ in such a way that the dipole moment $p = el$ remains finite: the expressions given then describe the field for all values of r (except zero). The infinitesimal dipole is used so freely in theoretical work that the qualification 'infinitesimal' is generally left understood.

If an infinitesimal dipole is placed in an external field **E** the external forces on the individual charges of the dipole are $\pm e\mathbf{E}$, which constitutes a couple whose vector moment about the point where the dipole is located is

$$\mathbf{G} = \mathbf{p} \wedge \mathbf{E}. \tag{2.32}$$

2.1.6 *The potential equation*

It has been seen that (2.1) implies that **E** satisfies the partial differential equations (2.4) and (2.14), namely

$$\operatorname{curl}\mathbf{E} = 0, \tag{2.33}$$

$$\operatorname{div}\mathbf{E} = \rho/\epsilon_0. \tag{2.34}$$

Moreover (2.33) can be replaced by $\mathbf{E} = -\operatorname{grad}\phi$, so in the potential representation the equations are equivalent to

$$\nabla^2\phi \equiv \operatorname{div}\operatorname{grad}\phi = -\rho/\epsilon_0, \tag{2.35}$$

known as *Poisson's equation.*

In any region in which ρ is zero (2.35) becomes

$$\nabla^2\phi = 0, \tag{2.36}$$

known as *Laplace's equation.*

These equations for the potential are very important because they make available methods of treating problems in which the charge density is not prescribed.

It is worth verifying directly that $1/r$ and $\cos\theta/r^2$, proportional respectively to pole and dipole potentials, do satisfy Laplace's equation (2.36) everywhere except at $r = 0$. The verification for $1/r$ is in effect contained in (2.20); the other is left as an exercise (problem 2.5).

Can it be verified directly that (2.10) satisfies Poisson's equation (2.35)? This is, in fact, quite difficult. If the field point P at which

$$\int \frac{\rho}{R}\,\mathrm{d}\tau$$

is evaluated lies outside the region in which ρ is non-zero, then the ∇^2 operator can be taken under the integral sign, giving zero and thereby confirming that the expression is a solution of (2.36). But if P is within the region in which ρ is non-zero then it is not legitimate to take the ∇^2 operator under the integral sign, since a double differentiation of the integrand makes the integral divergent (cf. the remarks in §2.1.1).

The difficulty in the latter case can be surmounted in the following way. Consider, first, a situation in which the charge density is ρ', where ρ' is everywhere the same as the actual density ρ except within some sphere Σ centred on P; and throughout Σ ρ' is uniform, having the value ρ_0 that ρ takes at P. Then at P

$$\nabla^2 \int \frac{\rho'}{R}\,\mathrm{d}\tau = -4\pi\rho_0, \tag{2.37}$$

because the charge outside Σ contributes nothing to the left hand side, and the charge inside Σ contributes $-4\pi\rho_0$, as follows from the known result for the special case of a sphere of uniform charge density discussed in §2.1.4 (see problem 2.3). The required result is therefore established if it can be shown that at P

$$\nabla^2 \int \frac{\rho-\rho'}{R}\,\mathrm{d}\tau = 0. \tag{2.38}$$

Now $\rho-\rho'$ vanishes at P, and if it does so sufficiently rapidly, as P is approached, the singularity of the integrand in (2.38) becomes weak enough to validate taking the ∇^2 operator under the integral sign, which gives zero. The details of the argument would be out of place here, but it is useful to be aware of the issues involved.

From what has been said it might appear that the direct verification that (2.10) satisfies (2.35) is appreciably more cumbersome than the demonstration of Gauss' theorem given in §2.1.3. But this is not really so, because for that demonstration to be made rigorous it would be necessary to have an equally detailed investigation into the circumstances in which **E** defined by (2.1) is sufficiently well behaved for the divergence theorem to be applied to it.

The expression (2.10) is a solution of (2.35). Moreover, as shown later in §3.3.2, it is the only solution that tends to zero, as the field point recedes to infinity, at least as fast as the inverse distance. This indicates the sense in which the argument that led from the integral relation (2.1) to the differential relations (2.33) and (2.34) is reversible. In summary, the latter relations yield Poisson's equation, and the solution of Poisson's equation that vanishes at infinity at least as fast as the inverse distance is the integral (2.10).

2.2 The field of steady current

2.2.1 *Steady current flow*

In §2.1 the situation was considered in which the field was purely electrostatic. The case now to be treated is that in which the field is purely magnetostatic, being produced by steady (that is, time independent) current that flows without giving rise to any charge density.

The discussion is not at this stage concerned with how such current may be generated. It is simply assumed that a current density **J**, bounded in extent, is present in what is otherwise a vacuum; and the aim is to relate **J** to the associated field **B**.

The first point to note is that, with $\partial\rho/\partial t = 0$, the equation of charge conservation merely expresses the fact that the flux of **J** out of any closed surface is zero; (1.9) becomes

$$\int \mathbf{J}.\mathrm{d}\mathbf{S} = 0, \qquad (2.39)$$

and the equivalent differential form is, from (1.10),

$$\operatorname{div}\mathbf{J} = 0. \qquad (2.40)$$

Thus **J** is what is called a *solenoidal* vector.

This characteristic of **J** has a simple geometrical interpretation. Lines of current flow are envisaged that at each point have tangent in the direction of **J** at that point. Then a *tube* of current flow, which is the region bounded by the surface containing all the lines of flow intersecting some simple closed curve (see figure 2.6), has the property, deduced immediately from (2.39), that the flux of **J** across a cross-section of the tube is independent of the position of the cross-section. Each tube of flow is therefore closed (or, in idealized models, emerges from and proceeds to infinity).

Steady current therefore consists of closed tubes of flow; and if it is desired to break the distribution down into infinitesimal elements of steady flow, each element can itself be a tube of flow of infinitesimal cross-section. Such a tube is equivalent to current $I = J\mathrm{d}S$ flowing in a (thin) wire loop, where J is the current density at any point along the tube and $\mathrm{d}S$ is the area of normal cross-section of the tube at that point.

Moreover, current I in a wire loop of any size or shape is equivalent, as regards the associated magnetic field, to current I circulating in the same

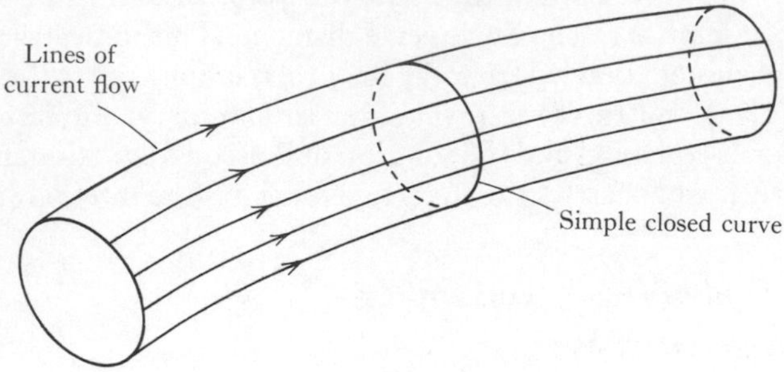

Figure 2.6

sense in each mesh of a net whose rim is the given loop (see figure 2.7); for the currents in the common portion of adjacent meshes are oppositely directed, and therefore cancel out. Each mesh can evidently be arbitrarily small and effectively plane.

An experimentally determined law for the magnetic field produced by a steady current in a small plane loop is therefore a sufficient basis from which to develop the theory. Just such a law was established by Ampère (*c.* 1825), and is now introduced. It will be seen to lead to the same results as the application of (1.6), and can be regarded as justification for the claim that inaccuracies in (1.6) are self-compensating in the case of steady current flow.

2.2.2 *Ampère's dipole law*

Consider first the couple on steady current I flowing in a plane wire loop situated in a uniform external magnetic field $\mathbf{B}_0$. The force due to $\mathbf{B}_0$ on each vector element $\mathrm{d}\mathbf{s}$ of the loop is $I\,\mathrm{d}\mathbf{s}\wedge\mathbf{B}_0$, which evidently produces a couple whose magnitude is proportional both to B_0 and to the area of the loop. Specifically, the couple about some origin O is

$$\mathbf{G} = I\oint \mathbf{r}\wedge(\mathrm{d}\mathbf{s}\wedge\mathbf{B}_0), \tag{2.41}$$

where $\mathbf{r}$ is the position vector of $\mathrm{d}\mathbf{s}$, and the integration is round the loop. The integral in (2.41) is

$$\oint(\mathbf{r}\,.\,\mathbf{B}_0)\,\mathrm{d}\mathbf{s} - \mathbf{B}_0\oint \mathbf{r}\,.\,\mathrm{d}\mathbf{s},$$

and if the line integrals are transformed to surface integrals (see (A.21) and (A.22)) it is seen that the second vanishes, since $\operatorname{curl}\mathbf{r} = 0$, and the first is

$$\int \mathrm{d}\mathbf{S}\wedge\operatorname{grad}(\mathbf{r}\,.\,\mathbf{B}_0) = \int \mathrm{d}\mathbf{S}\wedge\mathbf{B}_0$$

taken over any surface spanning the loop. The couple is therefore

$$\mathbf{G} = I\mathbf{S}\wedge\mathbf{B}_0, \tag{2.42}$$

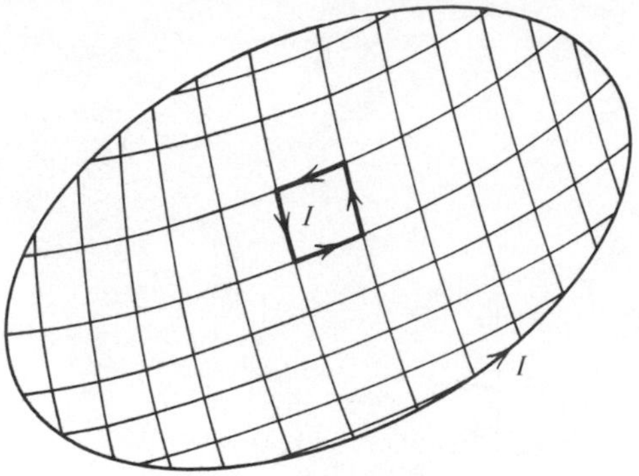

Figure 2.7

where **S** is the vector whose magnitude S is the area of the loop and whose direction is normal to the plane of the loop in the sense of a right-handed screw relative to the direction of current flow.

For an infinitesimal loop located at O, defined by letting the maximum chord of the loop tend to zero and I tend to infinity with IS remaining finite, the uniformity of the magnetic field is immaterial to the argument. Thus

$$\mathbf{G} = \mathbf{m} \wedge \mathbf{B}, \tag{2.43}$$

where

$$\mathbf{m} = I\mathbf{S}, \tag{2.44}$$

and **B** is the field at the loop.

The formal similarity between (2.43) and (2.32), the couple on an electric dipole in an electric field, is a clue to the law in question. This may be summarized in the statement that an infinitesimal current loop has the character of a *magnetic dipole* of vector moment **m**, both with regard to the forces acting on it and with regard to the field it generates. The latter feature is now accepted as an additional experimental result on which further theory is based. The specific statement of the result is that the non-zero components of the field of the current loop are

$$B_r = \frac{\mu_0 m}{4\pi} \frac{2\cos\theta}{r^3}, \quad B_\theta = \frac{\mu_0 m}{4\pi} \frac{\sin\theta}{r^3}, \tag{2.45}$$

where $m = IS$, and r, θ are spherical polar coordinates referred to the position of the loop as origin and the normal to the loop as axis (see figure 2.8).

The expressions (2.45) have precisely the same form as (2.29), (2.30), and whereas in practice they hold only for values of r much greater than the maximum chord of the loop, the concept of an 'infinitesimal dipole' validates them for all values of r, as in the electric case.

Since a product of **B** with velocity gives a force per unit charge the mks unit of **B** is that of **E** times m^{-1} sec, that is, volt sec m^{-2}. A volt sec is called a *weber*, and the unit of **B** is therefore weber m^{-2} (for which *tesla* has been

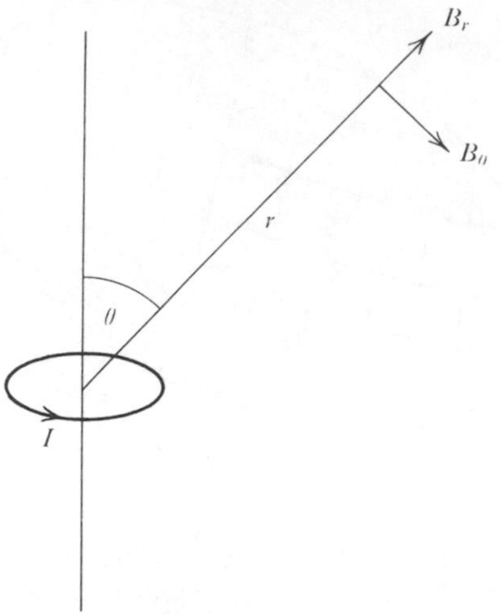

Figure 2.8

suggested). Using (1.17) it appears from (2.45) that the field 1 m along the normal from a loop of area $10^{-3}\,\text{m}^2$ carrying a 10 amp current is

$$2 \times 10^{-9}\,\text{weber m}^{-2}.$$

Fields exceeding 1 weber m^{-2} are in common use in the laboratory.

By exact analogy with the mathematics of § 2.1.5 the magnetic field (2.45) can be expressed in terms of a scalar potential function

$$\mathbf{B} = -\mu_0 \operatorname{grad} \phi, \tag{2.46}$$

where (cf. (2.27))

$$\phi = \frac{\mathbf{m}.\mathbf{r}}{4\pi r^3}. \tag{2.47}$$

It is merely a matter of convention that the factor μ_0 is included in (2.46) rather than (2.47); it gives the magnetic scalar potential the same dimensions as current.

This dipole law for current flowing in an infinitesimal plane loop can be immediately applied to current flowing in a loop L of arbitrary size and shape by the construction indicated in figure 2.7. It is imagined that L is spanned by a surface S on which is a net of infinitesimal mesh, and it is recognized that current I circulating in the same sense in every mesh is equivalent to current I in L. Each infinitesimal mesh of vector area $\mathrm{d}\mathbf{S}$ is a magnetic dipole of vector moment $I\,\mathrm{d}\mathbf{S}$, so that their combined potential is

$$\phi = \frac{I}{4\pi}\int_S \frac{\mathbf{R}.\mathrm{d}\mathbf{S}}{R^3}, \tag{2.48}$$

where the integration is over the surface S, and $\mathbf{R}$ is the vector distance from the surface element $\mathrm{d}\mathbf{S}$ to the field point P (see figure 2.9).

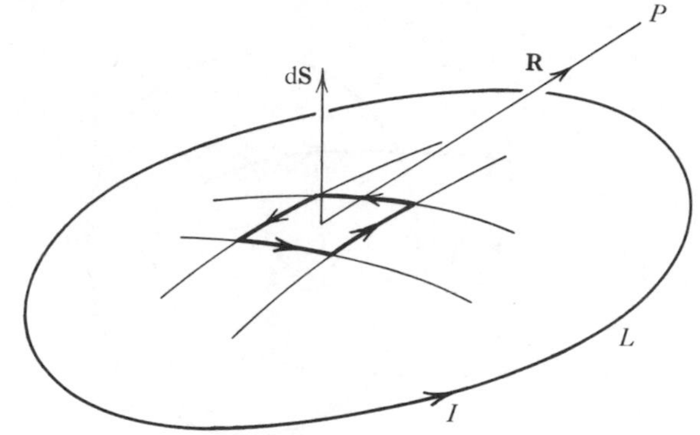

Figure 2.9

It is possible to express (2.48) in a formally very simple way. In an electrostatic context the expression would emerge as the flux across S of the field of a point charge at P. Reference to the second paragraph of §2.1.3 thus shows that

$$\phi = \frac{I\Omega}{4\pi}, \tag{2.49}$$

where Ω is the solid angle subtended at P by the surface spanning L.

2.2.3 *Ampère's circuital law*

The simplicity of the readily remembered formula (2.49) is somewhat deceptive. The solid angle Ω is defined by the integral in (2.48), and there is in general no short cut to its calculation. However, (2.49) does make transparent one important feature of ϕ, namely that it jumps by the amount I as the field point crosses S. Thus ϕ is continuous throughout the whole of space, provided S is treated as a barrier which the field point is not allowed to cross.

Why the calculation of the field **B** is independent of the choice of surface spanning the current loop is no puzzle. If two such surfaces, S_1 and S_2, are considered, then the corresponding potentials ϕ_1 and ϕ_2 will be the same everywhere, except in the region between S_1 and S_2, where they will differ by I; but this difference, being a constant, will not affect B calculated from (2.46).

However, the barrier associated with S only appears as an adjunct to the method of analysis, having no root in the actual physical problem; and it is therefore natural to adopt a potential function that is continuous throughout the whole of space (except on the loop itself) without qualification. The only modification this entails is that ϕ must be regarded as a multi-valued function, in the following sense: that the value of ϕ increases by I as the field point, starting from an arbitrary position P_0, subsequently returns to P_0 after tracing out a contour C that links the current loop L once (see figure 2.10).

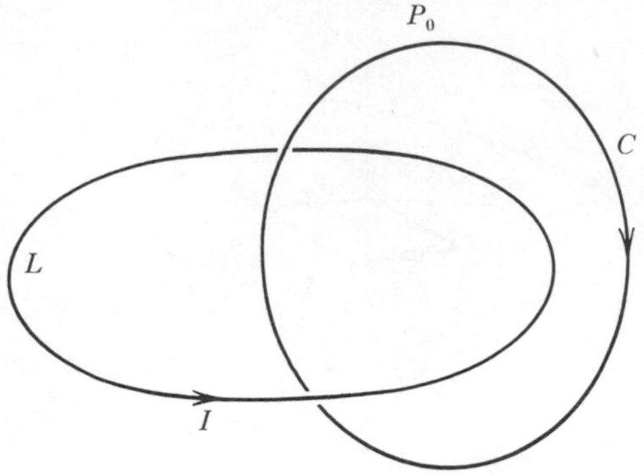

Figure 2.10

More precisely, if C links L n times ($n = 0, 1, 2, \ldots$) then the final value of ϕ at P_0 exceeds its initial value there by nI. In making these statements it is understood that the senses of circulation of the current in L and of the field point round C are related by right-handed screw motions, as indicated in figure 2.10; if the sense of one of these circulations were reversed ϕ would decrease rather than increase.

That the continuous magnetostatic potential is not single-valued is in sharp contrast to the electrostatic potential. The magnetostatic field is not conservative since the result analogous to (2.2) in the present problem is evidently

$$\oint_C \mathbf{B} \cdot \mathrm{d}\mathbf{s} = \mu_0 n I. \tag{2.50}$$

The relation (2.50) can be applied to the general case of a steady current density **J**, giving a most important result. The current distribution is regarded as consisting of tubes of flow of infinitesimal cross-section. The field contributed by each tube satisfies (2.50), where I is the current in the tube. The complete field **B** is obtained by superposition, so that

$$\oint_C \mathbf{B} \cdot \mathrm{d}\mathbf{s}$$

is the sum of the currents in all the tubes that link C, a tube that links C n times being counted n-fold. This sum may legitimately be called the total current flowing through C: it is conveniently expressed as the flux of **J** across any surface spanning C, since this description makes automatic allowance for the current tubes that link C more than once.

The result may therefore be summarized in the mathematical form

$$\oint \mathbf{B} \cdot \mathrm{d}\mathbf{s} = \mu_0 \int \mathbf{J} \cdot \mathrm{d}\mathbf{S}, \tag{2.51}$$

where the surface integral is over any surface spanning the arbitrary closed circuit round which the line integral is taken. This is Ampère's circuital law.

Once again it is the case that an integral relation has an equivalent differential form. Stokes' theorem (A. 21) applied to the left hand side of (2.51) gives

$$\int (\operatorname{curl} \mathbf{B} - \mu_0 \mathbf{J}) . \mathrm{d}\mathbf{S} = 0,$$

for an arbitrary surface of integration. Hence (cf. (2.2) and (2.4))

$$\operatorname{curl} \mathbf{B} = \mu_0 \mathbf{J}. \tag{2.52}$$

This is the important differential form of the circuital law. It shows conclusively that **B** can only be expressed as the gradient of a potential function in regions throughout which **J** is zero.

It should also be remembered that, even if curl **B** = 0 throughout some region, the potential function may not be single-valued throughout that region, as the case of a current loop demonstrates. The argument from Stokes' theorem clearly requires the further condition that the region be *simply connected* in the sense that any closed path lying wholly within the region can be spanned by a surface lying wholly within the region.

2.2.4 *The Biot–Savart law*

From the dipole law (2.45) has been deduced in turn the corresponding potential (2.47); the potential of a finite current loop (2.48), or equivalently (2.49); and finally the circuital law, (2.50) for a current loop, and (2.51) for a general steady current, with the differential form (2.52).

There is another chain of deduction that leads by a different route from (2.48) to (2.52), and establishes further equally important relations. The new approach is primarily concerned with the derivation of an explicit expression for **B** as an integral over the current distribution; this is known as the Biot–Savart law, and corresponds to the application of (1.6) to steady currents as discussed in § 1.3.1.

Consider, first, current I flowing in a wire loop of arbitrary size and shape. Then there is available the potential formula (2.48), from which **B** can be obtained by taking the gradient. If the gradient operation is performed under the integral sign, **B** is given as a surface integral. The same form would be obtained by superposing dipole fields directly, without introducing potentials.

However, this surface integral for **B** is comparatively complicated. Much more useful is the expression of **B** as a line integral taken round the current loop, which is indeed the result being sought. This line integral can be obtained by a transformation of the surface integral, but a quicker derivation is possible.

The idea is to use (2.48) to calculate the first order increment $\delta\phi$ in ϕ corresponding to an arbitrary small displacement $\delta\mathbf{r}$ of the field point, and then to obtain **B** from

$$\mathbf{B} . \delta\mathbf{r} = -\mu_0 \delta\mathbf{r} . \operatorname{grad} \phi = -\mu_0 \delta\phi. \tag{2.53}$$

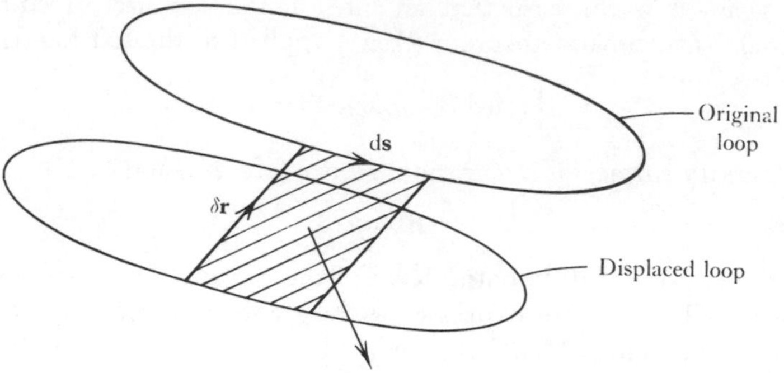

Figure 2.11

Now the increment in ϕ resulting from a displacement $\delta\mathbf{r}$ of the field point, keeping the current loop fixed, is the same as the increment resulting from a bodily displacement $-\delta\mathbf{r}$ of the current loop, keeping the field point fixed. The latter viewpoint has the advantage of giving a neat calculation of $\delta\phi$, since the effect of the displacement of the current loop is accounted for simply by extending the surface of integration in (2.48) to cover the region swept out by the loop in the course of its displacement. The surface thus added is a ribbon whose edges are the original loop and the displaced loop respectively, and the vector element of area of the ribbon is clearly $\mathrm{d}\mathbf{s} \wedge \delta\mathbf{r}$, where $\mathrm{d}\mathbf{s}$ is the element of arc of the loop (see figure 2.11). Thus

$$\delta\phi = \frac{I}{4\pi}\oint \frac{\mathbf{R}.(\mathrm{d}\mathbf{s} \wedge \delta\mathbf{r})}{R^3} = \delta\mathbf{r}.\frac{I}{4\pi}\oint \frac{\mathbf{R} \wedge \mathrm{d}\mathbf{s}}{\mathbf{R}^3}, \tag{2.54}$$

where now, of course, the integral is a line integral round the current loop, and $\mathbf{R}$ is the vector displacement from the arc element $\mathrm{d}\mathbf{s}$ to the field point.

Since the direction of $\delta\mathbf{r}$ is arbitrary a comparison of (2.54) with (2.53) gives

$$\mathbf{B} = \frac{\mu_0 I}{4\pi}\oint \frac{\mathrm{d}\mathbf{s} \wedge \mathbf{R}}{R^3}. \tag{2.55}$$

This is the Biot–Savart law for the magnetostatic field produced by a current loop I.

The extension of (2.55) to the case of general steady current flow is immediate. The general flow is regarded as a combination of tubes of flow of infinitesimal cross-section; and the field contributed by each tube is obtained from (2.55) with $I\mathrm{d}\mathbf{s}$ replaced by $\mathbf{J}\mathrm{d}S\,\mathrm{d}s$, where $\mathrm{d}S$ is the area of the normal cross-section and $\mathrm{d}s$ the element of arc length of the tube. Since $\mathrm{d}S\mathrm{d}s$ is a volume element $\mathrm{d}\tau$ superposition gives the complete field as the volume integral

$$\mathbf{B} = \frac{\mu_0}{4\pi}\int \frac{\mathbf{J} \wedge \mathbf{R}}{R^3}\mathrm{d}\tau, \tag{2.56}$$

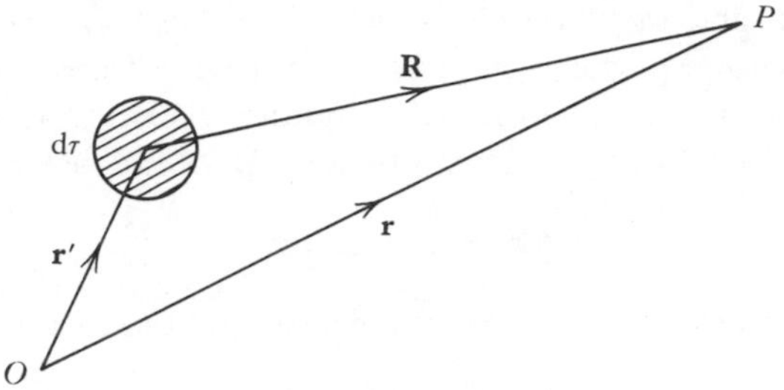

Figure 2.12

where **R** is now the vector displacement from dτ to the field point, and the integration is over the entire current distribution. This is the Biot–Savart law for the magnetostatic field of steady current of density **J**.

The expression (2.56) gives a prescription for the magnetostatic field in much the same way as (2.1) gives a prescription for the electrostatic field. Just as the integral in (2.1) remains convergent when the field point is inside the charge distribution, so that in (2.56) remains convergent when the field point is inside the current distribution. Again, just as the acceptance of the inverse square law (1.5) led to (2.1), so the acceptance of (1.6) would lead to (2.56). However, in the former case the converse holds, whereas in the latter it does not; (1.6) cannot be deduced from (2.56), since (2.56) only refers to steady currents, and a single moving charged particle does not constitute a steady current. In fact, as already remarked, (1.6) is at best an approximation; (2.56) is exact.

Although the magnitude of the integrand of (2.56) falls off with increasing R as $1/R^2$, the field of any spatially bounded steady current distribution is $O(1/r^3)$ as the distance r from some origin tends to infinity. That this must be the case is evident from the Ampère dipole law. Specifically, the dominant part of the distant field of a current loop I is evidently the field of a dipole whose vector moment is the product of I with the vector area of any surface spanning the loop; and since this can be written

$$\tfrac{1}{2}I\oint \mathbf{r}' \wedge \mathrm{d}\mathbf{s},$$

where $\mathbf{r}'$ is the position vector of the element $\mathrm{d}\mathbf{s}$ of the loop, it follows, by the argument leading from (2.55) to (2.56), that the dominant part of the distant field of a general steady current distribution is that of a dipole of vector moment

$$\frac{1}{2}\int \mathbf{r}' \wedge \mathbf{J}\,\mathrm{d}\tau, \tag{2.57}$$

where now $\mathbf{r}'$ is the position vector of the volume element dτ (see figure 2.12). This result is established by an alternative method in §6.1.2.

An interesting point about the forces between steady currents is conveniently made here: for two steady current distributions the net interaction forces are equal and opposite (Newton's third law), in contrast to those between two charged particles in motion (see (1.26), (1.27) *et seq.*).

To prove this, consider two spatially bounded distributions $\mathbf{J}$ and $\mathbf{J}'$. The force on the former due to the magnetic field $\mathbf{B}'$ of the latter is

$$\mathbf{F} = \int \mathbf{J} \wedge \mathbf{B}' \, d\tau, \tag{2.58}$$

where the integral can be taken over all space; and the formula (2.56) for $\mathbf{B}'$ then gives

$$\mathbf{F} = \frac{\mu_0}{4\pi} \iint \frac{\mathbf{J} \wedge (\mathbf{J}' \wedge \mathbf{R})}{R^3} \, d\tau' \, d\tau, \tag{2.59}$$

where in the present context $\mathbf{R}$ is the vector displacement from $d\tau'$ to $d\tau$. Now the integrand in (2.59) is

$$\frac{(\mathbf{J}.\mathbf{R})}{R^3} \mathbf{J}' - (\mathbf{J}.\mathbf{J}') \frac{\mathbf{R}}{R^3}.$$

But $\operatorname{div} \mathbf{J} = 0$ implies $$\frac{\mathbf{J}.\mathbf{R}}{R^3} = -\operatorname{div}\left(\frac{\mathbf{J}}{R}\right)$$

for each specific position of $d\tau'$. Hence

$$\int \frac{\mathbf{J}.\mathbf{R}}{R^3} \, d\tau = -\int \frac{\mathbf{J}.d\mathbf{S}}{R},$$

and is zero since $\mathbf{J}$ is spatially bounded. Therefore

$$\mathbf{F} = -\frac{\mu_0}{4\pi} \iint (\mathbf{J}.\mathbf{J}') \frac{\mathbf{R}}{R^3} \, d\tau \, d\tau'. \tag{2.60}$$

The expression (2.60) makes evident the validity of Newton's third law in this context, in the sense that the total force on the distribution $\mathbf{J}$ due to the field of $\mathbf{J}'$ is equal and opposite to that on $\mathbf{J}'$ due to the field of $\mathbf{J}$. In particular, it shows, on identifying $\mathbf{J}'$ with $\mathbf{J}$, that no steady current distribution can exert a total force on itself. A similar analysis shows likewise that no steady current distribution can exert a couple on itself (see problem 2.18 and also § 4.45).

2.2.5 *The vector potential*

It has been noted that, in view of (2.52), $\mathbf{B}$ cannot in general be expressed as the gradient of a scalar potential. However, as will be shown in a moment, it appears that (2.56) always admits expression as the curl of a vector function, known as the *vector potential*. Admittedly this representation does not have the particular advantage attached to the replacement of a vector field by a scalar field, but it does facilitate the analysis in other important ways.

To introduce the vector potential write (2.56) as

$$\mathbf{B} = -\frac{\mu_0}{4\pi} \int \mathbf{J} \wedge \operatorname{grad} \frac{1}{R} \, d\tau, \tag{2.61}$$

where the grad operation signifies differentiation with respect to the coordinates of the position vector $\mathbf{r}$ of the field point, not the position vector $\mathbf{r}'$ of the volume element $d\tau$ (see figure 2.12). Now the integrand depends on $\mathbf{r}$ only through $R = |\mathbf{r}-\mathbf{r}'|$; in particular, $\mathbf{J}$ there is solely a function of $\mathbf{r}'$, so from (A. 14) the integrand can be written $\operatorname{curl}(\mathbf{J}/R)$.

Thus
$$\mathbf{B} = \operatorname{curl}\mathbf{A}, \tag{2.62}$$

where
$$\mathbf{A} = \frac{\mu_0}{4\pi}\int \frac{\mathbf{J}}{R}\,d\tau. \tag{2.63}$$

$\mathbf{A}$ is the vector potential.

A noteworthy consequence of (2.62) is that, everywhere,

$$\operatorname{div}\mathbf{B} = 0. \tag{2.64}$$

For a given field $\mathbf{B}$ the relation (2.62) alone does not specify $\mathbf{A}$ uniquely. To (2.63) could be added any vector function whose curl is zero; that is to say, a vector grad ψ, where ψ is any scalar function. This arbitrariness in the relation between field and potential is reminiscent of that in electrostatics between $\mathbf{E}$ and ϕ, though less trivial. There is no call here to depart from the natural form (2.63), which is patently analogous to the expression (2.10) for ϕ.

Since each cartesian component of (2.63) is of identical mathematical form to (2.10), which, as emphasized in § 2.1.6, satisfies Poisson's equation (2.35), it follows that the x component of $\mathbf{A}$, for example, satisfies

$$\nabla^2 A_x = -\mu_0 J_x;$$

and with the notation

$$\nabla^2\mathbf{A} \equiv (\nabla^2 A_x, \nabla^2 A_y, \nabla^2 A_z)$$

the scalar equations can be combined into the vector form

$$\nabla^2\mathbf{A} = -\mu_0\mathbf{J}. \tag{2.65}$$

In fact equation (2.65) for the vector potential is another version of (2.52). This follows from the identity

$$\nabla^2\mathbf{A} = \operatorname{grad}\operatorname{div}\mathbf{A} - \operatorname{curl}\operatorname{curl}\mathbf{A},$$

provided it can be shown that $\operatorname{div}\mathbf{A} = 0$.

Now, from (2.63),

$$\operatorname{div}\mathbf{A} = \frac{\mu_0}{4\pi}\int \operatorname{div}(\mathbf{J}/R)\,d\tau$$

where the divergence operation taken under the integral sign signifies differentiation with respect to the coordinates of the field point $\mathbf{r}$. But here $\operatorname{div}\mathbf{J} = 0$, $\mathbf{J}$ being solely a function of $\mathbf{r}'$, the position vector of the volume element $d\tau$. Also $\operatorname{div}'\mathbf{J} = 0$, being the charge conservation relation, where the dash signifies differentiation with respect to the coordinates of $\mathbf{r}'$; and $\operatorname{grad}(1/R) = -\operatorname{grad}'(1/R)$. Therefore, using (A. 13),

$$\operatorname{div}\mathbf{A} = -\frac{\mu_0}{4\pi}\int \operatorname{div}'(\mathbf{J}/R)\,d\tau.$$

This can be transformed to a surface integral, which vanishes because the current is spatially bounded. Thus

$$\operatorname{div}\mathbf{A} = 0. \tag{2.66}$$

At the end of §2.1.6 it was in effect recognized that the specification of the curl and the divergence of $\mathbf{E}$, through (2.33) and (2.34), coupled with the assumption that the field vanish rapidly enough at infinity, was equivalent to the inverse square law relation (2.1). The equations satisfied by $\mathbf{B}$ exhibit a like feature.

It has been seen that the Biot–Savart law implies

$$\operatorname{curl}\mathbf{B} = \mu_0\mathbf{J}, \tag{2.52}$$

$$\operatorname{div}\mathbf{B} = 0. \tag{2.64}$$

To examine the converse it is noted that (2.64) ensures the existence of a vector $\mathbf{A}$ such that $\mathbf{B} = \operatorname{curl}\mathbf{A}$ (see § A. 2). If $\mathbf{A}$ is solenoidal, this representation in (2.52) implies $\nabla^2\mathbf{A} = -\mu_0\mathbf{J}$; and the solution of this equation that vanishes at infinity at least as fast as the inverse distance, being unique (see § 3.3.2), is (2.63).

As an example of the vector potential consider the infinitesimal dipole of vector moment $\mathbf{m}$ whose scalar potential is (2.47). Its vector potential is comparably simple, and is perhaps most quickly derived by noting that $\nabla^2(\mathbf{m}/r) = 0$ implies, using (A. 5) and (A. 13),

$$\operatorname{curl}\operatorname{curl}\left(\frac{\mathbf{m}}{r}\right) = \operatorname{grad}\operatorname{div}\left(\frac{\mathbf{m}}{r}\right) = -\operatorname{grad}\left(\frac{\mathbf{m}.\mathbf{r}}{r^3}\right).$$

Hence, if $\mathbf{B}$ is the dipole field, (2.62) is satisfied by

$$\mathbf{A} = \frac{\mu_0}{4\pi}\operatorname{curl}\frac{\mathbf{m}}{r} = -\frac{\mu_0}{4\pi}\mathbf{m}\wedge\operatorname{grad}\frac{1}{r} = \frac{\mu_0}{4\pi}\frac{\mathbf{m}\wedge\mathbf{r}}{r^3}; \tag{2.67}$$

and since, from the first form, (2.67) is patently solenoidal, it is the required vector potential.

2.3 Faraday's law of induction

2.3.1 *The voltage in a current loop*

An unvarying charge distribution gives rise to a purely electrostatic field. One that varies in time generates a varying electric field, and must also, in general, be associated with a magnetic field, since charge conservation demands that $\mathbf{J}$ be non-zero if $\dot{\rho}$ is non-zero.

On the other hand it is perfectly possible for current flow to vary in time in the absence of charge density; for $\rho = 0$ only imposes on $\mathbf{J}$ the restriction that $\operatorname{div} \mathbf{J} = 0$. It might then be thought that, just as in § 2.2 there was a magnetostatic field without electric field, so in this case there would be a varying magnetic field without electric field. However, this is not so. Faraday made the key discovery that there cannot be a varying magnetic field without an associated electric field.

The quantitative statement of this discovery must be accepted as a new fact that makes an essential contribution to the completion of electromagnetic theory: it cannot be proved from any of the results so far introduced. Nevertheless, its *plausibility* can be demonstrated, and this is the aim of the following discussion.

Consider the flow of current in a wire, taking the flow to be uniform (that is, the same at each point of the wire) but not necessarily constant. The current consists of the average motion of conduction electrons along the wire, and this motion naturally experiences some resistance; the resistive mechanism, analogous to dynamic friction, is discussed in § 6.2.2, but is immaterial to the present argument. In order to drive the conduction electrons along the wire work must be done on them by some force with which the resistance comes into balance. Let $\mathbf{F}$ denote the force per unit charge that performs this function; the wire may not be uniform, and $\mathbf{F}$ will in general depend on the position of the electron in the wire, and on time.

The situation discussed in § 2.2 was that in which the wire formed a closed loop L. What are the requirements on $\mathbf{F}$ to achieve a uniform current in this case? Evidently $\mathbf{F}$ must have, at each point of the loop, a component tangential to the loop, directed in the sense of circulation of the current. Thus

$$\oint_L \mathbf{F} . \mathbf{ds} \tag{2.68}$$

must be non-zero, so that the force must be non-conservative. Conversely it can be argued that if the wire loop is placed in a force field for which (2.68) is non-zero, then, in suitably restricted circumstances, uniform current will circulate in the loop.

The stated converse may at first sight seem surprising. For if, at any instant, the tangential component of $\mathbf{F}$ can vary almost arbitrarily along the wire, being restricted only by the requirement that (2.68) be non-zero, how does it come about that the current is uniform? The reason is that any tendency to non-uniformity in the current gives rise, by charge conservation, to a charge distribution in the wire that continuously adjusts itself so that the electric field associated with it, $\mathbf{E}'$ say, makes a contribution to the force field of precisely the amount that ensures that the current due to the total force field $\mathbf{F}+\mathbf{E}'$ is uniform. The feedback is very fast, and can be taken as instantaneous, for a good conductor, if the time scale of the variation of $\mathbf{F}$ is not too short.

Two further points may be made in connection with this argument. First, that the charge density adjusts itself suitably not only to control the tangential component of $\mathbf{F}+\mathbf{E}'$, but also to make the normal component zero; for until this is achieved charge is built up by flow across the wire. Secondly, that

$$\oint_L \mathbf{E}'.\mathrm{d}\mathbf{s}$$

is negligible compared with (2.68). The validity of this latter claim is again dependent on circumstances. Pains were taken in §2.1 to point out that the integral would be strictly zero if the source of $\mathbf{E}'$ were time independent charge density, so it is enough to assume that the time scale of the variation of the current is sufficiently long. In fact, since the only fundamental dimension in electromagnetic theory is the speed of light c, it may be anticipated that the criterion is that the time scale of variation is much greater than $1/c$ times the maximum chord of the loop.

It is admittedly impossible at this stage fully to discuss the issues just raised. However, they are only mentioned in order to clarify the importance of the quantity (2.68) in relation to uniform current circulating in a loop. It can now be appreciated that, under the conditions indicated, the non-vanishing of (2.68) is both a necessary and a sufficient condition for such current to flow.

Traditionally, if somewhat confusingly, (2.68) is called the 'electromotive force' (emf for short) in the loop. But here the alternative term *voltage* is preferred, the quantity having, of course, the same dimensions as potential.

What might $\mathbf{F}$ be? Natural candidates are the force per unit charge associated with an electric or a magnetic field.

If $\mathbf{F}$ is an electric field $\mathbf{E}$, (2.68) is

$$\oint_L \mathbf{E}.\mathrm{d}\mathbf{s}. \tag{2.69}$$

Now, as has just been emphasized, for an electrostatic field produced by a time independent charge density, the line integral of $\mathbf{E}$ round any closed path is zero; so the possibility of easily measurable current flow round a loop being sustained by an electric field may well be questioned. That it can be done, implying that curl $\mathbf{E}$ can be significantly different from zero, is an integral part of Faraday's law, as shown in §2.3.4.

Before pursuing the matter further it is instructive to contrast current flow in a closed loop of wire with the perhaps more familiar situation in which current is made to flow in a wire by means of a battery. For present purposes a battery may be ideally conceived as a device that maintains a constant potential difference V between two terminals A and B, so that it creates an electrostatic field for which

$$\int_A^B \mathbf{E}.\mathrm{d}\mathbf{s} = \phi_A - \phi_B = V, \tag{2.70}$$

where the integral is along any path from A to B. If one end of a piece of wire is connected to A and the other to B a steady current will flow in the wire. Conduction electrons in the wire are driven by the electric field, the work done per unit charge on an electron that passes from B to A along the wire being V. The effective transfer of electrons from B to A by their movement along the wire does, of course, tend to reduce the potential difference between A and B. It is the function of the battery to keep the balance by restoring electrons to B, and this is done within the cells of the battery by a chemical reaction that gives rises to a force just sufficient to overcome the oppositely directed force of the electrostatic field in the cells. Current thus produced therefore relies on forces other than electric forces. The wire itself is not a closed loop, but the flow of charge within the battery completes a current circuit in which the voltage (2.68) is V, since $\mathbf{F}$ is zero inside the battery, where the chemical force cancels the electric force, and is $\mathbf{E}$ in the wire.

2.3.2 *Motion of a wire loop in a magnetostatic field*

Continuing with consideration of current in a closed wire loop, the alternative possibility that $\mathbf{F}$ is the force associated with a magnetic rather than an electric field is now examined. That the magnetic force can give rise to a voltage in a loop may be seen as follows.

Suppose that a wire loop L, not necessarily rigid, is moved in an arbitrary fashion in a field that is purely magnetostatic. Then the magnetic force per unit charge on a conduction electron is, at any time,

$$(\mathbf{v}+\mathbf{u}) \wedge \mathbf{B}, \qquad (2.71)$$

where $\mathbf{B}$ and $\mathbf{v}$ are the magnetic field and velocity of the wire, respectively, at the location of the electron at the time in question, and $\mathbf{u}$ is the velocity of the electron relative to the wire. The corresponding voltage in the loop is

$$\oint_L [(\mathbf{v}+\mathbf{u}) \wedge \mathbf{B}].\mathrm{d}\mathbf{s} = -\oint_L \mathbf{B}.[(\mathbf{v}+\mathbf{u}) \wedge \mathrm{d}\mathbf{s}]$$

$$= -\oint_L \mathbf{B}.(\mathbf{v} \wedge \mathrm{d}\mathbf{s}), \qquad (2.72)$$

where the last step is made on the presumption that the thermal velocities of the electrons can be discounted, so that $\mathbf{u}$ is the local macroscopic (or average) velocity, and is therefore parallel to $\mathrm{d}\mathbf{s}$, as argued, in effect, in § 2.3.1.

That (2.72) can differ from zero is put beyond doubt by a most significant transformation of the line integral into the rate of change of a surface integral.

The deduction of the transformation is mathematically closely analogous to a reversal of the argument immediately preceding (2.54). Consider two

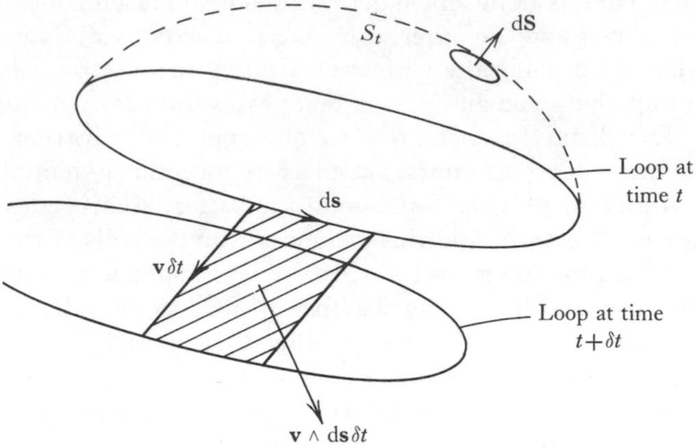

Figure 2.13

positions of the loop at neighbouring times t and $t+\delta t$, as illustrated in figure 2.13 (cf. figure 2.11). It is evident that

$$\oint_L \mathbf{B}.(\mathbf{v}\wedge \mathrm{d}\mathbf{s})\,\delta t \tag{2.73}$$

is precisely the flux of **B** across the surface swept out by the loop in the course of its displacement during the time interval δt. But the flux of **B** out of any closed surface is zero. Hence if S_t and $S_{t+\delta t}$ are any surfaces spanning the loop at times t and $t+\delta t$, respectively, (2.73) is the same as

$$\int_{S_{t+\delta t}} \mathbf{B}.\mathrm{d}\mathbf{S} - \int_{S_t} \mathbf{B}.\mathrm{d}\mathbf{S}, \tag{2.74}$$

where it is understood that the sense of direction of the normal to the surface in each integral is related to the sense of circulation in the line integral (2.73) by a right-handed screw.

Now (2.73) is $-\delta t$ times (2.72). It therefore follows, on dividing by δt and then letting $\delta t \to 0$, that the voltage in the wire loop is

$$-\frac{\mathrm{d}}{\mathrm{d}t}\int \mathbf{B}.\mathrm{d}\mathbf{S}, \tag{2.75}$$

where the integration is over any surface spanning the loop.

The magnitude of the voltage induced by the movement of the loop is thus the rate of change of magnetic flux through the loop; if the loop is moved in such a way that the flux does change then a current will circulate in it.

This result, here deduced from the expression for the force on a charged particle moving in a magnetic field, was established experimentally by Faraday. It is an intrinsic part, but only a part, of Faraday's law of induction.

The general statement, now to be discussed, must be accepted as an experimental result, though it is made plausible by this prior derivation of the restricted result.

2.3.3 *Faraday's law*

In the situation just envisaged consider the special case in which the wire loop is rigid and has uniform translation through the magnetostatic field. Provided the field is inhomogeneous (2.75) will in general be non-zero and a current will flow in the loop.

Now imagine an observer moving with the loop. His frame of reference merely has constant velocity relative to the 'laboratory' frame so far understood, and it is reasonable to anticipate that his measurement of the voltage driving the current will also equate it to the rate of decrease of magnetic flux through the loop. Since to this observer the loop is stationary and the magnetic field time varying, the voltage is expressed as

$$-\frac{\mathrm{d}}{\mathrm{d}t}\int \mathbf{B}.\mathrm{d}\mathbf{S} = -\int \dot{\mathbf{B}}.\mathrm{d}\mathbf{S} \tag{2.76}$$

where the integration is over a fixed surface spanning the loop, and the dot signifies partial differentiation with respect to time t, that is $\dot{\mathbf{B}} \equiv \partial \mathbf{B}/\partial t$ is the rate of change of $\mathbf{B}$ at each point.

That indeed there is induced in a fixed loop by any varying magnetic field a voltage that is the rate of change of magnetic flux through the loop is again one of the results established experimentally by Faraday; and here its acceptance must ultimately rest on the basis of experiment, since the argument just given fails to constitute a proof in the absence of precise knowledge of the effect on observations of the uniform motion of the frame of reference. The implication in this result, that there is no ambiguity in specifying the flux of $\mathbf{B}$ through the loop, should be noted: it is equivalent to the assertion that the expressions in (2.76) are independent of the particular surface chosen to span the loop; thus the flux of $\dot{\mathbf{B}}$ out of any closed surface is zero, and hence, by the divergence theorem (cf. (2.40))

$$\operatorname{div} \dot{\mathbf{B}} = 0, \tag{2.77}$$

that is

$$\frac{\partial}{\partial t}(\operatorname{div} \mathbf{B}) = 0, \tag{2.78}$$

so that $\operatorname{div} \mathbf{B}$ is independent of time. The part played by this condition in the present discussion of time varying fields is, of course, entirely analogous to that played by the condition (2.64) for static fields, which was invoked in the argument leading to (2.75). In fact there are no grounds for supposing that magnetic fields can exist for which the constant value of $\operatorname{div} \mathbf{B}$ differs from zero, and hence

$$\operatorname{div} \mathbf{B} = 0 \tag{2.79}$$

is accepted as an equation satisfied by arbitrary fields.

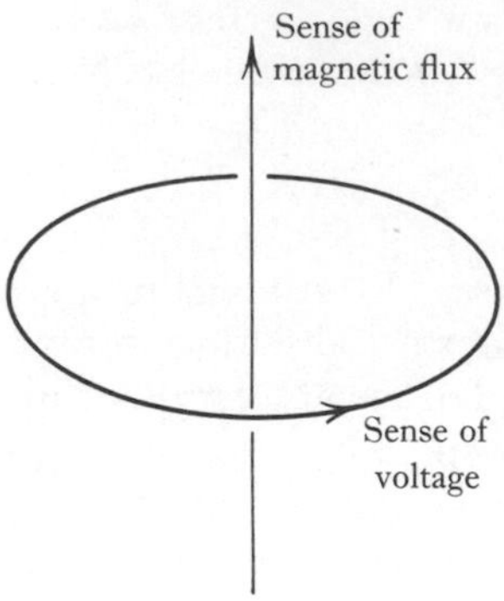

Figure 2.14

The complete statement of Faraday's law of induction is but a natural generalization of preceding results, namely that *a magnetic field induces in a closed wire loop a voltage that is equal to the rate of decrease of the flux of magnetic induction through the loop, where the sense of the voltage and the direction of the flux are related by a right-handed screw as indicated in figure* 2.14.

After the exposure of the two special cases the general statement is likely to be acceptable with little further comment: the point is simply that it is based on experiments in which the magnetic flux through the wire loop is made to vary in a variety of ways, and the finding is that the current flow is determined solely by the rate of change of flux, being independent of the mechanism by which the rate of change is produced.

2.3.4 *Maxwell's equation*

The physics revealed by Faraday's law is, of course, of tremendous practical consequence. The most straightforward applications, perhaps, are to the dynamo and electric motor, which essentially involve flux changes arising from the rotation of a loop in a magnetostatic field. However the immediate concern here is with the vital contribution that the law makes to fundamental theory, and this is pinpointed by the special case in which the loop is fixed and the magnetic field is time varying. That the voltage is then given by (2.76) implies a relation between magnetic and electric fields, obtained by the following argument.

For a fixed loop of wire the mean motion of the conduction electrons can only be along the wire. The magnetic force on the electrons can therefore only be at right angles to the wire and can make no contribution to the voltage. What force, then, gives rise to the voltage (2.76)? The sole remaining candidate in this context is an electric force, and the conclusion is that with a time varying magnetic field **B** there is inevitably associated an electric field **E** such that

$$\oint \mathbf{E}\,.\,d\mathbf{s} = -\int \dot{\mathbf{B}}\,.\,d\mathbf{S}, \tag{2.80}$$

where the line integral is round an arbitrary loop and the surface integral is over any surface spanning the loop. This is the result promised in § 2.3.1.

The mathematical form of the integral relation (2.80) is a repetition of that of (2.51). Thus, by the same argument that leads from (2.51) to (2.52), it follows that

$$\operatorname{curl}\mathbf{E} = -\dot{\mathbf{B}}. \tag{2.81}$$

Equation (2.81) is one of the two main vector partial differential equations of the set known as Maxwell's equations. All experience is consistent with the supposition that it is satisfied by every electromagnetic field. It can be regarded as a generalization of the electrostatic equation $\operatorname{curl}\mathbf{E} = 0$, and shows that in general the electric field cannot be expressed as the gradient of a scalar potential.

It is, of course, a mathematical consequence of (2.81) that $\operatorname{div}\dot{\mathbf{B}} = 0$, that is, $\operatorname{div}\mathbf{B}$ is independent of time; in fact, as already asserted in (2.79),

$$\operatorname{div}\mathbf{B} = 0 \tag{2.82}$$

holds with complete generality. This particular relation, deduced originally for magnetostatic fields, remains intact in the full theory.

One further point may be made in connection with (2.81). The plausibility argument leading to it was framed in the context of Faraday's law, because final appeal was made to the experimental basis of that law. But it may be helpful to state the essence of the argument somewhat more baldly in the following terms. Suppose, by some agency, a unit point charge is moved with constant velocity $\mathbf{v}$ through an inhomogeneous magnetostatic field $\mathbf{B}$. The agency must exert a force $-\mathbf{v}\wedge\mathbf{B}$ to overcome the magnetic force. To an observer moving with the point charge, however, the force is necessarily exerted to keep the electron at rest. The force to be overcome can therefore only be an electric force, and the observer thus detects an electric field that may plausibly be expected to be

$$\mathbf{E} = \mathbf{v}\wedge\mathbf{B}. \tag{2.83}$$

The explicit appearance of $\mathbf{v}$ in (2.83) can be eliminated by taking the curl. This gives, from (A. 17),

$$\operatorname{curl}\mathbf{E} = \mathbf{v}\operatorname{div}\mathbf{B} - (\mathbf{v}.\operatorname{grad})\mathbf{B}, \tag{2.84}$$

which is in agreement with (2.81) because the first term on the right hand side is zero, and the second term is the rate of decrease of $\mathbf{B}$ as seen by the observer moving with the point charge.

2.4 Maxwell's 'displacement' current

2.4.1 *The generalization of Ampère's circuital law: Maxwell's equation*

Of the four equations governing time independent electric and magnetic fields, two, namely (2.33) and (2.64), have now been extended, by an appeal to Faraday's law, to cover arbitrary electromagnetic fields. It is natural to attempt to reach a complete theory by generalizing the remaining static equations (2.34) and (2.52). This was achieved by Maxwell, not on the basis of experimental results, but rather by theoretical reasoning that led to a self-consistent formulation. Subsequently, all experience with electromagnetic phenomena has confirmed the validity of Maxwell's equations.

The essence of the argument is as follows. It is evident that (2.52), which is the differential form of Ampère's circuital law for steady current flow, cannot hold for general time varying currents; for it requires the conservation relation $\operatorname{div}\mathbf{J} = 0$, special to steady current flow, whereas the general charge conservation relation (1.10) is

$$\operatorname{div}\mathbf{J} + \dot{\rho} = 0. \tag{2.85}$$

If, however, (2.34) is accepted as it stands, (2.85) can be written

$$\operatorname{div}(\epsilon_0\dot{\mathbf{E}} + \mathbf{J}) = 0, \tag{2.86}$$

and the inconsistency in (2.52) can be removed by replacing $\mathbf{J}$ on the right hand side by

$$\epsilon_0\dot{\mathbf{E}} + \mathbf{J}. \tag{2.87}$$

Thus it may be conjectured that the general equations are

$$\frac{1}{\mu_0}\operatorname{curl}\mathbf{B} = \epsilon_0\dot{\mathbf{E}} + \mathbf{J}, \tag{2.88}$$

$$\epsilon_0 \operatorname{div}\mathbf{E} = \rho. \tag{2.89}$$

Equation (2.88) is the other main Maxwell equation, to be taken in conjunction with (2.81). Its acceptance of course implies (2.86), which together with the charge conservation relation (2.85) in turn implies that the time derivative of (2.89) is satisfied. Thus (2.89) is related to (2.88) in much the same way as (2.82) is related to (2.81).

It is sometimes claimed that the integral relations rather than the corresponding differential equations give a more illuminating picture of the physics. That corresponding to (2.88) is

$$\frac{1}{\mu_0}\oint \mathbf{B}.\mathrm{d}\mathbf{s} = \int (\epsilon_0\dot{\mathbf{E}} + \mathbf{J}).\mathrm{d}\mathbf{S}, \tag{2.90}$$

and the reader may find it instructive to recast the argument in terms of integral relations to reach this form directly. Equation (2.90) expresses the fact that the line integral of $\mathbf{B}/\mu_0$ round any closed circuit is equal to the flux of (2.87) through the circuit; and with the idea much in mind that this is a generalization of Ampère's circuital law for steady currents, in which the right hand side would be the current flowing through the circuit, (2.87) is sometimes called the *total current* density. Furthermore, Maxwell gave the name *displacement current* density to the contribution

$$\epsilon_0 \dot{\mathbf{E}}; \tag{2.91}$$

this designation is still used, although it only arose from the additional consideration of the displacement of charge in dielectric media (see § 6.1), and must therefore be regarded as no more than a historical label, since it fails to indicate the essential point that (2.91) is independent of the local presence of charge.

2.4.2 *The vectors* **D** *and* **H**

In addition to **E** and **B** Maxwell introduced two further field vectors, **D** and **H**. The chief purpose of working with four (or more) vectors is to facilitate the description of the electromagnetic properties of material media, and for the vacuum the distinction between **D** and **E**, and between **H** and **B**, is mathematically trivial, since in fact

$$\mathbf{D} = \epsilon_0 \mathbf{E}, \tag{2.92}$$

$$\mathbf{H} = \frac{1}{\mu_0} \mathbf{B}. \tag{2.93}$$

However, in mks units there is here a physical distinction in dimensions; and it may well be instructive at this stage to indicate the mode of thought behind the associated concepts by presenting the argument leading to (2.88), (2.89) in a slightly different way, which is indeed perhaps closer to Maxwell's original reasoning.

First, introduce a field vector **D** whose 'source' is electric charge in the sense that the flux of **D** out of any closed surface is the total charge within the surface: that is

$$\int \mathbf{D} \,.\, \mathrm{d}\mathbf{S} = \int \rho \, \mathrm{d}\tau; \tag{2.94}$$

or equivalently,

$$\operatorname{div} \mathbf{D} = \rho. \tag{2.95}$$

No matter what space or time variation ρ may have there certainly exists a vector **D** satisfying (2.95), and of course it is not uniquely specified by this equation. Then it follows from the charge conservation relation (2.85) that the divergence of

$$\dot{\mathbf{D}} + \mathbf{J} \tag{2.96}$$

is identically zero, and therefore (§ A. 2) there exists a vector **H** such that

$$\operatorname{curl} \mathbf{H} = \dot{\mathbf{D}} + \mathbf{J}. \tag{2.97}$$

These statements are mathematically impeccable, but only lead to closed equations for the electromagnetic field if they are supplemented by two further relations. What might these relations be? In the case of vacuum

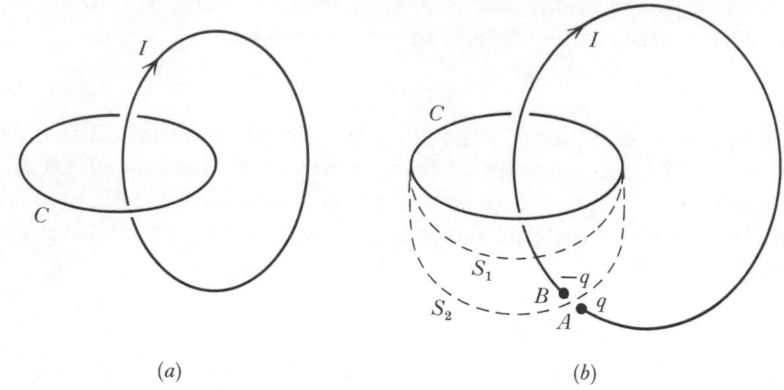

Figure 2.15

electrostatics the development from Coulomb's law has shown that $\mathbf{D} = \epsilon_0 \mathbf{E}$ is appropriate, since then (2.95) is the same as (2.34); whereas in the case of vacuum magnetostatics of steady currents $\mathbf{H} = \mathbf{B}/\mu_0$ is appropriate, since then (2.97) is the same as (2.52). In the absence of evidence to the contrary it is therefore natural to suppose that the relations (2.92) and (2.93) apply to arbitrary vacuum fields, and in this way (2.88) and (2.89) are recovered from (2.97) and (2.95) respectively.

It will be seen later how **D** and **H** can be defined physically and how the electromagnetic properties of material media can often be described phenomenologically in terms of other relatively simple relations between **D**, **H**, **E** and **B**. At present it is merely observed that, $\dot{\mathbf{D}}$ being the displacement current, the dimensions of **D** are charge per unit area, so that the unit is coulomb m^{-2}; and that those of **H** are current per unit length, with unit amp m^{-1}.

2.4.3 *A simple example*

The general mathematical reasoning leading to Maxwell's equation (2.88) is well illustrated by a simple example.

Consider a closed loop of wire round which flows a steady current I. Imagine a path C linking the wire, as shown in figure 2.15 (*a*). Then the line integral of $\mathbf{B}/\mu_0$ round C is the current flowing through C (more precisely, the current density flux across any surface spanning C), namely I.

Now suppose the wire is cut at some point, leaving two ends A and B with a small separation. Then the current I can still be presumed to flow in the wire provided the ends A and B carry charges q and $-q$, respectively, where

$$I = \mathrm{d}q/\mathrm{d}t. \tag{2.98}$$

It would be expected that the magnetic field was sensibly unaffected by the cut in the wire, so that the line integral of $\mathbf{B}/\mu_0$ round C would still be I.

This is, of course, the 'current flowing through C' if the current density flux is evaluated across a surface spanning C such as S_1 that intersects the wire, as indicated in figure 2.15(*b*). But what if the flux is evaluated across a surface S_2 that avoids intersecting the wire by squeezing between A and B? No current crosses S_2.

If, however, the displacement current density $\dot{\mathbf{D}}$ is also included, the fluxes across S_1 and S_2 are seen to agree: for the flux of $\mathbf{D}$ across S_1 is negligible (taking the moment of the dipole AB to be negligible), and that across S_2 is q; hence the flux of $\dot{\mathbf{D}}$ across S_1 is negligible, whereas that across S_2 is $\mathrm{d}q/\mathrm{d}t = I$.

2.4.4 *Connection with the force law*

It has been seen that Maxwell's equation (2.81) is intimately related to the fact that $\mathbf{E}$ is the electric force and $\mathbf{v}\wedge\mathbf{B}$ the magnetic force acting on a moving unit point charge. It is also interesting to observe that the displacement current term in Maxwell's equation (2.88) follows directly from the expressions (1.5) and (1.6) for the electric and magnetic fields produced by a moving point charge. The demonstration of this result does not constitute a complete proof of the displacement current term, if only because (1.5) and (1.6) are approximations of restricted validity: but the establishment of the displacement current term even for restricted circumstances is, of course, a significant step, and might logically have come, though it did not historically, before the argument of §2.4.1.

The result is obtained at once from the relation (cf. (2.84))

$$\operatorname{curl}\left(\frac{\mathbf{v}\wedge\mathbf{r}}{r^3}\right) = \mathbf{v}\operatorname{div}\frac{\mathbf{r}}{r^3} - (\mathbf{v}\,.\operatorname{grad})\frac{\mathbf{r}}{r^3}, \tag{2.99}$$

where $\mathbf{v}$ is the velocity of the point charge, and $\mathbf{r}$ is the vector displacement from it to the field point. For the first term on the right hand side is zero, as shown in (2.20); and the second term is the rate of change of $\mathbf{r}/r^3$, the change being solely due to the motion of the point from which $\mathbf{r}$ is measured. Thus the use of (1.5) and (1.6) implies

$$\frac{1}{\mu_0}\operatorname{curl}\mathbf{B} = \epsilon_0\dot{\mathbf{E}}. \tag{2.100}$$

Equation (2.100) is indeed the form that (2.88) takes in any region where the current density $\mathbf{J}$ is zero.

An alternative proof, which arrives at the equivalent integral form

$$\frac{1}{\mu_0}\oint\mathbf{B}\,.\,\mathrm{d}\mathbf{s} = \epsilon_0\int\dot{\mathbf{E}}\,.\,\mathrm{d}\mathbf{S} \tag{2.101}$$

is given by essentially the same mathematical procedure as that adopted in the derivation of (2.55). To the approximation (1.5), the flux of **E** across any fixed, open surface S is

$$\int \mathbf{E}.\mathrm{d}\mathbf{S} = \frac{e}{4\pi\epsilon_0}\int \frac{\mathbf{r}.\mathrm{d}\mathbf{S}}{r^3}, \tag{2.102}$$

where **r** is the vector from the point charge to the element d**S** of S. Now in an infinitesimal time interval δt the flux changes because **r** changes by the negative of the displacement $\mathbf{v}\delta t$ through which the charge moves during the interval. However, the same change in **r** would arise were the charge held fixed and the surface S translated bodily through the displacement $-\mathbf{v}\delta t$. Thus, by precisely the argument leading to (2.54), the flux increment is

$$\delta \int \mathbf{E}.\mathrm{d}\mathbf{S} = \frac{e}{4\pi\epsilon_0}\oint \frac{\mathbf{r}.(\mathrm{d}\mathbf{s} \wedge \mathbf{v}\delta t)}{r^3}.$$

Division by δt, and procedure to the limit $\delta t \to 0$, gives

$$\frac{\mathrm{d}}{\mathrm{d}t}\int \mathbf{E}.\mathrm{d}\mathbf{S} = \frac{e}{4\pi\epsilon_0}\oint \frac{\mathbf{v}\wedge\mathbf{r}}{r^3}.\mathrm{d}\mathbf{s},$$

which, invoking (1.6), is indeed (2.101).

It may perhaps be questioned whether (2.81) cannot also be deduced from (1.5) and (1.6). The answer is no, since the curl of (1.5) vanishes. However there is no inconsistency. As the rate of change of (1.6) indicates, it is simply that terms smaller by the factors v^2/c^2 or $r\,|\dot{\mathbf{v}}|/c^2$ are not represented in the approximation.

2.5 Maxwell's equations

2.5.1 *A summary*

It is as well as this stage to summarize the differential equations governing the electromagnetic field, known collectively as Maxwell's equations. In conjunction with the Lorentz force law they state the whole basis of the classical (that is, non-quantum) theory of electromagnetism; and, moreover, in a way that has generally been found most useful for the development and application of the theory.

The relevant equations are the charge conservation relation

$$\operatorname{div}\mathbf{J} + \dot{\rho} = 0; \tag{2.103}$$

the Faraday–Maxwell equation

$$\operatorname{curl}\mathbf{E} = -\dot{\mathbf{B}}, \tag{2.104}$$

together with

$$\operatorname{div}\mathbf{B} = 0; \tag{2.105}$$

and the Ampère–Maxwell equation

$$\frac{1}{\mu_0}\operatorname{curl}\mathbf{B} = \epsilon_0\dot{\mathbf{E}} + \mathbf{J}, \tag{2.106}$$

together with
$$\operatorname{div}\mathbf{E} = \rho/\epsilon_0. \tag{2.107}$$

To these must be added the force density exerted by the field on charge and current, namely
$$\rho\mathbf{E}+\mathbf{J}\wedge\mathbf{B}. \tag{2.108}$$

Equation (2.104) implies that div **B** is constant (that is, time independent), and (2.106) with (2.103) that div $\mathbf{E}-\rho/\epsilon_0$ is constant. Equations (2.105) and (2.107) therefore give additional information only about static fields: they are, of course, essential to the description of static fields, but are otherwise redundant, as for example in the discussion of time harmonic fields.

When there is no time variation Maxwell's equations separate into two independent groups. On the one hand there are the equations of electrostatics,
$$\operatorname{curl}\mathbf{E} = 0, \quad \operatorname{div}\mathbf{E} = \rho/\epsilon_0, \tag{2.109}$$
which may be equivalently represented in terms of a scalar potential ϕ in the form
$$\mathbf{E} = -\operatorname{grad}\phi, \quad \nabla^2\phi = -\rho/\epsilon_0. \tag{2.110}$$
On the other hand there are the equations of magnetostatics
$$\operatorname{curl}\mathbf{B} = \mu_0\mathbf{J}, \quad \operatorname{div}\mathbf{B} = 0, \tag{2.111}$$
which in terms of a vector potential **A** admit the equivalent representation
$$\mathbf{B} = \operatorname{curl}\mathbf{A}, \quad \operatorname{div}\mathbf{A} = 0, \quad \nabla^2\mathbf{A} = -\mu_0\mathbf{J}. \tag{2.112}$$

When there is a time variation the electric and magnetic fields must coexist. The most striking single deduction from the equations is then that electromagnetic disturbances are propagated with the speed
$$c = (\epsilon_0\mu_0)^{-\frac{1}{2}} = 2.998\times 10^8\,\text{m}\,\text{sec}^{-1}. \tag{2.113}$$
A preliminary investigation of this aspect of the equations is now undertaken.

2.5.2 *The wave equation*

Evidently **E** can be eliminated between (2.104) and (2.106) by taking the curl of the latter equation and substituting for curl **E** from the former. If (A. 5) is invoked, in conjunction with (2.105), the result is
$$\nabla^2\mathbf{B}-\frac{1}{c^2}\ddot{\mathbf{B}} = -\mu_0\operatorname{curl}\mathbf{J}. \tag{2.114}$$

In a similar way the elimination of **B**, by taking the curl of (2.104), and using (2.107), gives
$$\nabla^2\mathbf{E}-\frac{1}{c^2}\ddot{\mathbf{E}} = \frac{1}{\epsilon_0}\operatorname{grad}\rho+\mu_0\dot{\mathbf{J}}. \tag{2.115}$$

Thus each cartesian component of the vectors **B** and **E** satisfies an *inhomogeneous wave equation* of the form

$$\nabla^2\chi - \frac{1}{c^2}\ddot{\chi} = -s, \tag{2.116}$$

in which

$$\nabla^2 - \frac{1}{c^2}\frac{\partial^2}{\partial t^2} \tag{2.117}$$

may be called the *wave operator*, and s (in general a function of position and time) the *source* term. In this sense μ_0 curl **J** is the source of **B**; and a current density whose curl is everywhere zero does not give rise to a magnetic field.

However the derivation from the general inhomogeneous wave equations of explicit expressions for the field in terms of the sources, expressions of the kind already introduced in the time independent cases, is a comparatively advanced piece of work which it is not appropriate to embark on at this stage. On the other hand the common physical context is that in which the charge-current source is confined to a highly restricted region of space; and since elsewhere the cartesian components of **B** and **E** each satisfy the *homogeneous* wave equation

$$\nabla^2\chi = \frac{1}{c^2}\frac{\partial^2\chi}{\partial t^2}, \tag{2.118}$$

solutions of this equation are of fundamental importance. Such solutions lead to an understanding of the nature of the propagation of electromagnetic waves, free from the details of their relation to the specific distribution of charge and current by which they are generated.

Consider an idealized situation in which the field depends spatially only on one cartesian coordinate, say x. Then (2.118) is

$$\frac{\partial^2\chi}{\partial x^2} = \frac{1}{c^2}\frac{\partial^2\chi}{\partial t^2}, \tag{2.119}$$

with the general solution

$$\chi = f_1(t - x/c) + f_2(t + x/c), \tag{2.120}$$

where f_1 and f_2 are arbitrary functions. The first term in (2.120) represents a disturbance travelling in the positive x direction, the second term one travelling in the negative x direction, each with speed c. For ease of description, and obviously without significant loss of generality since fields can be superposed, the ensuing discussion assumes that only the first type of disturbance is present.

It is known, then, that each cartesian component of **B** and **E** must be a function of $t - x/c$; it is, however, necessary to return to Maxwell's equations to obtain the relations between these functions. The x component of the equations

$$\text{curl}\,\mathbf{E} = -\dot{\mathbf{B}}, \tag{2.121}$$

$$\text{curl}\,\mathbf{B} = \frac{1}{c^2}\dot{\mathbf{E}}, \tag{2.122}$$

gives $\dot{B}_x = \dot{E}_x = 0$, so that B_x and E_x are independent of time, and therefore, by virtue of their form, of position also. Thus in effect

$$B_x = E_x = 0, \tag{2.123}$$

since uniform static fields are irrelevant in the present context. This conclusion ensures the satisfaction of

$$\operatorname{div}\mathbf{B} = \operatorname{div}\mathbf{E} = 0, \tag{2.124}$$

from which alternatively it might have been deduced.

The y and z components of (2.121) and (2.122) are

$$\frac{\partial E_z}{\partial x} = \dot{B}_y, \quad \frac{\partial E_y}{\partial x} = -\dot{B}_z \tag{2.125}$$

and

$$\frac{\partial B_z}{\partial x} = -\frac{1}{c^2}\dot{E}_y, \quad \frac{\partial B_y}{\partial x} = \frac{1}{c^2}\dot{E}_z. \tag{2.126}$$

The first equation of the pair (2.125) and the second of (2.126) involve only E_z and B_y, whereas the remaining two equations involve only E_y and B_z. Again, then, there is no significant loss of generality in supposing E_z and B_y, say, to be zero; in which case the identification

$$E_y = F_1(t-x/c), \quad B_z = \frac{1}{c}F_2(t-x/c) \tag{2.127}$$

yields a solution of Maxwell's equations providing only that the first derivatives of F_1 and F_2 are equal. In effect, then, F_1 and F_2 themselves are equal, since a difference between them of a constant other than zero only relates to static uniform fields.

The field thus arrived at is of the form

$$E_y = cB_z = F(t-x/c), \tag{2.128}$$

with the remaining components of **E** and **B** identically zero. The simple field geometry is indicated in figure 2.16.

A very direct interpretation of the speed c is obtained by relating the field (2.128) in $x > 0$ to its value at $x = 0$, where it is given by

$$E_y = cB_z = F(t). \tag{2.129}$$

If $F(t)$ starts at $t = 0$, being zero previously, the field arrives at $x = a$ at time $t = a/c$.

Note that the disturbance is *transverse*, in the sense that the field vectors are at right angles to the direction of propagation. Note also that **E** and **B** are mutually orthogonal, and that the ratio of their magnitudes is the constant c. These features are common to any disturbance whose space–time dependence is only through $t-x/c$; but it is a particular feature of the selected disturbance that the directions of **E** and **B** are unvarying, for which

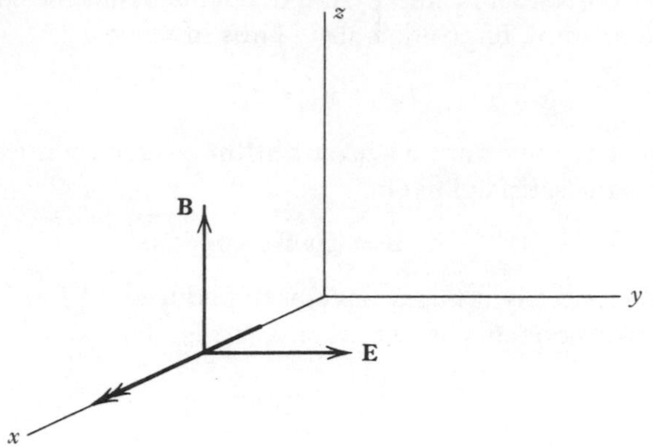

Figure 2.16

reason the disturbance is said to be *linearly polarized*. The superposition of a disturbance whose only non-zero field components are

$$E_z = -cB_y = G(t - x/c) \tag{2.130}$$

results in the field

$$\mathbf{E} = (0, F, G), \quad c\mathbf{B} = (0, -G, F), \tag{2.131}$$

in which the direction of each vector in general changes with position and time.

2.5.3 *Time harmonic fields*

A case of special importance for both theory and practice is that in which the field varies harmonically. For example (2.128) might be

$$E_y = cB_z = E_0 \cos[\omega(t - x/c)]. \tag{2.132}$$

This represents a travelling wave of frequency

$$f = \frac{\omega}{2\pi}, \tag{2.133}$$

and wavelength

$$\lambda = \frac{2\pi c}{\omega}, \tag{2.134}$$

with

$$f\lambda = c. \tag{2.135}$$

In the mathematical treatment it is more convenient to use the angular frequency ω rather than f, and also to introduce the *wave number*

$$k_0 = \omega/c = 2\pi/\lambda, \tag{2.136}$$

so that (2.132) is commonly written

$$E_y = cB_z = E_0 \cos(\omega t - k_0 x). \tag{2.137}$$

The range of frequencies, and correspondingly of wavelengths, that are encountered in electromagnetic phenomena is enormous, since all forms of radiation are embraced by the theory. The accompanying table indicates the physical significance of different regions of the frequency spectrum.

Frequency f (sec^{-1})	Designation	Wavelength λ (m)
10^{21}	gamma rays	3×10^{-13}
10^{19}	X-rays	3×10^{-11}
10^{17}	ultraviolet	3×10^{-9}
10^{15}	visible light	3×10^{-7}
10^{13}	heat	3×10^{-5}
10^{11}	radio waves: short	3×10^{-3}
10^{6}	radio waves: medium	3×10^{2}
10^{3}	radio waves: long	3×10^{5}

2.5.4 *Intimations of relativity*

The establishment of Maxwell's equations leaves the way open for the pursuit of their application to specific situations. In such applications they are treated as a set of partial differential equations whose solution presents a challenge to the mathematician and computer of broadly the same type as, say, that presented by the equations of hydrodynamics. However, it is also of great significance to the development of fundamental physical concepts that the acceptance of Maxwell's equations implies a breakaway from previous ideas of space and time. The new ideas crystallized in the theory first of 'special' then of 'general' relativity (1905, 1915), which are now in turn accepted as integral parts of physical knowledge.

The features of electromagnetic theory indicative of relativistic rather than Newtonian space–time concepts are the velocity dependence of the Lorentz force on a charged particle $e(\mathbf{E}+\mathbf{v}\wedge\mathbf{B})$, and the appearance of a velocity c of wave propagation. But are these novel? Are there not analogous features in Newtonian mechanics, such as a velocity dependent frictional force and the velocity of propagation of sound waves? There is a vital difference, namely that in the mechanical cases the velocities are relative to the medium in which the body or waves are travelling, whereas in the electromagnetic case the velocities are relative to the observer. The speed of a sound wave is different to observers in relative motion to each other; the speed of an electromagnetic wave is not.

Suppose that according to an observer O in a frame of reference S a charged particle e is moved with uniform velocity $\mathbf{u}$ in an electromagnetic field $\mathbf{E}$, $\mathbf{B}$. Then the electromagnetic force on it is

$$e(\mathbf{E}+\mathbf{u}\wedge\mathbf{B}). \qquad (2.138)$$

However, according to an observer O' in a frame of reference S' in which the particle is at rest, the corresponding force is

$$e'\mathbf{E}'. \tag{2.139}$$

If it is the case that measurements by O' give both the charge and the force the same values as measurements by O, then $e' = e$ and

$$\mathbf{E}' = \mathbf{E}+\mathbf{u}\wedge\mathbf{B}, \tag{2.140}$$

so that the fields measured by the two observers must differ. Apart possibly from a factor dependent only on $\mathbf{u}$, it can be argued that (2.140) is the only reasonable interpretation; for if u were very small the values of charge and force could not differ significantly between O and O', but $\mathbf{u}\wedge\mathbf{B}$ could nevertheless be comparable in magnitude to $\mathbf{E}$.

If the result corresponding to (2.140) for a charge with arbitrary velocity $\mathbf{v}$ in S is taken to be

$$\mathbf{E}+\mathbf{v}\wedge\mathbf{B} = \mathbf{E}'+(\mathbf{v}-\mathbf{u})\wedge\mathbf{B}',$$

it follows by substitution from (2.140) that $(\mathbf{v}-\mathbf{u})\wedge(\mathbf{B}'-\mathbf{B}) = 0$, or, since $\mathbf{v}$ is arbitrary,

$$\mathbf{B}' = \mathbf{B}. \tag{2.141}$$

The relation (2.140) is in fact approximately correct when $u^2/c^2 \ll 1$. Evidently, however, the reasoning presented does not preclude the right hand side having an extra factor which depends on $\mathbf{u}$ and tends to unity as $u \to 0$; and such a factor would in turn modify (2.141). To see that this is indeed likely to be the case, consider a point charge e at rest in S. To O its field is

$$\mathbf{E} = \frac{e}{4\pi\epsilon_0}\frac{\mathbf{r}}{r^3}, \quad \mathbf{B} = 0. \tag{2.142}$$

But relative to O' the charge has velocity $-\mathbf{u}$, and so (1.6) implies

$$\mathbf{B}' = -\frac{\mu_0 e}{4\pi}\frac{\mathbf{u}\wedge\mathbf{r}}{r^3} = -\frac{1}{c^2}\mathbf{u}\wedge\mathbf{E}. \tag{2.143}$$

This suggests that an improvement on (2.141) should be

$$\mathbf{B}' = \mathbf{B}-\frac{1}{c^2}\mathbf{u}\wedge\mathbf{E}. \tag{2.144}$$

The relations (2.140) and (2.144) are correct to within a factor whose difference from unity is of order u^2/c^2. To get them right requires a deeper investigation intimately linked with the fact that a wave travelling in vacuum appears to both O and O' to have the same speed c. This was essentially the result established experimentally by Michelson and Morley (*c.* 1880). It leads to a conclusion that can be expressed as follows.

Let S and S' be frames of reference in uniform relative motion with, for convenience, the specific choice that the velocity of S' relative to S is

$\mathbf{u} = (u, 0, 0)$, and the axes of the two frames coincide at $t = 0$. Then if an observer in S locates an 'event' at (x, y, z) at time t, an observer in S' locates the same event at (x', y', z') at time t', where

$$x' = \gamma(x - ut), \quad y' = y, \quad z' = z, \quad t' = \gamma(t - ux/c^2), \qquad (2.145)$$

with

$$\gamma = 1/(1 - u^2/c^2)^{\frac{1}{2}}. \qquad (2.146)$$

The relations (2.145) are known as the Lorentz transformation. The most striking feature is the inequality of t and t', which can be significant (for large enough x) even when $u/c \ll 1$.

The reasoning that leads to the transformation, and the development of its consequences, belong to the theory of special relativity, which is not pursued in this book. The one further point to be made here, again without proof, is that Maxwell's equations are invariant in form under the Lorentz transformation together with the transformations

$$J_x' = \gamma(J_x - \rho u), \quad J_y' = J_y, \quad J_z' = J_z, \quad \rho' = \gamma(\rho - uJ_x/c^2), \quad (2.147)$$

and

$$E_x' = E_x, \quad B_x' = B_x,$$

$$\mathbf{E}_\perp' = \gamma(\mathbf{E}_\perp + \mathbf{u} \wedge \mathbf{B}_\perp), \quad \mathbf{B}_\perp' = \gamma(\mathbf{B}_\perp - \mathbf{u} \wedge \mathbf{E}_\perp/c^2). \qquad (2.148)$$

The suffix $\perp$ indicates the component perpendicular to $\mathbf{u}$ (here, that in the y, z plane), and (2.148) evidently yields (2.140), (2.144) as approximations when $u^2/c^2 \ll 1$.

That Maxwell's equations accommodate the Lorentz transformation, without amendment, is further evidence of their correctness. In this respect they succeed where Newton's dynamical equations of motion fail. Thus any exact solution of Maxwell's equations is 'relativistic'. Moreover, any exact solution referred to one frame of reference can be referred to a frame in uniform relative motion by means of the transformations given (see problems 2.33 to 2.36).

Problems 2

2.1 N particles, each of charge e, are uniformly spaced on a circle with centre O and radius a. Find the potential ϕ and field $\mathbf{E}$ at points on the line through O normal to the plane of the circle.

2.2 Show that the field of a line charge (obtained from the inverse square law in problem 1.11) can be deduced from Gauss' theorem and symmetry considerations (see §3.1.5, (3.31)).

2.3 Verify directly that the potential (2.21), (2.22) of uniform charge density ρ confined to the sphere $r < a$ satisfies Poisson's equation $\nabla^2\phi = -\rho/\epsilon_0$. Sketch the variation of potential with r.

Where do the potential and field reach their maxima? Find the values of these maxima when $a = 10^{-15}$ m and the total charge is that of a proton.

2.4 Use Gauss' theorem to obtain the field, everywhere, of charge of uniform density ρ occupying the region $a < r < b$, where r is distance from the origin. Find the potential.

Consider the limit $b \to a$, $\rho \to \infty$, with $(b-a)\rho = \sigma$ remaining finite. Show that in this limit the electric field suffers a discontinuity of amount σ/ϵ_0 in crossing the layer of charge. What happens to the potential?

2.5 Verify directly that the dipole potential (2.26) satisfies Laplace's equation $\nabla^2\phi = 0$ $(r \neq 0)$.

2.6 Find the couple on an infinitesimal electric dipole of moment $\mathbf{p}$, of fixed magnitude but variable direction, moving with velocity $\mathbf{v}$ in a magnetic field $\mathbf{B}$.

Show that in circumstances in which (1.6) is valid the magnetic field generated by the moving dipole is

$$\frac{\mu_0}{4\pi r^3}\left[\frac{d\mathbf{p}}{dt} \wedge \mathbf{r} + \mathbf{p} \wedge \mathbf{v} + \frac{3(\mathbf{p}\cdot\mathbf{r})(\mathbf{v} \wedge \mathbf{r})}{r^2}\right].$$

2.7 An infinitesimal electric dipole of moment $\mathbf{p} = (0, 0, p)$ is located at the origin O, and another, of moment $-\mathbf{p}$, at $(0, 0, a)$. Show that at points P whose distance r from O is much greater than a the potential is

$$\frac{pa}{4\pi\epsilon_0}\frac{1-3\cos^2\theta}{r^3},$$

where θ is the angle OP makes with the z axis.

2.8 Two parallel line charges, $\pm e$ per unit length, vector distance $\mathbf{l}$ apart, where $l \to 0$, $e \to \infty$, with $\mathbf{p} = e\mathbf{l}$ remaining finite, constitute a *line dipole* of moment $\mathbf{p}$ per unit length. Find the potential and field.

2.9 Show that the electric field due to a charge distribution of density $\rho = \rho_0 e^{-k|z|}$, where ρ_0 and k are positive constants, is $\mathbf{E} = (0, 0, E)$, where

$$E = \frac{\rho_0}{\epsilon_0 k}\frac{|z|}{z}(1-e^{-k|z|}).$$

2.10 Show that current densities proportional to $(-y, x, 0)$ and $(x, y, -2z)$, respectively, represent steady current flow. Sketch the lines of flow in each case.

Find the most general form of the function $F(x, y, z)$ for which current density $(-yF, xF, 0)$ represents steady current flow.

2.11 At points on and above the earth's surface the earth's magnetic field is approximately the same as the field that would be produced (in vacuum) by a magnetic dipole of moment 0.81×10^{23} amp m^2 located at the centre of the earth and directed towards magnetic south. On this basis calculate the magnitude of the earth's field at the magnetic poles, at the magnetic equator and at magnetic latitude 45° N. Give also, in each case, the angle the field makes with the horizontal (known as the *dip* angle), and calculate the couple exerted on a horizontal wire loop of area 1 m^2 carrying a 10 amp current.

2.12 Charge q is uniformly distributed over the surface of a small sphere of radius a and rotates about a fixed diameter of the sphere with constant angular velocity ω. Show that the magnetic dipole moment is $\frac{1}{3}qa^2\omega$ directed along the fixed diameter.

2.13 Demonstrate explicitly that the direct application of the Biot–Savart law (2.55) to steady current I flowing in an infinitesimal plane loop gives a dipole field.

2.14 Show that the solid angle subtended at the vertex by the base of a right circular cone of semi-vertical angle α is $2\pi(1-\cos\alpha)$. Hence write down the magnetic scalar potential due to steady current I flowing in circular loop of radius a, at a point on the axis of the loop distance z from its centre. Derive the field at the centre of the loop, and confirm that the same result is given by the Biot–Savart law.

2.15 Use the Biot–Savart law to obtain the field at the centre of a wire loop, carrying steady current I, in the shape of a square of side a.

2.16 By using Ampère's circuital law in conjunction with symmetry arguments show that the current density $\mathbf{J} = (J, 0, 0)$, where $J = -J_0 e^{-k|z|}$ and k (>0) and J_0 are constants, gives rise to the magnetic field $\mathbf{B} = (0, B, 0)$, where

$$B = \frac{\mu_0 J_0}{k}\frac{|z|}{z}(1-e^{-k|z|}).$$

2.17 Confirm that the application of the Biot–Savart law to steady current I flowing along an infinite straight line is essentially the same analysis as that given in §1.3.1. Verify also that the result agrees with Ampère's circuital law (2.51) applied to a circle normal to and centred on the line current.

Use the circuital law to find the field of uniform current density $(0, 0, J)$ confined to the cylinder $r < a$, where r is the distance from the z axis (see §4.2.1, (4.18), (4.19)).

2.18 Show that the net couple on a spatially bounded steady current distribution $\mathbf{J}$ due to the field of a second distribution $\mathbf{J}'$ is

$$\mathbf{G} = \frac{\mu_0}{4\pi}\int\int\left(\frac{\mathbf{r}\wedge\mathbf{r}'}{R^3}\mathbf{J}.\mathbf{J}' + \frac{\mathbf{J}\wedge\mathbf{J}'}{R}\right)d\tau\, d\tau',$$

where R is the distance between volume element $d\tau$ at $\mathbf{r}$ and volume element $d\tau'$ at $\mathbf{r}'$.

Deduce that no steady current distribution exerts a net couple on itself.

2.19 Show that, for a magnetostatic field $\mathbf{B}$ with vector potential $\mathbf{A}$,

$$\int_S \mathbf{B}.d\mathbf{S} = \oint_C \mathbf{A}.d\mathbf{s},$$

where S is a surface spanning the loop C. Find a solenoidal vector potential ($\text{div}\,\mathbf{A} = 0$) for the uniform field $\mathbf{B} = (0, 0, B)$ confined to the cylinder $r < a$, where r is distance from the z axis.

2.20 A plane loop of wire, of area A, is rotated with constant angular velocity ω about a fixed line l in the plane of the loop. If there is a uniform magnetic field B perpendicular to l find the voltage V induced in the loop, neglecting the effect of the additional field due to the associated current I in the loop. If I is proportional to V find also the couple required to keep the loop rotating, and show that its rate of working is IV.

Evaluate the amplitude of V when $A = 10^{-2}$ m^2, $\omega = 10^2$ sec^{-1} and $B = 10^{-1}$ weber m^{-2}.

2.21 A uniform magnetic field $\mathbf{B} = (0, 0, B)$ is confined to the cylinder $r < a$, where r is distance from the z axis. The field is encircled by two wire loops in the plane $z = 0$, L_1 given by $r = b_1$ and L_2 given by $r = b_2$, where $b_2 > b_1 > a$. With L_1 fixed, L_2 is displaced in $z = 0$ until a point A on it makes contact with L_1. L_2 is then cut at A, and the two ends taken round L_1 in opposite senses. When the ends meet again they are rejoined, after which L_2 is further displaced so that it loses contact with L_1. Expose the fallacy in the following argument: 'In the initial state there is a magnetic flux $\pi a^2 B$ through L_2, and in the final state none. Thus a closed circuit is maintained throughout, in which there is a net change of magnetic flux. There must be an associated rate of change of flux, hence by Faraday's law a voltage, and hence some current flow.'

2.22 Two perfectly conducting metal plates in the form of equilateral triangles ABC and DEF, of side $2a$, lie without overlapping in a horizontal plane in such a way that $CBDF$ forms a rectangle of shorter side $b = a\sqrt{3}$, so that $AE = 3b$. The vertices A and E are joined (in the sense $ACFE$) by a semi-circular wire, of resistance R, in the same horizontal plane; and a circuit is completed by a straight wire, of length b and negligible resistance, that has one end L on CB, and the other end M on FD. Show that if a uniform magnetic field $B = B_0 \sin(\omega t)$, where B_0 and ω are constants, acts vertically downward through the system, and if the wire LM is moved with uniform velocity v so that, at time t, $CL = FM = a + vt$, for $-a/v \leqslant t \leqslant a/v$, then the current through the semi-circular wire is approximately

$$\frac{B_0 b}{R}\left[\left(\frac{9\pi}{8}b + 2vt\right)\omega \cos(\omega t) + v \sin(\omega t)\right] \quad (-a/v \leqslant t \leqslant a/v).$$

[Neglect the magnetic field of the current, and assume that current flow through the plates takes place along straight line paths.]

2.23 A conducting circular disc of radius a rotates about its axis with constant angular velocity ω. Initially the conduction electrons move outwards under the centrifugal force, and a steady state is reached when they are so distributed that they give rise to a counterbalancing electrostatic field. Show that the potential drop from the centre to the rim of the disc is then $m\omega^2 a^2/(2e)$, and find its value when $a = 10^{-1}$ m, $\omega = 10^2$ sec^{-1}.

2.24 A conducting circular disc of radius a rotates about its axis with angular velocity ω in a uniform magnetic field B parallel to the axis. Show that in the steady state the presence of the field gives rise to a potential difference $\frac{1}{2}a^2\omega B$ between the centre of the disc and the rim, and find its value when $a = 10^{-1}$ m, $\omega = 10^2$ sec^{-1}, $B = 10^{-1}$ weber m^{-2}.

2.25 A conducting rod $\mathbf{l}$ is translated with constant velocity $\mathbf{v}$ in a constant, uniform magnetic field $\mathbf{B}$. Show that an electric field is created, the line integral of which along the rod, when the steady state obtains, is $-\mathbf{l}.(\mathbf{v} \wedge \mathbf{B})$. What assumption is made in calling this quantity the 'potential drop' along the rod?

A car travels at 100 km hour^{-1} over flat ground in the northern hemisphere at a place where the earth's magnetic field is 0.3 gauss and the angle of dip 66°. What is the magnitude and direction of the potential drop along a bumper of length 2 m?

2.26 Show that for the general motion (including deformation) of a simple closed curve C spanned by a surface S,

$$-\frac{d}{dt}\int_S \mathbf{B}.d\mathbf{S} = \oint_C (\mathbf{E} + \mathbf{v} \wedge \mathbf{B}).d\mathbf{s}$$

where $\mathbf{v}$ is the velocity of the element $d\mathbf{s}$ of C.

2.27 A charged particle moves under the influence of a field for which

$$\mathbf{B} = [0, 0, B(r, t)],$$

where r denotes distance from the z axis. Show that the particle can be held in a circular orbit of fixed radius, in which its speed is increased by the action of the electric field associated with the varying magnetic field, provided the average value of the magnetic field inside the orbit is twice the field at the orbit.

2.28 Show that, for an electromagnetic field in a region free of charge and current, if E_z and B_z are identically zero, then the remaining cartesian components E_x, E_y, B_x and B_y each necessarily satisfy both the equations

$$\frac{\partial^2 \psi}{\partial z^2} = \frac{1}{c^2}\frac{\partial^2 \psi}{\partial t^2}, \quad \frac{\partial^2 \psi}{\partial x^2} + \frac{\partial^2 \psi}{\partial y^2} = 0.$$

2.29 Write down the charge conservation relation when current $I(z, t)$ flows along the z axis and $q(z, t)$ is the charge per unit length of the axis. If, in cylindrical polar coordinates r, θ, z, the only non-zero components of the corresponding field $\mathbf{E}$, $\mathbf{B}$ are E_r, B_θ, show from the integral form of Maxwell's equations (2.106) and (2.107) that

$$B_\theta = \frac{\mu_0 I}{2\pi r}, \quad E_r = \frac{q}{2\pi\epsilon_0 r}.$$

Deduce that all Maxwell's equations are satisfied by this field provided

$$\frac{1}{c^2}\frac{\partial I}{\partial t} + \frac{\partial q}{\partial z} = 0,$$

and that then q and I both satisfy the wave equation

$$\frac{\partial^2 \psi}{\partial z^2} = \frac{1}{c^2}\frac{\partial^2 \psi}{\partial t^2}.$$

2.30 Show that in general the field described in problem 2.29 is represented by

$$I = f(z-ct)+g(z+ct), \quad cq = f(z-ct)-g(z+ct),$$

where f and g are arbitrary functions; and that $E_r = cB_\theta$ when g is identically zero.

What are f and g in the particular cases (*a*) $I = I_0$, $q = 0$, (*b*) $I = 0$, $q = q_0$, where I_0 and q_0 are independent of z and t?

2.31 Show that for charge density $\rho(z, t)$ and current density $[0, 0, J(z, t)]$ all Maxwell's equations are satisfied by the purely electric field $\mathbf{E} = [0, 0, E(z, t)]$, where

$$\frac{\partial E}{\partial z} = \rho/\epsilon_0, \quad \frac{\partial E}{\partial t} = -J/\epsilon_0.$$

Deduce that to any time independent solution $\rho = f'(z)$, $J = 0$, $E = (1/\epsilon_0)f(z)$ there corresponds a solution $\rho = f'(z)\,g(t)$, $J = -f(z)\,g'(t)$, $E = (1/\epsilon_0)f(z)\,g(t)$, where the function g is arbitrary.

Verify that

$$J = v\rho = v\rho_0 \mathrm{e}^{-k|z-vt|}, \quad E = \frac{\rho_0}{\epsilon_0 k}\frac{|z-vt|}{z-vt}(1-\mathrm{e}^{-k|z-vt|}),$$

where k (> 0), v and ρ_0 are constants, is a solution.

2.32 Show that for current density $\mathbf{J} = [J(z, t), 0, 0]$, with zero charge density, all Maxwell's equations are satisfied by the field $\mathbf{B} = [0, B(z, t), 0]$, $\mathbf{E} = [E(z, t), 0, 0]$, where

$$\frac{\partial E}{\partial z} = -\frac{\partial B}{\partial t}, \quad -\frac{\partial B}{\partial z} = \frac{1}{c^2}\frac{\partial E}{\partial t} + \mu_0 J.$$

Verify that

$$J = -J_0 e^{-k|z-vt|}, \quad E = vB = \frac{\mu_0 J_0 v}{k(1-v^2/c^2)} \frac{|z-vt|}{z-vt}(1-e^{-k|z-vt|}),$$

where k (>0), v and J_0 are constants, is a solution.

2.33 Show that the displayed solutions to Maxwell's equations in problems (2.31) and (2.32) are given by the transformations (2.145), (2.147) and (2.148) applied to the solutions of problems (2.9) and (2.16), respectively.

2.34 (*a*) Show that (2.145) gives

$$x = \gamma(x' + ut'), \quad t = \gamma(t' + ux'/c^2).$$

(*b*) By considering uniform charge density in a rectangular box show that the measured net charge has the same value in S and S'.

2.35 Review problem 1.11 in the light of the relativistic transformations.

2.36 Use the relativistic transformations to obtain the exact field of a point charge e moving along the x axis with uniform speed v.

3

ELECTROSTATICS

3.1 Field and potential

3.1.1 *Field lines and equipotential surfaces*

The broad theory of electromagnetism has been set out in Chapter 2, and the task is now to develop various aspects of this theory. It is natural to begin with electrostatics, building on the foundations laid in § 2.1.

Much of the language of the theory is couched in terms of a pictorial representation of the field which it is convenient to explain at this stage. The picture is particularly simple in time independent situations. In electrostatics it consists essentially in envisaging space pervaded by a network of *field lines*, these being defined by the statement that at each point on a line the tangent is in the direction of the vector **E**. The alternative term *lines of force* is sometimes used, since **E** gives the force on a unit point charge.

A very simple example is afforded by a single point charge. Here the lines are radial from the charge, as shown in figure 3.1. Arrows can be used to indicate the sense of direction of **E**. The figure is drawn for a positive charge, with the arrows pointing outwards.

As another example, consider the field of an infinitesimal dipole. This is given analytically by (2.29) and (2.30). Since

$$E_r/E_\theta = 2\cos\theta/\sin\theta \tag{3.1}$$

the differential equation of the field lines in any plane containing the line of the dipole is

$$\frac{1}{r}\frac{\mathrm{d}r}{\mathrm{d}\theta} = 2\frac{\cos\theta}{\sin\theta} \tag{3.2}$$

which integrates to

$$r = a\sin^2\theta. \tag{3.3}$$

Here, a is the constant of integration, and each field line is given by assigning a particular value to a. The lines are shown in figure 3.2.

In general the electric field has a definite direction at each point of the region of space that it pervades, implying that field lines do not intersect or touch one another. However, the direction of **E** is ambiguous at the location of point charge singularities, such as those in figures 3.1 and 3.2; and also at *neutral points*, which are isolated points where the electric field is zero. Figure 3.3 shows the field lines for two equal positive point charges: a neutral point is located midway between the two charges.

There is a neat equation for the field lines of a set of collinear point charges $e_1, e_2, \ldots, e_n$. Consider a segment d**s** of field line through any point P.

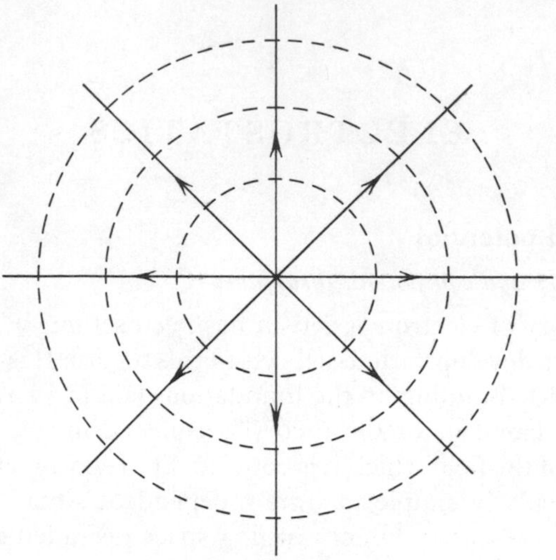

Figure 3.1

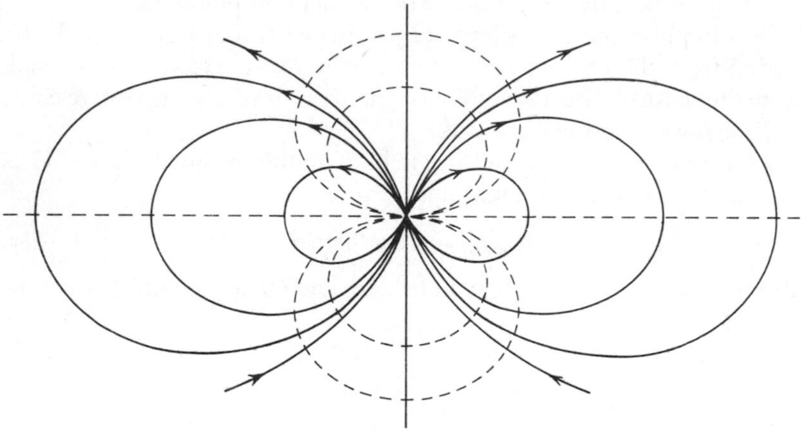

Figure 3.2

Then the expression of the fact that **E** at P has no component normal to d**s** is, from the inverse square law,

$$\sum_{i=1}^{n} \frac{e_i}{r_i^2} \sin \phi_i = 0, \tag{3.4}$$

where r_i is the distance from e_i to P, and ϕ_i is the angle that the radius vector from e_i to P makes with d**s** (figure 3.4). But

$$\frac{1}{r_i^2} \sin \phi_i = \frac{1}{r_i} \frac{d\theta_i}{ds} = \frac{\sin \theta_i}{p} \frac{d\theta_i}{ds}, \tag{3.5}$$

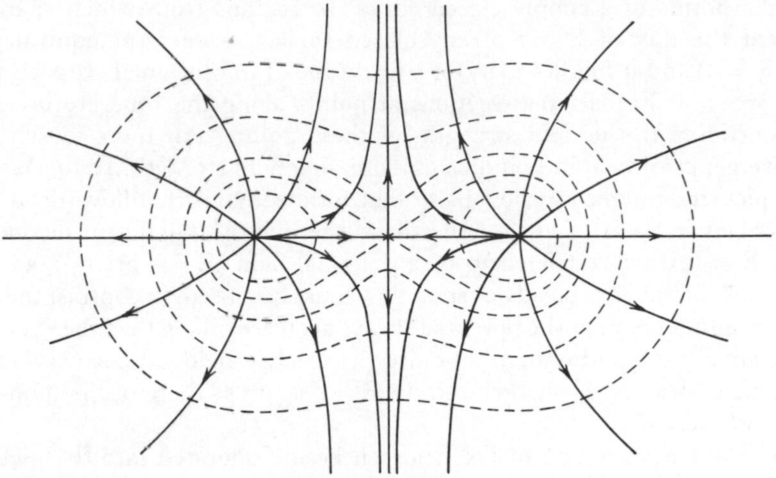

Figure 3.3

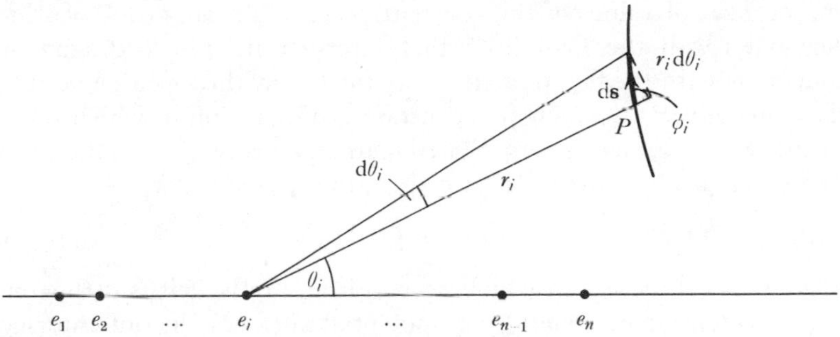

Figure 3.4

where p is the perpendicular distance from P to the line of charges. Hence (3.4) is

$$\sum_{i=1}^{n} e_i \sin\theta_i \, d\theta_i = 0, \tag{3.6}$$

which integrates to

$$\sum_{i=1}^{n} e_i \cos\theta_i = \text{constant}. \tag{3.7}$$

A specific application of (3.7) is the determination of the relation between the direction in which a field line leaves one of the point charges and the direction to which it tends asymptotically at infinity. At a particular charge e_j only $\theta_j = \theta_{0j}$ is unknown, each other θ_i being either zero or π; whereas at infinity all the θ_i are equal, say to $\theta_{\infty j}$ for the field line through e_j. Then (3.7) gives the relation between θ_{0j} and $\theta_{\infty j}$.

A *field tube* (or *tube of force*) is the tube bounded by the field lines through all the points of a simple closed curve. In regions from which charge is absent the flux of **E** out of any closed surface is zero (or equivalently, $\operatorname{div}\mathbf{E} = 0$) and it follows that, for a field tube of infinitesimal cross-section, the areas of normal cross-sections at points along the tube are inversely proportional to the field strengths at these points. Put more loosely, the convergence of field lines indicates increasing field strength. In this respect the picture is indeed analogous to that for steady current flow ($\operatorname{div}\mathbf{J} = 0$) described in § 2.2.1. On the other hand there is the significant distinction that **E** is further restricted to be a potential field ($\mathbf{E} = -\operatorname{grad}\phi$), so that, whereas the tubes of steady current flow must be closed, in contrast the field tubes cannot be, since the potential always decreases along the tube. There is, of course, no paradox here, because the electric field cannot exist in the complete absence of charge, and $\operatorname{div}\mathbf{E} = \rho/\epsilon_0$ must therefore be non-zero somewhere.

The potential nature of the field can be incorporated into the pictorial representation by the recognition that the *equipotentials*

$$\phi = \text{constant}$$

form, for a set of values of the constant, a set of surfaces each of which is orthogonal to all the field lines that intersect it. The orthogonality is synonymous with the fact that, at any point P, any direction tangent to the surface through P on which ϕ is constant is a direction in which the component of $\mathbf{E} = -\operatorname{grad}\phi$ is zero. Stated mathematically, if $\delta\phi$ is the excess of the potential at $\mathbf{r}+\delta\mathbf{r}$ over that at $\mathbf{r}$, then, to the first order,

$$\delta\phi = \delta\mathbf{r}\,.\operatorname{grad}\phi, \tag{3.8}$$

so that $\delta\mathbf{r}$ is a direction in which $\delta\phi = 0$ if and only if it is orthogonal to $\operatorname{grad}\phi$. Evidently, in general, equipotential surfaces do not intersect or touch one another.

The equipotentials for a single point charge are, of course, spherical surfaces centred on the charge (see figure 3.1). For an infinitesimal dipole, for which the potential is (2.26), they are given by the equation

$$r^2 = k\cos\theta, \tag{3.9}$$

each surface being specified by a particular value of k. A plane section of the surfaces is shown in figure 3.2: it is easy to verify analytically the orthogonality of (3.3) and (3.9).

In the case of two equal point charges (figure 3.3) the role of the neutral point in the transition from equipotential surfaces that enclose one of the charges to those that enclose both may be noted.

The outcome of the present observations is a picture of the electrostatic field as a honeycomb of small cells, in which each cell is the portion of a narrow field tube confined between a pair of neighbouring equipotential surfaces (see figure 3.5). Pictures of this kind are, of course, only adjuncts

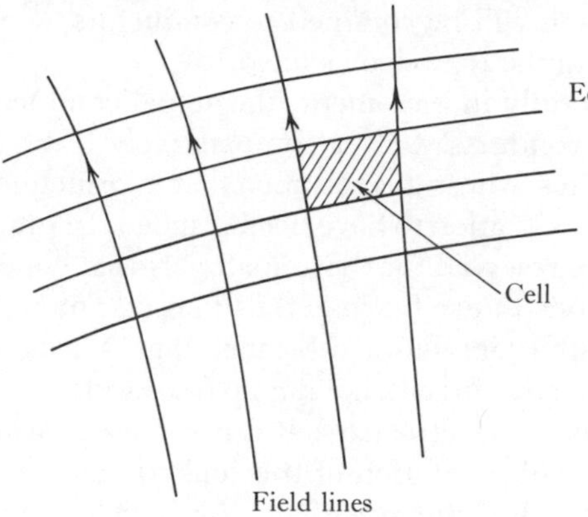

Figure 3.5

to the mathematical analysis; nevertheless, as will be seen, they may well help towards an understanding and interpretation of the physical concepts. The next step is to show the application of the theory in a more realistic context, in which the charges involved are density distributions on conductors.

3.1.2 *Conductors*

In Chapter 1 it was pointed out that a vast quantity of electricity exists in matter in the form of electrons and protons. For the most part, however, the positive and negative charges are so closely locked that they fail to give rise to macroscopic electromagnetic effects. For some materials, though, the atomic structure is such that a significant fraction of the electrons can move comparatively freely. By means of a battery a steady current could be made to flow through such material, which is called a conductor.

As well as providing a flow of current the free electrons in a conductor can also readily give rise to static charge distributions. Moreover, it is easy to arrange for a conductor to be charged, that is, carry a net charge: either electrons are conducted away from the conductor, leaving it positively charged, or conducted to it, leaving it negatively charged. The physics of electrostatics is essentially concerned with the relation between static charge distributions on conductors, and the associated electric field. It is not possible for a charge distribution to be in static equilibrium under Coulomb forces alone, and achievement of equilibrium is due to the fact that atomic binding forces are operative within the conducting medium to maintain the relative positions of the protons and non-conduction electrons. This is not to say that there could not be some deformation of the conductor. There certainly are circumstances of electrostatic equilibrium in which a non-rigid conductor may be distorted by the electric force. However this sort of complication involves considerations outside the basic theory of electricity, and here the discussion will be confined to conductors, such as a block of metal, whose shape can be regarded as fixed.

As is argued more fully in a moment, the actual conducting ability of a material, whether it conducts well or comparatively feebly, is not strictly relevant to electrostatics, where it is assumed that in principle sufficient time has elapsed for a static situation to have been reached. In practice, of course, the term conductor is reserved for those materials that conduct reasonably well. Other descriptions in use for materials that do not are *semi-conductor* and *insulator*. The latter signifies a substance that is sensibly without the ability to conduct a steady current, having no free electrons. However it too plays an important role in electrostatics if it has marked *polarizability*; that is to say, if the degree of separation of the constituent positive and bound negative charges under the influence of an electric field is sufficient itself to give rise to a not negligible field. A simple treatment of such *dielectrics* is given in § 3.4, but the rest of this chapter is confined solely to charges on conductors in what is otherwise in effect a vacuum.

In claiming that the following theory does give an adequate account of experience the further point should also be made that the electric fields are assumed not to be so strong that they could either drag electrons out of the conductors, or cause an electrical breakdown of the surrounding medium. If this medium were air at normal temperature and pressure, which is not significantly different from a vacuum as regards ordinary electrical effects, then to avoid breakdown the electric fields would have not to exceed about 3×10^6 volt m^{-1}. With regard to the emission of electrons from conductors, only thermionic and photoelectric processes are of general practical significance: they are not discussed in this book, apart from an oblique reference in § 6.2.5.

Suppose a conducting body is placed in a given electrostatic field, whose sources are so far distant that they remain unaffected. Initially the field exerts a force on the protons and electrons in the conductor, and the conduction electrons move under its influence. The motion of the conduction electrons sets up a charge density distribution, negative in regions where the electrons accumulate and positive where they leave a deficit. The charge density itself gives rise to an electric field, superposed on the given field, and the motion of the conduction electrons is modified accordingly. Very soon (indeed, in about 10^{-18} sec for a good conductor) a macroscopic equilibrium state is reached in which the charge density and the electric field are static, and in which there is necessarily (see § 2.3.1) no current flow. It is this state that is analysed in the theory of electrostatics.

Now in the equilibrium state, since there is no current flow, the *electric field must be zero throughout the body of the conductor*. This is the key statement that furnishes the mathematical description of the effect of a conductor. It is equivalent to saying that all points of a conductor are at the same potential; in particular, the *boundary of a conductor is an equipotential surface*.

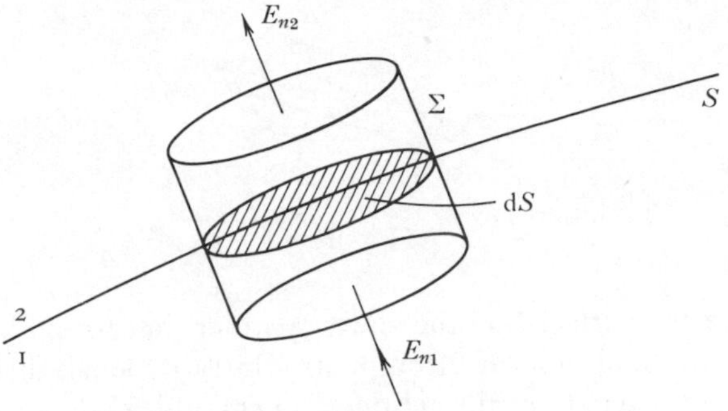

Figure 3.6

What are the consequences of this statement? One is at first sight paradoxical. If the total field throughout the conductor is zero then so also is the charge density, by virtue of the equation $\operatorname{div}\mathbf{E} = \rho/\epsilon_0$. Where then is the induced charge density, arising from the redistribution of conduction electrons, necessary to create the field that, throughout the conductor, precisely annuls the given field in which the conductor is placed?

The answer to the apparent paradox is that the *charge density is confined solely to the surface of the conductor*.

The concept of surface charge density has not previously been introduced. It is a mathematical idealization, implying an infinite volume charge density, associated in the present context with the idealized model of a conductor. In practice, though, the charge layer in the neighbourhood of the surface of a conductor is indeed only a few molecules thick; so the adoption of the idealized concept is entirely justified in a macroscopic theory. In view of the importance of the concept it is worth interrupting the discussion of conductors in order to explain the effect of the physical singularity that is implied.

3.1.3 *Surface charge*

Suppose that charge is distributed over a surface S, and that each infinitesimal element of S, of area $\mathrm{d}S$, carries a charge $\sigma\,\mathrm{d}S$. Then σ is the surface charge density. In general it is a function of position on the surface, and of time; but in the present context it is assumed to be time independent. Naturally, similar considerations to those of § 1.2.1. apply equally to the concept of surface density, and need not be repeated here.

Since it is supposed that a finite charge occupies zero volume, the volume charge density is infinite. The effect of this singularity on the behaviour of the electric field in the vicinity of the surface can be diagnosed from the integral form of Gauss' theorem and from the conservative nature of the field.

In figure 3.6 the effectively plane element $\mathrm{d}S$ of S is conceived to be

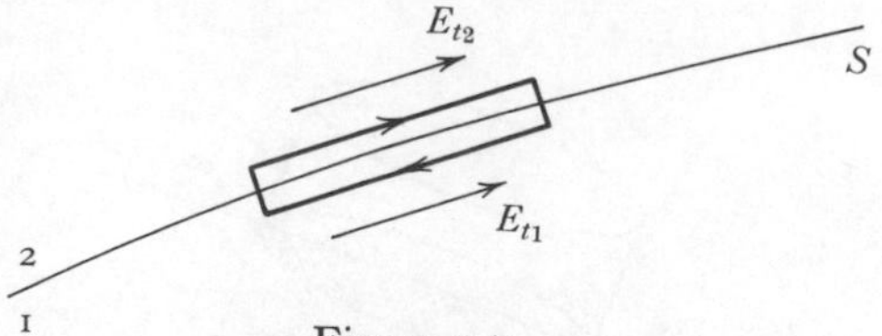

Figure 3.7

enclosed by the surface Σ of the short cylinder that would be generated by moving $\mathrm{d}S$ normal to itself through an arbitrarily small distance on either side of S. If the length of the cylinder is negligible compared with the ratio of the area $\mathrm{d}S$ to its perimeter, then the flux of $\epsilon_0\mathbf{E}$ out of Σ is in effect the flux across the end faces, namely

$$\epsilon_0(E_{n2}-E_{n1})\,\mathrm{d}S,$$

where the suffixes 1 and 2 denote evaluation just on one side or the other of $\mathrm{d}S$, and the suffix n signifies the component normal to S directed from 1 to 2. Identification of this flux with the total charge $\sigma\,\mathrm{d}S$ inside Σ gives

$$[\mathbf{n}.\mathbf{E}] \equiv E_{n2}-E_{n1} = \sigma/\epsilon_0, \tag{3.10}$$

using square brackets to signify the jump in the quantity embraced. Thus, in crossing a charged surface the normal component of $\epsilon_0\mathbf{E}$ jumps by the amount of the surface charge density.

The behaviour of the tangential component of $\mathbf{E}$ is obtained by considering an elementary rectangular circuit whose longer sides, of length $\mathrm{d}s$, are on either side of S and effectively parallel to it (figure 3.7). If the shorter sides of the rectangle are negligible compared with $\mathrm{d}s$ then the line integral of $\mathbf{E}$ round the rectangle is in effect

$$(E_{t2}-E_{t1})\,\mathrm{d}s,$$

where the suffix t signifies the component in the direction tangential to S that was arbitrarily selected. Since the line integral of $\mathbf{E}$ round any closed curve is zero it follows that the tangential components of $\mathbf{E}$ are continuous across a charged surface: vectorially,

$$\mathbf{E}_{t1} = \mathbf{E}_{t2} \quad \text{or} \quad [\mathbf{n}\wedge\mathbf{E}] = 0. \tag{3.11}$$

The combination of (3.10) and (3.11) can be expressed in the single statement

$$[\mathbf{E}] = \sigma\mathbf{n}/\epsilon_0. \tag{3.12}$$

It shows that field lines are refracted when they cross a charged surface: the field just on one side, $\mathbf{E}_1$, differs in direction from that just on the other side,

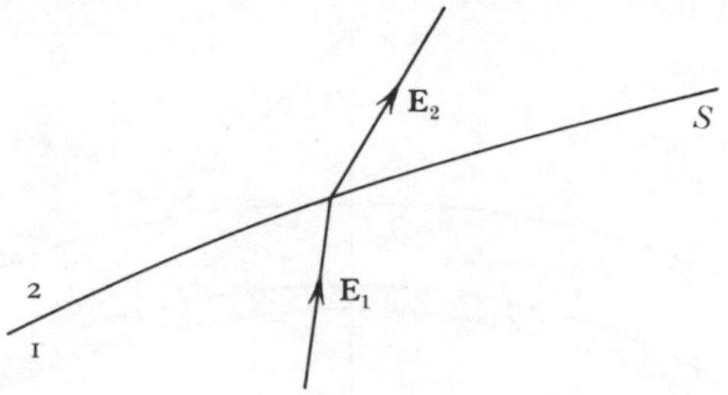

Figure 3.8

$\mathbf{E}_2$, as in figure 3.8. This poses an interesting and important problem. What is the force per unit area on a charged surface? To say that it is σ times the electric field at the surface is no help unless the ambiguity with regard to the field 'at the surface' is resolved. One way of obtaining the answer is as follows.

Consider an element Σ of S small enough to be treated as a plane circular element over which σ is uniform. Let $\mathbf{E}'$ be the field generated solely by the charge on Σ, and $\mathbf{E}''$ that generated by all other charge. Let $\mathbf{E}_1'$ and $\mathbf{E}_2'$ denote the limits of $\mathbf{E}'$ as the centre of Σ, P say, is approached from side 1 or side 2 of S respectively. Then, by symmetry,

$$\mathbf{E}_1' + \mathbf{E}_2' = 0, \tag{3.13}$$

and the force per unit area on an infinitesimal sub-element at P is

$$\sigma \mathbf{E}_P'', \tag{3.14}$$

where $\mathbf{E}_P''$ is the value of $\mathbf{E}''$ at P. Now the total field is

$$\mathbf{E} = \mathbf{E}' + \mathbf{E}'', \tag{3.15}$$

and its values on either side of the surface at P are

$$\mathbf{E}_1 = \mathbf{E}_1' + \mathbf{E}_P'', \quad \mathbf{E}_2 = \mathbf{E}_2' + \mathbf{E}_P''. \tag{3.16}$$

The application of (3.13) and (3.16) then shows (3.14) to be

$$\tfrac{1}{2}\sigma(\mathbf{E}_1 + \mathbf{E}_2). \tag{3.17}$$

Thus the average of $\mathbf{E}_1$ and $\mathbf{E}_2$ gives the force field on the surface charge.

General expressions for the field and potential due to a surface charge distribution can, of course, be written down. Analogous to (2.1),

$$\mathbf{E} = \frac{1}{4\pi\epsilon_0}\int \frac{\sigma \mathbf{R}}{R^3}\,\mathrm{d}S; \tag{3.18}$$

and analogous to (2.10),

$$\phi = \frac{1}{4\pi\epsilon_0}\int \frac{\sigma}{R}\,\mathrm{d}S; \tag{3.19}$$

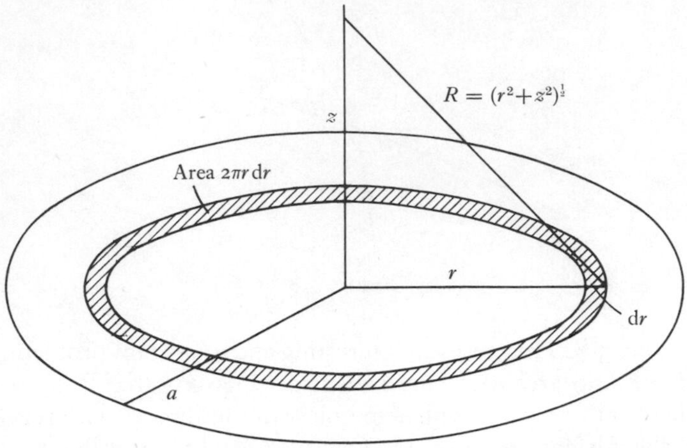

Figure 3.9

where in each case the integral is over the charged surface S, and $\mathbf{R}$, of magnitude R, is the vector from the element dS to the field point P.

The integral (3.19) can be evaluated at a point P in S. For let coordinates in the surface be adopted that are equivalent as $R \to 0$ to plane polar coordinates R, θ. Then $dS \sim R\,dR\,d\theta$ as $R \to 0$, so the singularity at $R = 0$ in the integrand of (3.19) is removed and ϕ has a unique value at P.

Thus the potential is continuous across a charged surface. This implies that the tangential components of $\mathbf{E}$ are continuous, as has already been found.

The points made in this discussion of surface charge can be instructively illustrated by a simple explicit calculation.

Suppose σ is uniform over a plane circular area S of radius a, so that, scale being of no significance, the field is in fact that written $\mathbf{E}'$ in the argument leading to (3.17). Then it is easy to evaluate the potential and field on the axis of S. For example, on the axis distance z from the centre on either side of S (3.19) gives (see figure 3.9)

$$\phi = \frac{1}{4\pi\epsilon_0}\int_0^a \frac{2\pi\sigma r}{(r^2+z^2)^{\frac{1}{2}}}\,dr; \tag{3.20}$$

that is

$$\phi = \frac{\sigma}{2\epsilon_0}[(a^2+z^2)^{\frac{1}{2}} - z]. \tag{3.21}$$

$\mathbf{E}$ at the same point is directed along the axis, away from S assuming $\sigma > 0$, and has magnitude

$$E = -\frac{\partial\phi}{\partial z} = \frac{\sigma}{2\epsilon_0}\left[1 - \frac{z}{(a^2+z^2)^{\frac{1}{2}}}\right]. \tag{3.22}$$

It may be noted that (3.22) is $\sigma/(4\pi\epsilon_0)$ times the solid angle subtended by S at the field point, in evident agreement with (3.18).

The continuity of ϕ at $z = 0$, and the jump in the normal component of $\epsilon_0 \mathbf{E}$ are shown explicitly in (3.21) and (3.22).

3.1.4 *More about conductors*

It was pointed out in § 3.1.2 that in a purely electrostatic situation there can be no current flow and hence no field within the body of a conductor: the conductor is at uniform potential, carries no volume charge distribution, but has a charge distribution on its surface. These conclusions evidently hold whether or not the conductor is charged. Of course, if it is uncharged there will be no field at all without an externally maintained field.

What can be said of the electrostatic field outside a conductor? Since the boundary of a conductor is an equipotential surface, **E** at the surface must be normal to it. This is palpably consistent with the facts that **E** is zero within the conductor and that the tangential components of **E** are continuous across the charged surface. Moreover, the corresponding discontinuity in the normal component means that, just outside the surface,

$$\mathbf{E} = (\sigma/\epsilon_0)\,\mathbf{n}, \tag{3.23}$$

where **n** is the unit vector along the outward normal (cf. 3.12)).

As will be seen later, it is commonly the case in the mathematical treatment of electrostatic problems that the primary calculation is of the potential or field, and not of the charge density. However, when the field is known the charge density on a conductor is found at once from the value of the field at the surface. Moreover, the force per unit area acting on the surface of a conductor is similarly available: the application of (3.17), with $\mathbf{E}_1 = 0$, $\mathbf{E}_2 = (\sigma/\epsilon_0)\,\mathbf{n}$ gives the force per unit area as

$$\tfrac{1}{2}\sigma^2/\epsilon_0 = \tfrac{1}{2}\epsilon_0 E^2 \tag{3.24}$$

along the outward normal. Since the charge is bound to the conductor this traction is exerted on the conductor, the total force on which is given by integration over the surface.

Turning now to the representation described in § 3.1, the picture of a conductor in an electrostatic field is illustrated in figure 3.10: (*a*) shows an uncharged conductor in an externally maintained field and (*b*) a positively charged conductor with no externally maintained field; the full lines are the field lines, the dashed lines indicate the equipotential surfaces, and ± signs indicate the sign and density of the surface charge.

The blank space within the boundary of the conductor emphasizes that it is immaterial to the field outside the boundary whether the physical region inside the boundary is occupied by conducting matter or is empty. The entire electric effect of the conductor is accounted for by the charge distribu-

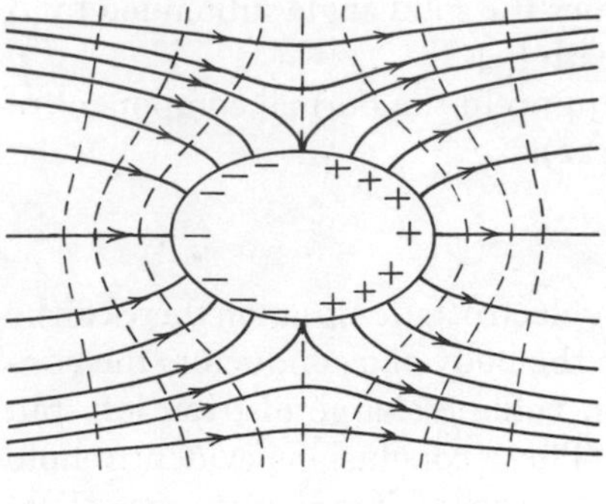

(a)

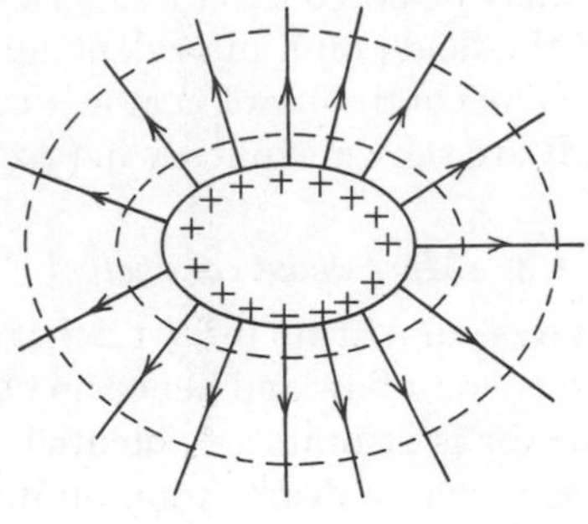

(b)

Figure 3.10

tion on its surface. This charge shields the interior from any electric field, and neutral matter in the interior will neither be affected nor produce any effect.

A hollow conductor consists of a *conducting sheet* in the form of a closed surface. A conducting sheet is conceived ideally as a conductor of negligible thickness which nevertheless retains its full shielding property; it is in effect a charged surface on which the charge is so distributed that the surface is an equipotential. In the calculation of an electrostatic problem the presence of a conductor is treated simply as the specification of an equipotential surface. Moreover, there is a fruitful converse to this statement, namely that any complete equipotential surface can be replaced by an uncharged conducting sheet without affecting the field. Thus in figure 3.10(*b*), for example, any of the dashed lines representing equipotential surfaces could be taken to be conducting sheets forming hollow conductors enclosing the original conductor.

It is worth examining this point in more detail. Suppose one of the equipotential surfaces in figure 3.10(*b*) is taken to be a conducting sheet. It is useful in this particular context to regard the sheet itself as having an exterior surface and an interior surface. The exterior surface, which the field lines leave along the outward normals, carries a positive charge, and the interior surface a negative charge. The situation is depicted in figure 3.11, where the thickness of the sheet is exaggerated for the sake of clarity. The positive and negative charge densities have the same magnitude at corresponding points on the two surfaces, and so cancel in combination. Nevertheless, their individual roles are most significant for the following reason. The field outside the conducting sheet on the one hand, and the field between the conducting sheet and the original conductor on the other, have entirely separate existences; the source of the external field is the positive charge density on the exterior surface of the conducting sheet, whereas that of the field in the bounded region is the negative charge density on the interior

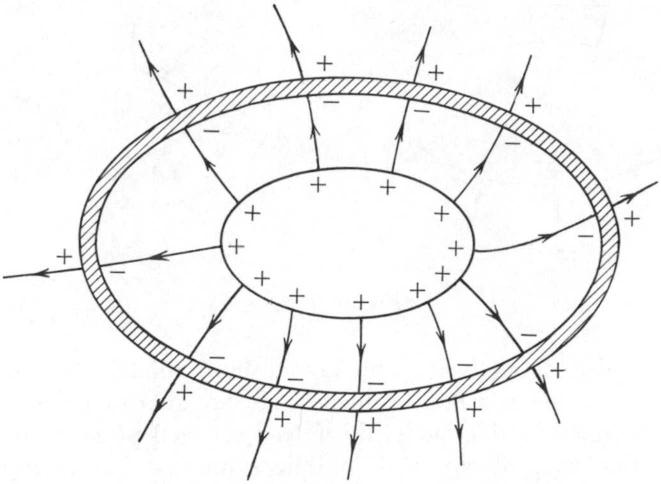

Figure 3.11

surface of the conducting sheet, together with the positive charge density on the surface of the original conductor.

Thus the field of figure 3.11 can be resolved as the separate fields depicted in figure 3.12, (*a*) and (*b*) respectively. Note that in figure 3.12(*b*) the total negative charge on the outer conductor is, of course, equal to the total positive charge on the inner conductor. More specifically, the corresponding charges on the terminal cross-sections of a narrow field tube proceeding from the inner to the outer conductor are equal, the charge densities being inversely proportional to the cross-section areas.

At the end of § 3.1.1 attention was drawn to the honeycomb picture of a region pervaded by an electrostatic field, illustrated in figure 3.5. It has now been seen that this picture can be supplemented by supposing that complete equipotential surfaces are conducting sheets, and that between each pair of adjacent sheets the field is generated independently by the charges on the mutually facing surfaces of the sheets.

All these statements are in conformity both with experiment and with deeper mathematical analysis. Experimentally it is found, for example, that there is no field inside an empty hollow conductor; indeed, testing this result is the essence of the indirect method that has established the inverse square law with such accuracy. Again, the independence of the fields shown in figure 3.11 could be confirmed by joining the conductors with a wire: this would have the effect of bringing the entire region inside the outermost conducting surface to a uniform potential, the disappearance of the field there being associated with the discharging process, which would leave the outer conductor with the positive charge on its outer surface.

Where theory is concerned the discussion cannot be faulted in so far as

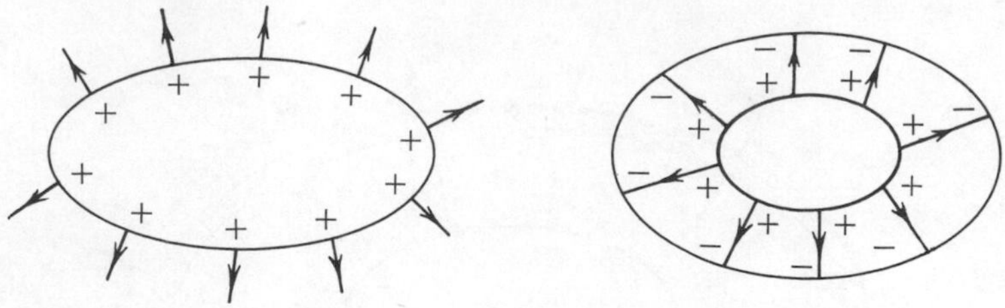

(a) (b)

Figure 3.12

it has proceeded deductively from an idealized model of a conductor. However, for the theory to be soundly based it should be established that the conditions implicit in this model do indeed set well posed mathematical problems; that is, problems with unique solutions. For example, with reference to the region between the two closed surfaces in figure 3.12(*b*), is there, to within an arbitrary constant, just one potential function satisfying $\nabla^2\phi = 0$ that is constant on each surface and consistent with the specified charge? The existence and uniqueness of solutions to such problems can be established. Proof of uniqueness is given in § 3.3, but the question of existence is a quite sophisticated part of the mathematics of potential theory, and well beyond the scope of this book.

3.1.5 *Capacitors*

Some simple examples serve to illustrate the foregoing ideas, and also to introduce the concept of *capacity* (or *capacitance*), which is the measure of the ability of a pair of conductors to store separated positive and negative electric charge.

Consider a uniform electric field; that is, one that has the same magnitude and direction throughout all space. This clearly satisfies the fundamental equations $\operatorname{div}\mathbf{E} = 0$ and $\operatorname{curl}\mathbf{E} = 0$. The equipotential surfaces are planes normal to $\mathbf{E}$, and if $\mathbf{E}$ is in the negative z direction

$$\phi = Ez + \phi_0$$

where ϕ_0 is an arbitrary constant.

Suppose that two of the equipotentials are taken to be conducting sheets. Then the situation is as depicted in figure 3.13, where the charge densities on the surface are, of course, uniform with magnitude $\sigma = \epsilon_0 E$. Isolating the inner region of figure 3.13 gives figure 3.14. Thus for two infinite, plane, parallel conducting sheets, carrying surface charge densities $\pm\sigma$ respectively, the electric field exists only between the sheets, where it is uniform, normal to the sheets, and of magnitude σ/ϵ_0. If the sheets are distance d apart their potential difference is

$$V = Ed = d\sigma/\epsilon_0. \tag{3.25}$$

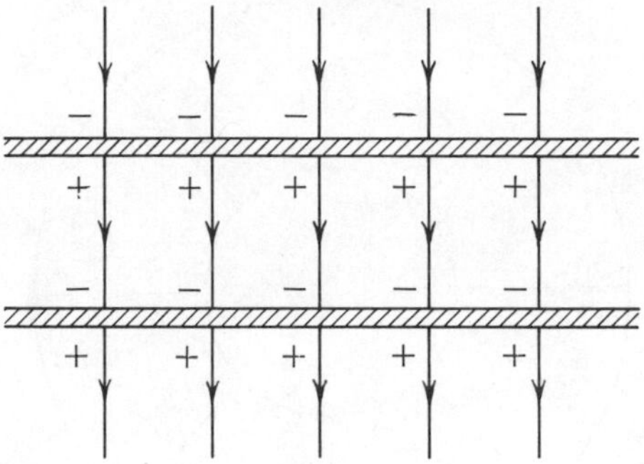

Figure 3.13

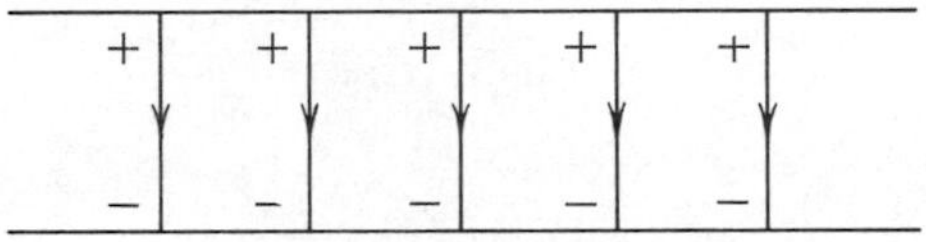

Figure 3.14

There are other slightly different ways of arriving at the situation depicted in figure 3.14. For example, the individual fields of each uniformly charged sheet, which will be mirror-symmetric about the sheet and of magnitude $\frac{1}{2}\sigma\epsilon_0$, can be superposed. Since the field of the positively charged sheet is directed away from it, and that of the negatively charged sheet towards it, the fields reinforce one another between the sheets and cancel out elsewhere. Another method is the direct application of Gauss' theorem, choosing for the closed surface a cylinder with plane ends whose generators are normal to the sheets.

Confinement of the electric field between parallel conducting plates is an important technique. The fact that in practice the area of the plates is finite is not of great significance. If the linear dimensions of the plates are large compared with their separation d the field between them is sensibly as described, with significant departure from uniformity arising only near the edges of the plates where the field spills into the surrounding space and becomes comparatively very weak. When edge effects are negligible the charges $\pm q$ on the respective plates are given by

$$q = A\sigma, \tag{3.26}$$

where A is the area of each plate. The device is known as a parallel plate *capacitor* (formerly, *condenser*). Its *capacity* (or *capacitance*) C is defined as the ratio of the quantity of charge on either plate to the potential difference between the plates. That is

$$C = q/V. \tag{3.27}$$

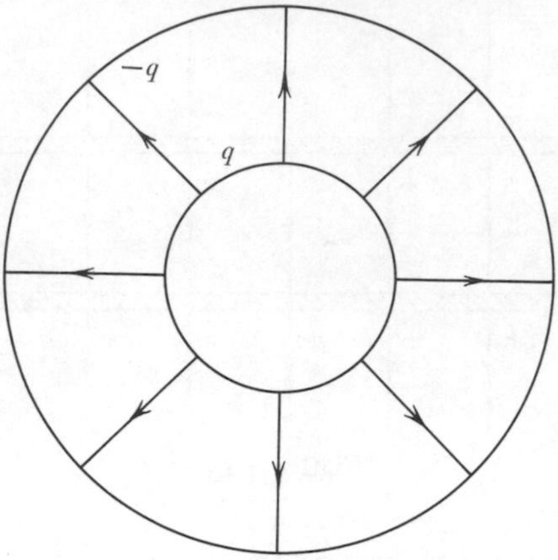

Figure 3.15

From (3.25) and (3.26),

$$C = \epsilon_0 A/d. \tag{3.28}$$

The spherical capacitor can be treated in a similar manner by isolating the region between two of the equipotentials of the field of a point charge q (see figure 3.15). If the conductors are $r = a$ and $r = b(> a)$ the potential difference between them is

$$V = \frac{q}{4\pi\epsilon_0}\left(\frac{1}{a}-\frac{1}{b}\right). \tag{3.29}$$

The charges on the conductors are, of course, $\pm q$ respectively, and hence the capacity is

$$C = 4\pi\epsilon_0 \frac{ab}{b-a}. \tag{3.30}$$

If $b = a+d$ this is
$$C = 4\pi a^2 \frac{\epsilon_0}{d}\left(1+\frac{d}{a}\right),$$

and it may be noted that when $d/a \ll 1$ the capacity per unit area of surface is approximately ϵ_0/d, which is what would have been expected from (3.28).

Another simple capacitor is that in which the conductors are coaxial circular cylinders. Consider, first, a *line charge* that is an infinite straight line l carrying a uniform charge distribution q per unit length. The corresponding field **E** is two-dimensional, being independent of distance parallel to l, and axially symmetric about l. Also, from the inverse square law, **E** clearly lies in the plane containing l, and is normal to l, so that it has only a radial component E_r, where r is distance from l. Application of Gauss' theorem to a cylinder $r =$ constant of finite length therefore gives

$$2\pi r E_r = q/\epsilon_0,$$

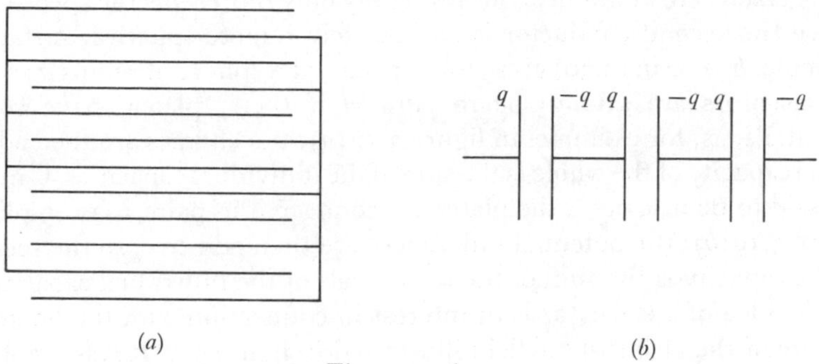

Figure 3.16

so that (cf. problems 1.11 and 2.2)

$$E_r = \frac{q}{2\pi\epsilon_0 r}. \tag{3.31}$$

The corresponding potential is evidently

$$\phi = -\frac{q}{2\pi\epsilon_0}\log r + \phi_0. \tag{3.32}$$

By isolating the region between the equipotentials $r = a$ and $r = b(> a)$ the cylindrical capacitor is seen to have capacity per unit length

$$2\pi\epsilon_0/\log(b/a). \tag{3.33}$$

As with the parallel plate capacitor this formula applies in practice with cylinders of finite length when edge effects are negligible.

The capacitors that have been described have simple geometries that lead to easily determined fields. Both the parallel plane and the circular cylindrical geometries are widely used in practice, often with a dielectric filling (see §3.4). In principle, for a given surface area, the capacity can be increased indefinitely by reducing the separation between the conductors; however, for a given potential difference the reduction is limited by the need to keep the field from becoming so large that it causes electrical breakdown. The standard way of conveniently obtaining a large surface area is to construct a uniform stack of parallel plates: the plates are connected together in two interleaved sets, as shown in figure 3.16(*a*), and the capacity is evidently the number of spaces times the capacity of a pair of adjacent plates.

In general a capacitor is formed by any two insulated conductors whose environment is such that when they carry charges $\pm q$, respectively, all the field lines run from one conductor to the other, which is certainly the case if no other conductors are present. If the potential difference between the conductors is then V the capacity is given by (3.27).

One sometimes talks of the capacity of a single conductor, meaning its charge to potential ratio when it is the only conductor present, the potential being taken zero at infinity. But this really only represents the special case in which the second conductor is an infinitely remote spherical surface. For example, $b \to \infty$ in (3.30) gives the capacity of a sphere of radius a as $4\pi\epsilon_0 a$.

Capacitors are said to be in *parallel* if their 'plates' have common potentials, as, for example, in figure 3.16(a): the charges are then additive, so the capacity of the whole is the sum of the individual capacities. Capacitors are said to be in *series* if the plates are connected in pairs, as exemplified in figure 3.16(b): the potential differences are then additive, so the reciprocal of the capacity is the sum of the reciprocals of the individual capacities.

The idea of a capacitor is of interest in connection with the honeycomb picture of the electrostatic field illustrated in figure 3.5. It was emphasized towards the end of §3.1.4 that complete equipotential surfaces can be regarded as conducting sheets. Now it is seen that a pair of adjacent sheets forms a capacitor. Moreover, by choosing the spacing between the sheets to be arbitrarily small, the elementary cell depicted in figure 3.5 is in effect a parallel plate capacitor (for the special case of spherical equipotentials this is implicit in the remark following (3.30)). In this sense an electrostatic field can be regarded as constructed from parallel plate capacitor fields.

The unit of capacity is called a *farad*, 1 farad being 1 coulomb volt^{-1}. The dimensions of ϵ_0 are often stated as farad m^{-1} (cf. (1.19)).

It was pointed out in §1.3.3 that a coulomb represents an enormous quantity of separated charge, and correspondingly one farad is a very large capacity. For example, the capacity of parallel plates of area 10^{-3} m^2 and separation 10^{-3} m is, from (3.28), 0.89×10^{-11} farad.

3.2 Energy

3.2.1 *Definition*

A static charge distribution has potential energy represented by the work that would have to be done against the electrostatic forces of interaction in building up the distribution from a state in which the charge is infinitely dispersed. Thus charge storage in a capacitor implies energy storage.

There is, of course, a close analogy with the potential energy of a mass distribution, the only difference being that the gravitational energy is always negative, since the force is attractive, whereas the electrostatic energy may be of either sign.

As a simple case, consider point charges e_1 and e_2, originally infinitely far apart, brought to a separation r. Clearly the work done against the Coulomb force is

$$\frac{e_1 e_2}{4\pi\epsilon_0 r}, \tag{3.34}$$

for if e_1 is held fixed the work done is e_2 times the potential due to e_1. Thus (3.34) is the potential energy.

The value of the energy concept in all branches of physics hardly needs stressing. It is, however, necessary to be clear about the sense in which the term energy is used in any particular context. For example, in the case of the two point charges just cited, no account was taken of the individual energies associated with each concentration of charge: these *self-energies* were implicitly assumed to be immutable (they are, in fact, infinite if calculated classically for charge located strictly at a point) and therefore not material to the issue, much as in gravitational theory the internal energy of massive bodies is often immaterial.

Particular expressions for the energy associated with various electrostatic situations are now obtained.

3.2.2 *Mutual energy of point charges*

The work done against the electric field in bringing a point charge e from infinity to a point where the potential is ϕ is $e\phi$. On the understanding that its self-energy is ignored, $e\phi$ is then the contribution of the point charge to the energy.

Consider, now, a system consisting solely of point charges $e_1, e_2, \ldots, e_n$ at specified positions in an unbounded vacuum. A first thought might be that the energy of the system is the sum of contributions $e_i \phi_i$, where ϕ_i is the potential at the location of e_i due to all the point charges other than e_i at their respective locations. However, this is not correct, simply because, apart from the last one, each point charge is brought into position when only a limited number of its companions are present so that the potential has not attained its final value.

Take for example the case of two point charges. No work is done in positioning the first, and the energy is therefore $e_2 \phi_2$, as displayed in (3·34). On the other hand the form $e_2 \phi_2$ lacks symmetry; it is preferable to write

$$\tfrac{1}{2}(e_1 \phi_1 + e_2 \phi_2), \tag{3·35}$$

and this symmetric form is easy to generalize, as the following argument shows.

With the point charges initially infinitely far apart, put in position successively $e_1, e_2, \ldots, e_i, \ldots$. The respective amounts of work done are

$$0, \quad \frac{e_1 e_2}{4\pi\epsilon_0 r_{12}}, \quad \ldots, \quad \frac{e_i}{4\pi\epsilon_0} \sum_{j=1}^{i-1} \frac{e_j}{r_{ij}}, \quad \ldots, \tag{3·36}$$

where r_{ij} is the distance between the given locations of e_i and e_j. The total work done can therefore be written

$$\frac{1}{4\pi\epsilon_0} \sum_{j<i} \frac{e_i e_j}{r_{ij}} \tag{3·37}$$

the sum being a double sum over i and j that comprises all terms for which $j < i$.

Now (3.37) is also

$$\frac{1}{4\pi\epsilon_0}\sum_{i<j}\frac{e_i e_j}{r_{ij}}, \tag{3.38}$$

and a further equivalent form can be obtained by taking half the sum of (3.37) and (3.38). This gives

$$\frac{1}{8\pi\epsilon_0}\sum_{i\neq j}\frac{e_i e_j}{r_{ij}}, \tag{3.39}$$

which is evidently

$$\frac{1}{2}\sum_{i=1}^{n} e_i\phi_i. \tag{3.40}$$

3.2.3 *Energy of a dipole in an external field: force and couple*

The concept of a dipole was explained in §2.1.5, and the dipole field was deduced from the Coulomb field of a point charge. For dipoles, as for point charges, there is an energy associated with their disposition that can be considered without regard to their respective self-energies.

A given dipole in a given electrostatic field has a potential energy measured by the work done against the field in bringing the dipole from infinity to its specified position. For an infinitesimal dipole of moment $\mathbf{p}$ this energy is easily written down by treating the dipole as point charges $\pm e$ separated by the displacement $\mathbf{l}$, in the limit $e\to\infty$, $l\to 0$ with $e\mathbf{l}=\mathbf{p}$. If the potential of the given field at the $-e$ charge is ϕ then that at the $+e$ charge is, to the first order in $\mathbf{l}$, $\phi+\mathbf{l}.\mathrm{grad}\,\phi$. The combined potential energy of the two charges is therefore

$$W = -e\phi + e(\phi+\mathbf{l}.\mathrm{grad}\,\phi) = e\mathbf{l}.\mathrm{grad}\,\phi;$$

that is

$$W = \mathbf{p}.\mathrm{grad}\,\phi = -\mathbf{p}.\mathbf{E}. \tag{3.41}$$

A dipole in an external field that is locally non-uniform experiences a net force. Since the line of action of this force depends on the field it is convenient to regard it as a force through the centre of the dipole together with a couple; and this representation has the added advantage of remaining valid in the special case, when the external field is locally uniform, for which the force vanishes.

If the force on the dipole is $\mathbf{F}$ the work done against this force in moving the dipole through an arbitrary infinitesimal displacement $\delta\mathbf{r}$, without change of orientation, is $-\mathbf{F}.\delta\mathbf{r}$. This, by definition, is the corresponding change δW in W. Hence

$$\mathbf{F} = -\mathrm{grad}\,W = \mathrm{grad}\,(\mathbf{p}.\mathbf{E}), \tag{3.42}$$

where the gradient is calculated with $\mathbf{p}$ kept constant.

On the other hand, a direct combination of the forces $-e\mathbf{E}$ and $e[\mathbf{E}+(\mathbf{l}.\mathrm{grad})\,\mathbf{E}]$ on the point charges $\pm e$ constituting the dipole clearly gives

$$\mathbf{F} = (\mathbf{p}.\mathrm{grad})\,\mathbf{E}. \tag{3.43}$$

At first sight (3.42) and (3.43) may appear to disagree. For example, the x component of (3.42) is $\mathbf{p}.\partial\mathbf{E}/\partial x$, which surprisingly involves all components of $\mathbf{E}$ in contrast to the x component of (3.43) which is $\mathbf{p}.\text{grad}\,E_x$. However, the apparent discrepancy is at once resolved by noting that $\partial\mathbf{E}/\partial x$ and $\text{grad}\,E_x$ are the same because $\text{curl}\,\mathbf{E} = 0$.

A similar analysis can be made of the couple on the dipole. As stated in (2.32) its vector moment is

$$\mathbf{G} = \mathbf{p}\wedge\mathbf{E}, \tag{3.44}$$

being the result of forces $\pm e\mathbf{E}$ acting at points vector distance $\mathbf{l}$ apart. The difference in the electric field at the two points makes no contribution in the limit $l \to 0$.

To deduce (3.44) from W, consider an arbitrary infinitesimal rotation of the dipole about its centre. If the rotation is specified by the vector $\delta\boldsymbol{\alpha}$ the work done against the couple is

$$\delta W = -\mathbf{G}.\delta\boldsymbol{\alpha}. \tag{3.45}$$

Now the change in dipole moment due to the rotation is

$$\delta\mathbf{p} = \delta\boldsymbol{\alpha}\wedge\mathbf{p}. \tag{3.46}$$

Hence, from (3.41),

$$\begin{aligned}\delta W &= -(\delta\boldsymbol{\alpha}\wedge\mathbf{p}).\mathbf{E}\\ &= -(\mathbf{p}\wedge\mathbf{E}).\delta\boldsymbol{\alpha}.\end{aligned} \tag{3.47}$$

A comparison of (3.47) with (3.45) leads to (3.44).

3.2.4 *Mutual energy of dipoles: forces of interaction*

The field of a dipole was derived in § 2.1.5, and expressed in the vector form (2.31). Using (3.41) in conjunction with (2.31) it appears that the mutual potential energy of a dipole of moment $\mathbf{p}_2$ at vector distance $\mathbf{r}$ from a dipole of moment $\mathbf{p}_1$ is

$$W = \frac{1}{4\pi\epsilon_0 r^3}\left[\mathbf{p}_1.\mathbf{p}_2 - 3\frac{(\mathbf{p}_1.\mathbf{r})(\mathbf{p}_2.\mathbf{r})}{r^2}\right]. \tag{3.48}$$

The force $\mathbf{F}_2$ on $\mathbf{p}_2$ is most easily obtained from (3.42). Since, for any constant $\mathbf{p}$,

$$\text{grad}\,(\mathbf{p}.\mathbf{r}) = \mathbf{p},$$

the negative of the gradient of (3.48) gives

$$\mathbf{F}_2 = \frac{3}{4\pi\epsilon_0 r^5}\left\{\left[\mathbf{p}_1.\mathbf{p}_2 - 5\frac{(\mathbf{p}_1.\mathbf{r})(\mathbf{p}_2.\mathbf{r})}{r^2}\right]\mathbf{r} + (\mathbf{p}_2.\mathbf{r})\,\mathbf{p}_1 + (\mathbf{p}_1.\mathbf{r})\,\mathbf{p}_2\right\}. \tag{3.49}$$

The couple $\mathbf{G}_2$ on $\mathbf{p}_2$, given by (3.44) and (2.31), is

$$\mathbf{G}_2 = \frac{1}{4\pi\epsilon_0 r^3}\left[\mathbf{p}_1\wedge\mathbf{p}_2 + 3\frac{(\mathbf{p}_1.\mathbf{r})(\mathbf{p}_2\wedge\mathbf{r})}{r^2}\right]. \tag{3.50}$$

What about the force $\mathbf{F}_1$ and couple $\mathbf{G}_1$ on $\mathbf{p}_1$? They must come from the expressions for $\mathbf{F}_2$ and $\mathbf{G}_2$ by interchanging $\mathbf{p}_1$ and $\mathbf{p}_2$ and replacing $\mathbf{r}$ by $-\mathbf{r}$. Evidently, from (3.49),

$$\mathbf{F}_1 = -\mathbf{F}_2. \tag{3.51}$$

And from (3.50)

$$\mathbf{G}_1 = \frac{1}{4\pi e_0 r^3}\left[\mathbf{p}_2 \wedge \mathbf{p}_1 + 3\,\frac{(\mathbf{p}_2.\mathbf{r})(\mathbf{p}_1 \wedge \mathbf{r})}{r^2}\right]. \tag{3.52}$$

Since the forces of interaction between dipoles derive from the inverse square law, (3.51) could have been anticipated. Neither is it unexpected that, in general, $\mathbf{G}_1$ is not equal to $-\mathbf{G}_2$; for $\mathbf{F}_1$ and $\mathbf{F}_2$ do not act along the line joining the dipoles, and there is therefore a couple $\mathbf{r} \wedge \mathbf{F}_2$ to be balanced. This balance implies

$$\mathbf{G}_1 + \mathbf{G}_2 + \mathbf{r} \wedge \mathbf{F}_2 = 0, \tag{3.53}$$

which is readily verified from (3.49), (3.50) and (3.52).

For a system of dipoles $\mathbf{p}_1, \mathbf{p}_2, \ldots, \mathbf{p}_n$ an argument entirely analogous to that of § 3.2.2 for point charges shows the energy to be

$$-\frac{1}{2}\sum_{i=1}^{n} \mathbf{p}_i.\mathbf{E}_i, \tag{3.54}$$

where $\mathbf{E}_i$ is the field at $\mathbf{p}_i$ due to all the dipoles other than $\mathbf{p}_i$ at their respective locations.

3.2.5 *Energy of distributed charge*

The energy of a charge distribution described by a density ρ that is spatially bounded and everywhere finite, but otherwise arbitrary, can be obtained as follows. Imagine the distribution built up, from the state in which it is infinitely dispersed, by successively putting in position infinitesimal multiplies of the total distribution. In this way the distribution at any intermediate stage has density $\lambda\rho$, where λ is a number between 0 and 1. Also, because the relation between charge and potential is linear, and there is no external field, the corresponding potential at the intermediate stage will be $\lambda\phi$, where ϕ is the final potential.

To find the work done against the forces of interaction it has only to be observed that the work done in increasing the density $\lambda\rho$ by the infinitesimal amount $\delta\lambda\rho$ is

$$\int \lambda\phi\,\delta\lambda\rho\,\mathrm{d}\tau,$$

where the volume integral is over any region enclosing all the charge. Then integration over λ from 0 to 1 yields a factor $\frac{1}{2}$, and the energy appears as

$$W = \frac{1}{2}\int \rho\phi\,\mathrm{d}\tau. \tag{3.55}$$

An important special case is when the charge resides solely on the surfaces of conductors. If there are n conductors, carrying charges $q_1, q_2, \ldots, q_n$ and at potentials $V_1, V_2, \ldots, V_n$, respectively, then evidently

$$W = \frac{1}{2}\sum_{i=1}^{n} q_i V_i. \tag{3.56}$$

The formal similarity between (3.56) and (3.40) is just what would be expected. Indeed (3.40) could have been obtained by the present argument, modified to the extent of ignoring the self-energies of the point charges.

As an example of (3.56), a capacitor holding charges $\pm q$ at a potential difference V stores energy

$$W = \tfrac{1}{2}qV. \tag{3.57}$$

In terms of the capacity $C = q/V$ this can be written in the alternative forms

$$W = \tfrac{1}{2}CV^2 = \tfrac{1}{2}q^2/C. \tag{3.58}$$

For a parallel plate capacitor, of plate area A and plate separation x, (3.28) gives $C = \epsilon_0 A/x$, and so

$$W = \tfrac{1}{2}\epsilon_0 AV^2/x = \tfrac{1}{2}q^2x/(\epsilon_0 A). \tag{3.59}$$

Each plate experiences a force of attraction towards the other. The magnitude F of this force can be obtained from

$$\delta W = F\,\delta x,$$

where the increment in W is that arising from the change in x when the charge on each plate is fixed, as it would be if the plates were insulated. Thus, from the second expression in (3.59),

$$F = \left(\frac{\mathrm{d}W}{\mathrm{d}x}\right)_{q\text{ constant}} = \tfrac{1}{2}q^2/(\epsilon_0 A). \tag{3.60}$$

This agrees with the force per unit area calculated from (3.24), since the charge density is q/A.

It is interesting to observe that the incremental energy for an increment in x is different if, instead of charge q, the voltage V across the plates is kept constant, say by a battery. Then, from the first expression in (3.59),

$$\left(\frac{\mathrm{d}W}{\mathrm{d}x}\right)_{V\text{ constant}} = -\tfrac{1}{2}\epsilon_0 AV^2/x^2, \tag{3.61}$$

which is equal in magnitude to (3.60), but opposite in sign. The interpretation is that, although mechanical work of amount (3.60) times δx is done *on* the system in increasing the plate separation, at the same time *twice* that amount of work is done *by* the system in giving up charge to the battery. This is confirmed by noting that, since $q = CV = \epsilon_0 AV/x$,

$$V\,\delta q/\delta x = -\epsilon_0 AV^2/x^2,$$

which is indeed the negative of twice (3.60).

The analogous result can be established for any number of arbitrary conductors. From (3.56), any infinitesimal displacement of the conductors in which their potentials are kept constant results in an energy increment

$$\delta W = \tfrac{1}{2}\Sigma V_i \delta q_i. \tag{3.62}$$

But in supplying charges δq_i to conductors at potentials V_i the 'electrical' work done on the system is

$$\Sigma V_i \delta q_i. \tag{3.63}$$

Hence the mechanical work done on the system, being necessarily (3.62) minus (3.63), is

$$-\tfrac{1}{2}\Sigma V_i \delta q_i; \tag{3.64}$$

and this is equal in magnitude to the change in energy of the system, but opposite in sign.

3.2.6 *Field energy*

The energy of a charge distribution, (3.55), can be alternatively expressed in terms of the field alone.

Since $\rho = \epsilon_0 \operatorname{div} \mathbf{E}$, the integrand in (3.55) can be written as ϵ_0 times

$$\operatorname{div}(\phi \mathbf{E}) - \mathbf{E} \, . \operatorname{grad} \phi.$$

The volume integral of the divergence can be transformed to the surface integral

$$\int \phi \mathbf{E} \, . \, \mathrm{d}\mathbf{S}. \tag{3.65}$$

This vanishes; for if the surface is taken to be a sphere whose radius r tends to infinity the integrand decreases at least as fast as $1/r^3$, since ϕ and $\mathbf{E}$ are $O(1/r)$ and $O(1/r^2)$ respectively (see §2.1.2), whereas the surface area only increases like r^2. Hence

$$W = \tfrac{1}{2}\epsilon_0 \int \mathbf{E}^2 \, \mathrm{d}\tau, \tag{3.66}$$

where the volume integral is over all space.

As an example, take the spherical capacitor of §3.1.5. Since $E = q/(4\pi\epsilon_0 r^2)$ for $a < r < b$, and is zero elsewhere, (3.66) gives

$$\frac{q^2}{8\pi\epsilon_0}\int_a^b \frac{\mathrm{d}r}{r^2} = \frac{q^2}{8\pi\epsilon_0}\left(\frac{1}{a} - \frac{1}{b}\right), \tag{3.67}$$

which, from (3.30), is indeed $\tfrac{1}{2}q^2/C$.

The parallel plate capacitor shows particularly clearly the origin of the field expression for the energy. If the plate area is A, and the plate separation x, the charge and potential are related to the field through the equations

$$q/A = \epsilon_0 E, \quad V/x = E. \tag{3.68}$$

Thus

$$\tfrac{1}{2}qV = \tfrac{1}{2}\epsilon_0 E^2 Ax, \tag{3.69}$$

in which the left hand side is (3.57), the expression first obtained for the energy of a capacitor, and the right hand side is $\frac{1}{2}\epsilon_0 E^2$ times the volume of the region between the plates.

This last result supplies a further feature of the honeycomb picture of the electric field, in which the cells are regarded as parallel plate capacitors; namely, that with each cell there is associated an energy

$$\tfrac{1}{2}\epsilon_0 \mathbf{E}^2 \tag{3.70}$$

per unit volume. It is certainly tempting to interpret (3.66) in this way, by conceiving the energy to be localized in the field with a volume density (3.70); and the concept of *field energy density* thus introduced, which can be extended to general electromagnetic fields in a way discussed later on, does play a useful role in electromagnetic theory.

3.3 Potential theory

3.3.1 *The problem*

The basic electrostatic problem is that in which there are conductors of specific shape and position, for each of which is given either its potential or the net charge it carries.

It may perhaps be felt that, from a physical viewpoint, the essence of the problem is to find the charge distribution on the surfaces of the conductors that makes these surfaces equipotentials that accord both with the given values of the potential and with the given charges. In a mathematical development along these lines the surface charge density σ would be the unknown, and the formulation would be in terms of integral equations. For example, for a single conductor bounded by the surface S and carrying total charge q, σ would have to satisfy the conditions that (3.19), namely the surface integral over S,

$$\frac{1}{4\pi\epsilon_0}\int \frac{\sigma\,\mathrm{d}S}{R}, \tag{3.71}$$

be the same at all points on S, and that

$$\int \sigma\,\mathrm{d}S = q. \tag{3.72}$$

Such a formulation has some merit. However, a more generally useful and less mathematically sophisticated approach is to treat the potential ϕ as the unknown; and the discussion here is confined solely to some of the comparatively elementary aspects of this formulation.

With ϕ as the unknown the problem just referred to would be posed in the following way: to find, in the region outside S, the solution of

$$\nabla^2\phi = 0 \tag{3.73}$$

that is $O(1/r)$ at infinity, is constant on S, and satisfies

$$\int \frac{\partial \phi}{\partial n}\, \mathrm{d}S = -q/\epsilon_0, \tag{3·74}$$

where $\partial/\partial n$ denotes differentiation along the outward normal to S.

Condition (3·74) expresses the fact that the total charge on S is q. It is the counterpart to (3·72), σ/ϵ_0 being equal to $-\partial\phi/\partial n$.

That ϕ is $O(1/r)$ at infinity is automatically achieved in the σ formulation, because ϕ is expressed in the form (3·71). But the condition has to be stated explicitly in the ϕ formulation, otherwise the solution would not be unique. The way in which imposed boundary conditions ensure that the solution (assuming it exists) is unique, is evidenced in the theorem of the next section. This also serves to categorize variants of the potential problems so far mentioned.

3·3·2 *Uniqueness theorem*

The mathematical result from which the uniqueness theorem is deduced is quickly obtained.

Let ψ be any suitably differentiable function defined throughout some region, and integrate the identity

$$\operatorname{div}(\psi \operatorname{grad}\psi) = \psi\nabla^2\psi + (\operatorname{grad}\psi)^2 \tag{3·75}$$

over the region, to get

$$\int (\operatorname{grad}\psi)^2\, \mathrm{d}\tau = \int \psi \frac{\partial \psi}{\partial n}\, \mathrm{d}S - \int \psi\nabla^2\psi\, \mathrm{d}\tau, \tag{3·76}$$

where the surface integral is over the boundary surface or surfaces of the region, including the sphere at infinity if the region is unbounded. If it happens that

$$\int \psi\nabla^2\psi\, \mathrm{d}\tau \tag{3·77}$$

and

$$\int \psi \frac{\partial \psi}{\partial n}\, \mathrm{d}S \tag{3·78}$$

are both zero, then the left hand side of (3·76) is zero, which requires that grad ψ vanishes throughout the region; that is, ψ is constant. For example, let ψ satisfy Laplace's equation throughout a bounded region and be zero over the boundary; then the result demonstrates that ψ can only be zero throughout the region.

The result is applied to any potential problem by supposing that functions ϕ_1 and ϕ_2 each satisfy the conditions of the problem, setting

$$\psi = \phi_1 - \phi_2, \tag{3·79}$$

and proving ψ to be constant by establishing that both (3·77) and (3·78) are zero.

Take the problem to be that in which the field is associated solely with charges on conducting surfaces, and pervades an empty region of space which may or may not extend to infinity. Then throughout this region ϕ_1 and ϕ_2 each satisfy Laplace's equation, so $\nabla^2\psi = 0$, and (3.77) is zero.

Moreover, for each conducting surface (3.78) is

$$(\phi_1-\phi_2)\int\left(\frac{\partial\phi_1}{\partial n}-\frac{\partial\phi_2}{\partial n}\right)\mathrm{d}S, \tag{3.80}$$

since ϕ_1 and ϕ_2 must be constant on such a surface; and (3.80) is zero, either because ϕ_1 and ϕ_2 have the same value on the surface (potential given), or because the integral of $\partial\phi_1/\partial n$ equals that of $\partial\phi_2/\partial n$ (total charge given).

Finally, for the sphere at infinity, (3.78) is also zero; for ψ is $O(1/r)$ and $\partial\psi/\partial n$ is $O(1/r^2)$ (cf. the vanishing of (3.65)).

Thus (3.78) vanishes, as well as (3.77), whence ϕ_1 and ϕ_2 can differ only by a constant. For potentials this does not, of course, represent a material difference, and it is therefore established that there is not more than one solution to the stated problem.

It is easy to see that this uniqueness theorem can at once be generalized to apply to the case when, in addition to the charges on the conductors, there is also a *specified* bounded charge distribution. For the only difference this makes is that ϕ_1 and ϕ_2 now satisfy Poisson's equation $\nabla^2\phi = -\rho/\epsilon_0$ rather than Laplace's equation; but if ρ is given $\nabla^2(\phi_1-\phi_2)$ remains zero, and the argument is unaffected.

Two slight variants on this last problem may be mentioned. Rather than a charge density ρ there may be specified either point charges, or an external field, such as a uniform field, that does not vanish at infinity. Both variants are really only limiting cases of the original formulation: in one, a finite quantity of charge shrinks to a point; in the other, an infinite quantity of charge recedes to infinity. If desired, however, direct analyses can be given with only slight modifications to the foregoing treatment.

Whilst attention has naturally been directed towards the potential problems of electrostatics it is worth mentioning that, from the mathematical point of view, the vanishing of the surface integral (3.78), with ψ given by (3.79), is most directly achieved by either of the conditions on the bounding surface (*a*) $\phi_1 = \phi_2$, (*b*) $\partial\phi_1/\partial n = \partial\phi_2/\partial n$, with also $O(1/r)$ behaviour at infinity when relevant. The problem in which the potential function is given on the bounding surface is known as a *Dirichlet* problem, and that in which the normal derivative of the potential function is given is known as a *Neumann* problem. The electrostatic problem with conductors at specific potentials is a special case of the Dirichlet problem, in which the potential function is a given constant on each conducting surface. The problem with conductors carrying specific total charges is *not* a Neumann problem, since the normal derivative of the potential is not given at each point of the conducting surface; only its integral over the surface is known, the additional information being that the potential itself is constant over the surface, though of unspecified value.

The electrostatic problem is well posed in so far as there is not more than one solution. The question remains whether there is any solution. This can be answered affirmatively with adequate generality, but the treatment is too sophisticated to be considered here. There are, though, many methods of finding solutions in particular cases, some of which manage without extensive analysis; a few simple examples are now given along the lines of those mentioned in §3.1.5.

3.3.3 *Images*

Consider the field of point charges $e(>0)$ and $-e$ (figure 3.17(*a*)). If the charges are situated on the z axis at $(0, 0, d)$, and $(0, 0, -d)$, respectively, the plane $z = 0$ is an equipotential, being in fact at zero potential. Take this equipotential to be a perfectly conducting sheet, which therefore carries a negative surface charge density on the upper $(z = 0+)$ face, and a corresponding positive surface charge density on the lower $(z = 0-)$ face. Now consider separately the field in the upper half-space, which is generated by the point charge e and the negative surface charge density on $z = 0+$, as indicated in figure 3.17(*b*).

Differently expressed, figure 3.17(*b*) represents the solution of the problem in which a point charge e is placed a distance d from an infinite plane conductor at zero potential. Evidently the charge density σ induced on the conductor generates in $z > 0$ precisely the same field as would a charge $-e$ placed at the point that is the 'image' in the plane of the location of the given charge e. In the problem as now conceived the *image* charge $-e$ only provides a convenient way of describing the field in $z > 0$; the lower half-space is screened from the given charge e by the conducting plane, the field in $z < 0$ due to the surface charge density σ being, of course, precisely the negative of that due to e.

Since the field is known σ is easily calculated from the value of $\mathbf{E}$ at $z = 0+$. From the inverse square law (see figure 3.17(*c*))

$$\sigma = -\frac{ed}{2\pi R^3} \qquad (3.81)$$

at distance R from e. The net charge on the plane is, of course, $-e$; and it is easy to check that the integral of (3.81) over the plane agrees with this.

The charge e experiences a force of attraction towards the plane. This is because it is situated in the field generated by σ. But this field is the same as that of the image charge. Hence the force on e is just that of attraction between itself and the image charge, namely

$$e^2/(16\pi\epsilon_0 d^2). \qquad (3.82)$$

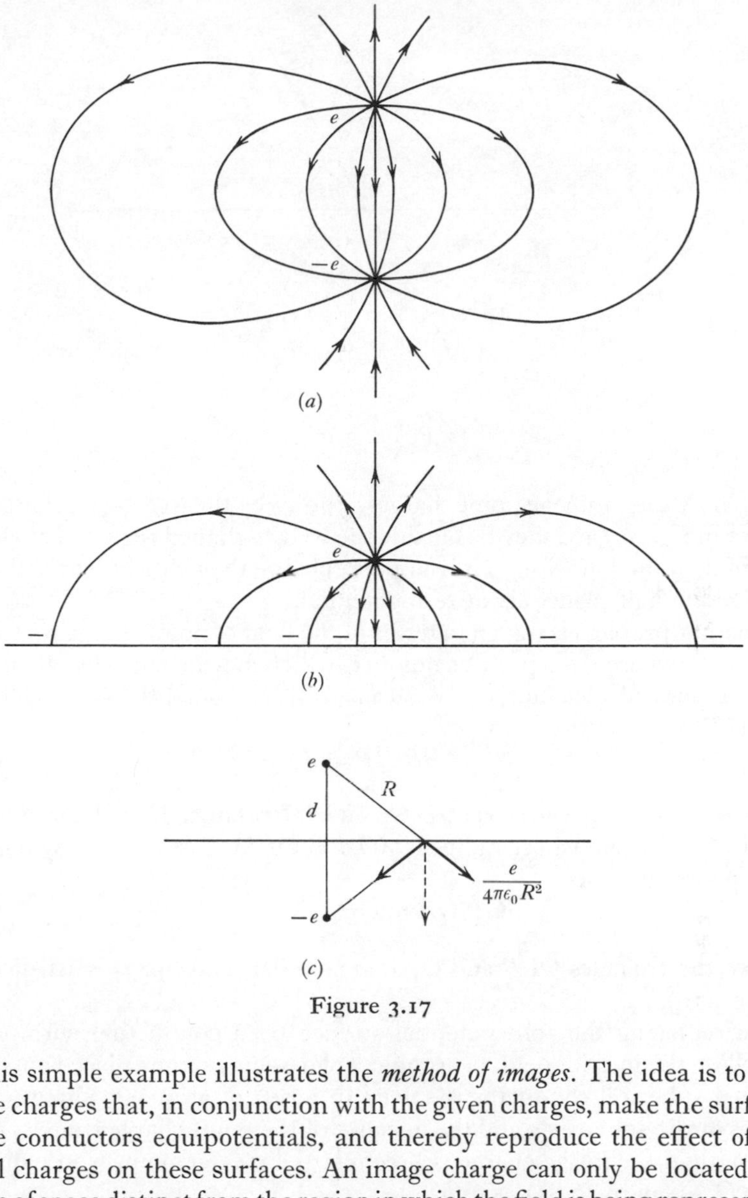

Figure 3.17

This simple example illustrates the *method of images*. The idea is to find image charges that, in conjunction with the given charges, make the surfaces of the conductors equipotentials, and thereby reproduce the effect of the actual charges on these surfaces. An image charge can only be located in a region of space distinct from the region in which the field is being represented.

The method is of quite limited applicability, but it does provide various elementary solutions, some of which are now described.

Closely related to the problem of a point charge in the presence of an infinite conducting plane is that of a point charge in the wedge-shaped region bounded by conducting half-planes that meet along a common edge at an angle π/n, where n is a positive integer. It is easy to see that the solution

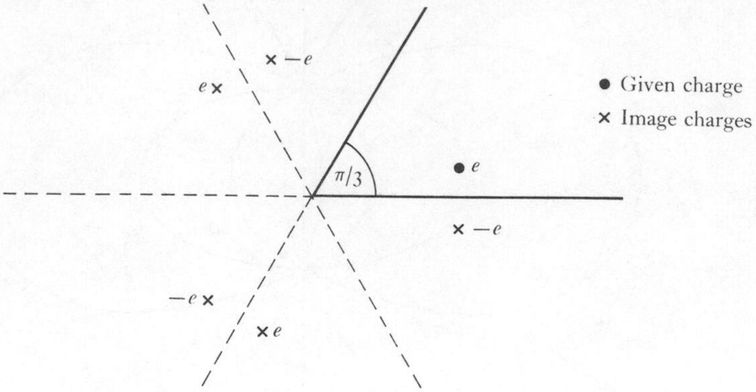

Figure 3.18

is given by $2n-1$ image point charges. For $n = 3$ the five image charges are shown in figure 3.18; they lie outside the wedge-shaped region of angle $\pi/3$, and in conjunction with the given point charge they clearly insure that the conducting half-planes are at zero potential.

Another problem is solved in terms of the field of point charges e at A and $-\lambda e$ at B, where λ is a positive number, which may be taken less than unity without loss of generality. The surface of zero potential is the locus of points P such that

$$BP/AP = \lambda. \tag{3.83}$$

This is a sphere, whose centre O, on AB produced, is determined by $OB/OA = \lambda^2$, and whose radius is $\lambda OA = OB/\lambda$. For if P is any point on this sphere,

$$OP/OA = OB/OP = \lambda, \tag{3.84}$$

so that the triangles OPB and OAP are similar, and (3.83) is satisfied (see figure 3.19).

On replacing the zero potential surface by a conducting surface, and annulling the interior field, the problem of a point charge e placed outside an earthed spherical conductor is solved in terms of an image charge. If the radius of the sphere is a and the distance of the point charge from its centre O is b, then the image charge is $-ea/b$ at the inverse point whose distance from O, given by (3.84), is a^2/b.

The total charge on the earthed sphere is $-ea/b$, and the force of attraction between it and the point charge is

$$\frac{e^2}{4\pi\epsilon_0}\frac{ab}{(b^2-a^2)^2}. \tag{3.85}$$

The charge density σ is calculated by finding the electric field at the surface

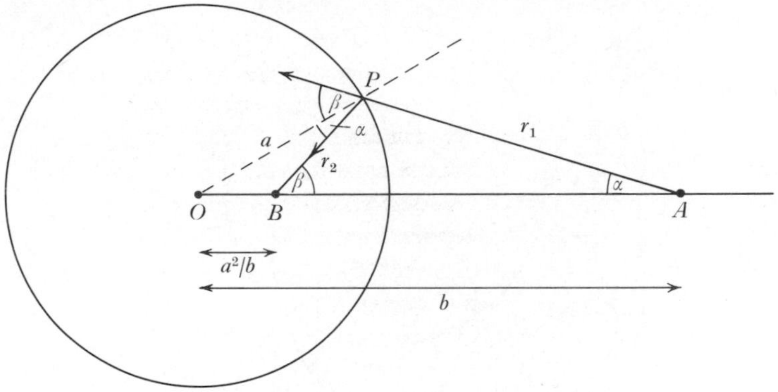

Figure 3.19

of the sphere, where it is, of course, normal to the surface. With reference to figure 3.19 the magnitude of the inward normal component is

$$\frac{e}{4\pi\epsilon_0}\left(\frac{1}{r_1^{\,2}}\cos\beta+\frac{a}{br_2^{\,2}}\cos\alpha\right),$$

and since $r_2 = \lambda r_1 = ar_1/b$ this is

$$\frac{e}{4\pi\epsilon_0}\frac{1}{r_1^{\,2}r_2}(r_2\cos\beta+r_1\cos\alpha) = \frac{e}{4\pi\epsilon_0}\frac{1}{r_1^{\,2}r_2}\left(b-\frac{a^2}{b}\right).$$

Hence $$\sigma = -\frac{e}{4\pi}\frac{b^2-a^2}{a}\frac{1}{r_1^{\,3}}. \tag{3.86}$$

Note that, if b and a are allowed to tend to infinity, with $b-a = d$ kept fixed, the results for a point charge e distance d from an infinite conducting plane are recaptured: evidently (3.85) tends to (3.82) and (3.86) to (3.81).

Suppose that, instead of being earthed, the sphere has either a given potential or a given charge. A trivial modification only is required; namely, to add an image charge e' at the centre of the sphere. The spherical surface then has potential $e'/(4\pi\epsilon_0 a)$ and total charge $e'-ea/b$; and e' can be chosen to meet one or other of the given conditions.

If the sphere is uncharged $e' = ea/b$. This leads to the interesting case of a spherical conductor in a uniform external field E_0, by taking

$$\frac{e}{4\pi\epsilon_0 b^2} = E_0$$

and letting e and b tend to infinity. The image is then evidently a dipole at the centre of the sphere, parallel to $\mathbf{E}_0$, of moment

$$\frac{ea}{b}\frac{a^2}{b} = 4\pi\epsilon_0 a^3 E_0. \tag{3.87}$$

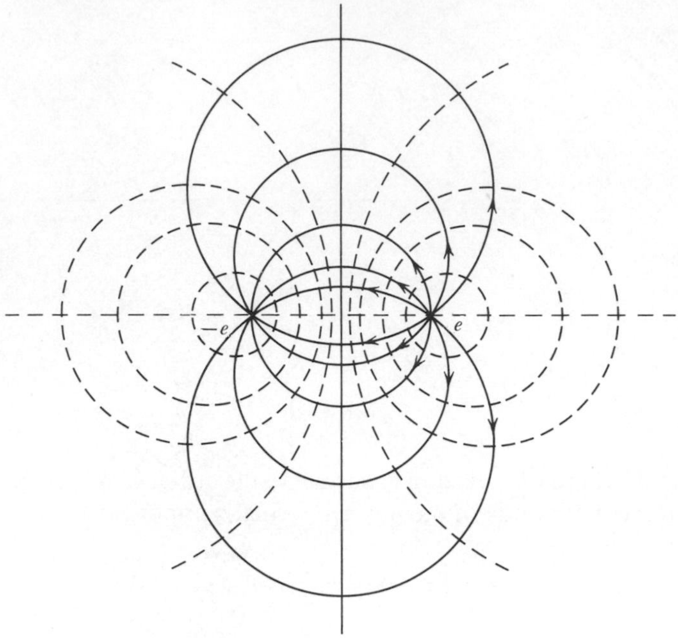

Figure 3.20

It is easy to confirm directly the correctness of this result; for if r, θ are spherical polar coordinates, with pole at the centre of the sphere and axis in the direction of $\mathbf{E}_0$, then the potential of the external field is

$$-E_0 r \cos\theta \tag{3.88}$$

and that of the image dipole is

$$a^3 E_0 \cos\theta / r^2, \tag{3.89}$$

so that the two cancel on $r = a$.

As a final problem consider the two-dimensional field of a pair of line charges, e and $-e$ per unit length. It is only necessary to describe the situation in some plane perpendicular to the charges. Let r_1 and r_2 denote distances from the line charges. Then the potential, from (3.32), is

$$\frac{e}{2\pi\epsilon_0}\log(r_2/r_1). \tag{3.90}$$

The equipotentials are therefore given by

$$r_2 = \lambda r_1, \tag{3.91}$$

where λ is a parametric constant; and the remarks following (3.83) show that these are circles, each of which encloses one or other of the charges (figure 3.20). In fact they form the coaxial system with the charges as limit points, and the lines of force are also circles (cf. problem 3.3). The problem

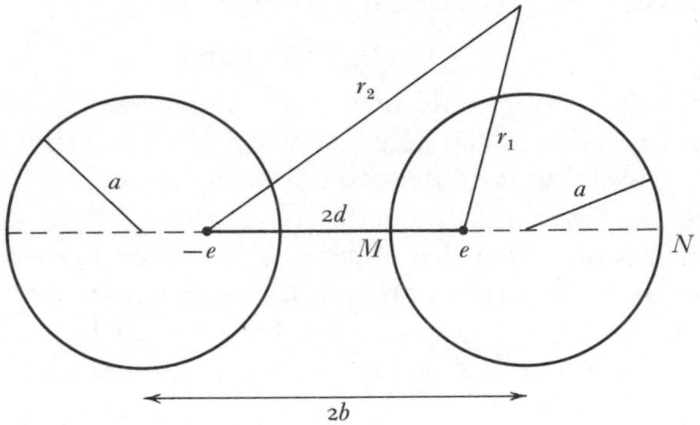

Figure 3.21

of a line charge e parallel to a conducting circular cylinder is therefore solved in terms of an image charge $-e$ along the inverse line.

Another interesting case is obtained by taking two of the equipotentials to be conducting surfaces. Both line charges can then play the role of image charges in representing the solution: if one conducting surface encloses the other the region pervaded by the field is that between them; otherwise, it is that external to them both.

A particular example is afforded by the calculation of the capacity per unit length of the twin wire line formed by a pair of parallel, conducting, circular cylinders. Suppose that the cylinders have radius a and that their axes are distance $2b$ apart (figure 3.21). Their cross-sections are given by (3.91) with λ taking the respective values $\lambda_0 (> 1)$ and $1/\lambda_0$, where

$$\lambda_0 = \frac{d+b-a}{d-b+a} = \frac{d+b+a}{a+b-d}. \tag{3.92}$$

In (3.92) d is the distance between the line charges, and the first expression for λ_0 follows from the values of r_1 and r_2 at M, the second from the values at N (figure 3.21). Simpler expressions, obtained by first summing and then differencing numerators and denominators, are

$$\lambda_0 = \frac{b+d}{a} = \frac{a}{b-d}. \tag{3.93}$$

Thus $b^2 - d^2 = a^2$, confirming that the positions of the line charges are inverse to one another; and in terms of the parameters a and b that specify the twin wire line

$$\lambda_0 = [b + (b^2 - a^2)^{\frac{1}{2}}]/a. \tag{3.94}$$

Now, from (3.90), the potential difference between the cylinders is $(e/\pi\epsilon_0) \log \lambda_0$; so the capacity per unit length is

$$C = \pi\epsilon_0 / \log \left[\frac{b + (b^2 - a^2)^{\frac{1}{2}}}{a}\right]. \tag{3.95}$$

The expression (3.95) is exact. If $a \ll b$ it gives

$$C \simeq \pi\epsilon_0/\log(2b/a). \tag{3.96}$$

This approximate result could have been anticipated by arguing that, if $a \ll b$, the image charges can be taken along the axes of the cylinders; for each charge placed thus contributes a uniform potential to the near cylinder and an almost uniform potential to the far cylinder.

The problems discussed that have involved given line or point charges can be extended by superposition to other given charge distributions. The treatment of dipoles, for example, can be achieved by taking equal and opposite point charges and then proceeding to the limit in the usual way.

3.3.4 *Minimum energy theorem*

It is possible to establish various general theorems concerning the energy of an electrostatic system. The most fundamental, perhaps, is Kelvin's theorem to the effect that the charge on conductors is so distributed that the energy is minimized. To state this more previsely, consider a system consisting of a number of fixed charged conductors; then the energy of the actual surface charge distribution on the conductors is less than the energy of any different (hypothetical) distribution in which the net charge on each conductor is unaltered. Only the actual charge distribution, of course, makes all the surfaces of the conductors equipotentials and the interior of the conductors correspondingly field free.

To prove the theorem let $\mathbf{E}$ and $\mathbf{E}'$ be the respective fields of the actual charge distribution and of any different distribution. Then the corresponding energies are

$$W = \tfrac{1}{2}\epsilon_0\int_V \mathbf{E}^2\,\mathrm{d}\tau, \quad W' = \tfrac{1}{2}\epsilon_0\int_{\text{all space}} \mathbf{E}'^2\,\mathrm{d}\tau,$$

where V is the region exterior to the conducting material. Thus

$$W' - W > \tfrac{1}{2}\epsilon_0\int_V (\mathbf{E}'^2 - \mathbf{E}^2)\,\mathrm{d}\tau$$

$$= \epsilon_0\int_V [\tfrac{1}{2}(\mathbf{E}' - \mathbf{E})^2 - \mathbf{E}\,.\,(\mathbf{E} - \mathbf{E}')]\,\mathrm{d}\tau. \tag{3.97}$$

But if ϕ is the potential of the actual charge distribution,

$$-\int_V \mathbf{E}\,.\,(\mathbf{E} - \mathbf{E}')\,\mathrm{d}\tau = \int_V (\mathbf{E}' - \mathbf{E})\,.\,\mathrm{grad}\,\phi\,\mathrm{d}\tau$$

$$= \int_V \mathrm{div}\,[\phi(\mathbf{E} - \mathbf{E}')]\,\mathrm{d}\tau, \tag{3.98}$$

since $\mathrm{div}\,\mathbf{E}$ and $\mathrm{div}\,\mathbf{E}'$ vanish throughout V. Now (3.98) can be written as the surface integral

$$\int \phi(\mathbf{E} - \mathbf{E}')\,.\,\mathrm{d}\mathbf{S}. \tag{3.99}$$

It is therefore zero: for over the sphere at infinity $\mathbf{E}$ and $\mathbf{E}'$ are $O(1/r^2)$, and

ϕ is $O(1/r)$; and over the surface of each conductor ϕ is constant, and $\int \mathbf{E}.\mathrm{d}\mathbf{S} = \int \mathbf{E}'.\mathrm{d}\mathbf{S}$ since each integral gives $-1/\epsilon_0$ times the net charge on the conductor.

Hence

$$W < W' - \tfrac{1}{2}\epsilon_0 \int_V (\mathbf{E}' - \mathbf{E})^2 \,\mathrm{d}\tau < W', \tag{3.100}$$

and the theorem is proved.

A closely related result is the theorem, that the introduction into any system of charged conductors of a further conductor that is either uncharged or kept at zero potential reduces the energy of the system.

The mathematics of the proof follows the lines of that just given, with $\mathbf{E}$ denoting the field after the introduction of the extra conductor and $\mathbf{E}'$ the field prior to its introduction. Then $\mathbf{E}$ pervades the region V exterior to all the conductors, and $\mathbf{E}'$ the region $V + V_0$, where V_0 is the region occupied by the extra conductor. Hence

$$W' - W = \tfrac{1}{2}\epsilon_0 \int_V (\mathbf{E}'^2 - \mathbf{E}^2)\,\mathrm{d}\tau + \tfrac{1}{2}\epsilon_0 \int_{V_0} \mathbf{E}'^2 \,\mathrm{d}\tau. \tag{3.101}$$

By virtually the same steps as before

$$\frac{1}{2}\int_V (\mathbf{E}'^2 - \mathbf{E}^2)\,\mathrm{d}\tau = \frac{1}{2}\int_V (\mathbf{E}' - \mathbf{E})^2 \,\mathrm{d}\tau - \phi \int_{S_0} \mathbf{E}.\mathrm{d}\mathbf{S}, \tag{3.102}$$

where $\mathbf{S}_0$ is the surface of the extra conductor, and the fact that

$$\int_{S_0} \mathbf{E}'.\mathrm{d}\mathbf{S} = 0$$

has been used.

Clearly, then, (3.101) establishes that $W < W'$ if the surface integral term in (3.102) vanishes; and this is the case either when the extra conductor is zero potential ($\phi = 0$), or when it is uncharged $\left(\int_{S_0} \mathbf{E}.\mathrm{d}\mathbf{S} = 0\right)$.

It is instructive to appreciate that in the case when the extra conductor is uncharged this latter theorem is no more than a special case of the minimum energy theorem, in which, with all the conductors present, the actual charge distribution is compared with a different distribution in which the charge density on $\mathbf{S}_0$ is everywhere zero.

Theorems such as these provide a means of getting approximate solutions to electrostatic problems. For example, the minimum energy theorem suggests that a 'trial solution' can be optimized by allocating to adjustable parameters those values that minimise the energy.

3.3.5 *Green's reciprocal theorem*

This is another useful general theorem that may be stated as follows. If charge densities ρ and ρ' separately give rise to respective potentials ϕ and ϕ', then

$$\int \rho\phi' \,\mathrm{d}\tau = \int \rho'\phi \,\mathrm{d}\tau, \tag{3.103}$$

where the integrals are over all space.

The proof is simple; for the left hand side of (3.103) is

$$\epsilon_0 \int \phi' \operatorname{div} \mathbf{E} \, d\tau = \epsilon_0 \int [\operatorname{div}(\phi' \mathbf{E}) - \mathbf{E} . \operatorname{grad} \phi'] \, d\tau$$

$$= \epsilon_0 \int \mathbf{E} . \mathbf{E}' \, d\tau, \tag{3.104}$$

which by symmetry is also the right hand side of (3.103).

A special case is when the charge resides solely on the surfaces of n fixed conductors. Then (cf. (3.55) and (3.56))

$$\sum_{i=1}^{n} q_i V_i' = \sum_{i=1}^{n} q_i' V_i, \tag{3.105}$$

where charges q_i on the conductors correspond to respective potentials V_i, and charges q_i' to potentials V_i'.

There is an immediate corollary, concerning the *coefficients of capacity* c_{ij} in terms of which the linear relation between charges and potentials is expressed in the form

$$q_i = \sum_{j=1}^{n} c_{ij} V_j \quad (i = 1, 2, \ldots, n). \tag{3.106}$$

For (3.105) gives

$$\sum_{i,j} c_{ij} V_i' V_j = \sum_{i,j} c_{ij} V_i V_j';$$

and this implies, by taking just one of the V_i and one of the V_i' to be non-zero, that

$$c_{ij} = c_{ji} \quad \text{for all } i, j. \tag{3.107}$$

It may also be noted that for point charges at given locations

$$\sum_i e_i \phi_i' = \frac{1}{4\pi\epsilon_0} \sum_{\substack{i,j \\ (i \neq j)}} \frac{e_i e_j'}{r_{ij}} = \sum_j e_j' \phi_j, \tag{3.108}$$

using the notation of § 3.2.2.

As a simple example of the use of the reciprocal theorem consider a point charge e placed outside an uncharged conducting sphere at a point P distance b from the centre of the sphere. What is the potential V of the sphere?

To apply the reciprocal theorem the alternative situation is envisaged in which the sphere carries unit charge and there is no charge at P, so that the potential at P is $1/(4\pi\epsilon_0 b)$. Then the reciprocal theorem states

$$V = e/(4\pi\epsilon_0 b).$$

This result obviously agrees with that given by the solution found in § 3.3.3.

3.4 Dielectrics

3.4.1 *Phenomenological theory*

So far the only material bodies introduced into the theory have been conductors. On grounds outlined in § 3.1.2 they have been treated on the basis of the simple hypothesis that no electrostatic field can exist in the interior of a conductor.

It is now appropriate to give a similar phenomenological description of another type of material body known as an *insulator*. In contrast to a conductor this is made of a substance that, ideally at least, contains no *free* charged particles; it is therefore not capable of yielding a steady flow of current when, say, under the influence of a potential gradient. Thus, for example, the region between the plates of a capacitor can be filled with an insulating substance without to any significant degree discharging the capacitor; the plates remain electrically isolated.

Many insulators can sustain appreciably stronger fields than air without breaking down and are obviously useful in keeping the plates of a capacitor in position. Moreover an insulating filling can also appreciably increase the capacity of a given plate configuration. This latter property arises from the possibility of creating a significant separation between the positive and negative charged particles in the insulator, which is therefore traditionally called a *dielectric*; the effect is readily demonstrated experimentally, and acceptance of the experimental findings suggests a simple description of dielectrics which is adequate for many theoretical calculations. This approach will be developed now, leaving until Chapter 6 some indication of that theory of the structure of a dielectric that aims to explain the origin of its characteristic electric properties.

Consider the parallel plate capacitor discussed in § 3.1.5. With the neglect of edge effects the field $\mathbf{E}$ between the plates is uniform, being normal to the plates and of magnitude σ/ϵ_0, where σ is the surface charge density on one plate and $-\sigma$ that on the other. The voltage between the plates is $Ed = \sigma d/\epsilon_0$, where d is the plate separation.

Now suppose the region between the plates to be filled with a homogeneous dielectric. If the dielectric is a good insulator the charge density on the plates is unaffected, but it is found that the voltage between the plates is reduced by a factor, say $1/\kappa$, the value of which is characteristic of the particular substance used. The voltage is therefore $\sigma d/\epsilon$, where

$$\epsilon = \epsilon_0 \kappa. \tag{3.109}$$

The capacity is, of course, increased by the factor κ.

In seeking to interpret this result in terms of field concepts it is recognized that, whatever the detailed mechanism, the effect can only arise through the presence of charged particles in the dielectric. If then the macroscopic field, as defined in § 1.2.1, is considered, the potential representation

$$\mathbf{E} = -\operatorname{grad} \phi \quad (\operatorname{curl} \mathbf{E} = 0) \tag{3.110}$$

holds, but the equation $\operatorname{div} \mathbf{E} = \rho/\epsilon_0$ fails if the contribution of the dielectric to the charge density is not explicitly allowed for.

In a phenomenological treatment the dielectric is regarded solely as a medium permeated by the field, and explicit reference to the charges it contains is avoided. Instead, following the reasoning of § 2.4.2, a vector **D** is introduced which satisfies

$$\operatorname{div} \mathbf{D} = \rho \tag{3.111}$$

and whose relation to **E** is determined so that the theory agrees with experiment.

When (3.111) is applied inside the parallel plate capacitor the charge density is simply that on the surface of the plates; with the neglect of edge effects **D** is therefore normal to the plates and of magnitude $D = \sigma$. The relation

$$\mathbf{D} = \epsilon \mathbf{E} \tag{3.112}$$

therefore meets the requirements; for if **E** is likewise uniform, of magnitude $E = \sigma/\epsilon$, (3.110) correctly predicts the voltage $\sigma d/\epsilon$ between the plates.

Equations (3.110), (3.111) and (3.112) contain no explicit reference to any particular geometry, and in the absence of evidence to the contrary it is natural to assume that they are the governing equations for any static disposition of charges, conductors and dielectrics. If the *permittivity* ϵ is uniform then the problem differs from the corresponding vacuum problem only in the replacement of ϵ_0 by ϵ: thus, for example, completely filling the region between the plates of a capacitor with a homogeneous dielectric increases the capacity by the factor $\kappa = \epsilon/\epsilon_0$ no matter what the shape of the capacitor. But ϵ may be a function of position, and then (3.112) is a point relation between the vectors **E** and **D**. The picture developed near the end of § 3.1.5, in which the electrostatic field is conceived as a honeycomb of parallel plate capacitor fields is modified only in that each capacitor is dielectric filled.

Provided the field is not too strong the electrostatic properties of many insulating substances, are, in fact, adequately characterized by a simple proportionality between the vectors **E** and **D** as expressed in (3.112). Such substances are called *isotropic*, in contrast to *anisotropic* substances for which **E** and **D**, whilst still linearly related, are no longer parallel.

Evidently ϵ has the same dimensions as ϵ_0. The dimensionless quantity $\kappa = \epsilon/\epsilon_0$ is called the *dielectric constant*, and its value depends on the molecular structure of the substance. It always exceeds unity, by little for gases, and commonly by about 4 or 5 for solids. Some values are given in the table below. The dielectric constant of distilled water is about 80, but water is not a good insulator.

Substance	Dielectric constant
Hydrogen	1.00026
Air	1.0006
Paraffin	2–2.5
Quartz	4–5
Mica	5–7
Glass	4–10
Porcelain	5–8

3.4.2 *Boundary conditions*

In §3.1.3 the idealized concept of surface charge density was introduced, and the nature of the corresponding field discontinuity was established, using Gauss' theorem and the fact that **E** is a potential field.

With the introduction of dielectrics an analogous problem arises. What happens at the interface between two dielectrics of different permittivities? The answer can be found in precisely the same way.

Application of Gauss' theorem (2.94) to the closed surface indicated in figure 3.6, where S is now the interface between two dielectrics, gives

$$D_{n2} = D_{n1}, \tag{3.113}$$

to be compared with (3.10); and a like interpretation of figure 3.7 again, of course, leads to (3.11), namely

$$\mathbf{E}_{t2} = \mathbf{E}_{t1}. \tag{3.114}$$

Thus both the normal component of **D** and the tangential components of **E** are continuous across the interface between two dielectrics. An evident corollary of this result is that the tangential components of **D** and the normal component of **E** are discontinuous. The field lines are therefore refracted in the sense that they suffer an abrupt change of direction in crossing the interface. Specifically, for dielectrics of permittivities ϵ_1 and ϵ_2, the continuity of D_n implies

$$\epsilon_2 E_{n2} = \epsilon_1 E_{n1}.$$

If, then, θ designates the angle that a field line makes with the normal to the interface, it follows at once from the continuity of $\mathbf{E}_t$ that

$$\epsilon_1 \tan\theta_2 = \epsilon_2 \tan\theta_1, \tag{3.115}$$

the line being bent away from the normal in the medium of greater permittivity. Moreover, the magnitude of **D** is greater, but that of **E** less, in the medium of greater permittivity. The situation is depicted in figure 3.22, where the thinner arrow represents **D** and the thicker **E**.

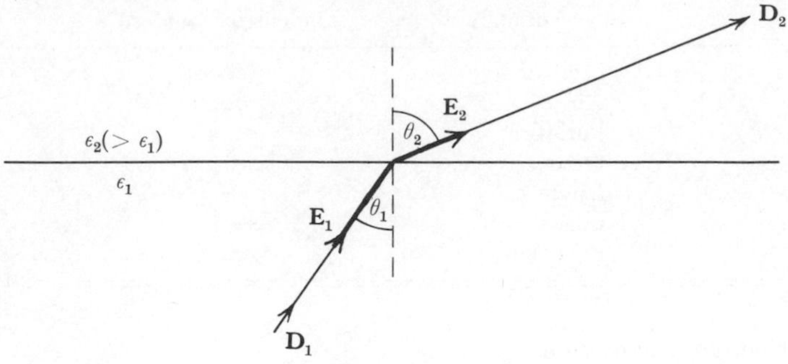

Figure 3.22

3.4.3 *Potential theory*

The formulation of electrostatics when dielectrics are present in terms of equations (3.110), (3.111) and (3.112) leads to easy generalizations of much of the potential theory described in § 3.3. If space is conceived as pervaded by an inhomogeneous dielectric characterized by a permittivity ϵ that is a function of position, then the generalization of Poisson's equation $\nabla^2\phi = -\rho/\epsilon_0$ is evidently

$$\operatorname{div}(\epsilon \operatorname{grad} \phi) = -\rho; \tag{3.116}$$

that is

$$\epsilon\nabla^2\phi + (\operatorname{grad}\epsilon).\operatorname{grad}\phi = -\rho. \tag{3.117}$$

In any homogeneous region throughout which ρ is zero this reduces to Laplace's equation $\nabla^2\phi = 0$.

The boundary conditions are that ϕ and $\epsilon\,\partial\phi/\partial n$ (being $-D_n$) are continuous across the interface between two dielectrics, and that ϕ is constant over the surface of a conductor.

The uniqueness theorem is a simple generalization of § 3.3.2. In the notation of that section, $\operatorname{div}(\epsilon \operatorname{grad}\psi) = 0$ in the region between conductors, and hence

$$\int \epsilon(\operatorname{grad}\psi)^2\,\mathrm{d}\tau = \int \epsilon\psi\frac{\partial\psi}{\partial n}\,\mathrm{d}S. \tag{3.118}$$

For each conducting surface the right hand side of (3.118) vanishes for just the reasons stated after (3.80). For the sphere at infinity it vanishes provided $\epsilon\,\partial\psi/\partial r$ tends to zero, as $r \to \infty$, faster than $1/r^2$. Thus, ϵ being positive, $\operatorname{grad}\psi$ is zero.

Two points should be made. First, that the proof outlined requires that ϵ be everywhere at least a differentiable function of position; strictly, therefore, the proof is not applicable to the commonly considered case in which ϵ has surfaces of discontinuity. This objection has no weight, since such surfaces can be represented by regions in which the change in ϵ is arbitrarily rapid but mathematically sufficiently smooth; or, more formally, such surfaces can be allowed for on the right hand side of (3.118), and are seen to have no effect by virtue of the boundary conditions.

The second point is of more interest. It concerns the condition on the behaviour of ϕ at infinity. In the vacuum case the behaviour is known from the inverse square law; but in the presence of inhomogeneous dielectrics no such simple presciption is available. The stated behaviour may be accepted as emerging from the consideration of uniqueness. Additionally, however, it is worth noting that, in the physically appropriate case in which the dielectric medium is bounded, ϕ is $O(1/r)$ as $r \to \infty$, since the field arises from a bounded distribution of charge: indeed, for that part of the field contributed solely by the dielectric charge density, ϕ is $O(1/r^2)$, since the net dielectric charge is zero.

The minimum energy theorem and Green's reciprocal theorem are also easily generalized.

The energy that represents the external work done in assembling charges in the presence of a prescribed dielectric medium is given by (3.55), namely

$$W = \frac{1}{2}\int \rho\phi \, d\tau.$$

Following the treatment of § 3.2.6, with (3.111) replacing $\rho = \epsilon_0 \operatorname{div} \mathbf{E}$, this can be transformed into

$$W = \frac{1}{2}\int \mathbf{E}.\mathbf{D} \, d\tau = \frac{1}{2}\int \epsilon \mathbf{E}^2 \, d\tau, \tag{3.119}$$

which reduces to (3.66) when ϵ is everywhere ϵ_0.

It is evident, then, that the analysis of § 3.3.4 goes through with appropriate minor modifications. In particular, the result corresponding to (3.100) is

$$W < W' - \frac{1}{2}\int \epsilon(\mathbf{E}' - \mathbf{E})^2 \, d\tau.$$

Moreover, it is obvious that Green's reciprocal theorem (3.103) still holds, since

$$\int \rho\phi' \, d\tau = \int \epsilon \mathbf{E}.\mathbf{E}' \, d\tau.$$

3.4.4 *Examples: generalizations from vacuum solutions*

If the solution is known for some disposition of charged conductors in a vacuum then it can immediately be generalized to the case in which the entire region between two equipotential surfaces is filled with a homogeneous dielectric. To make the generalization it need only be observed that in this context the presence of the dielectric does not affect $\mathbf{D}$. Thus $\mathbf{D}$ is known, being the same as in the vacuum case; and in consequence $\mathbf{E}$ is known, being $\mathbf{D}/\epsilon$ in the dielectric and $\mathbf{D}/\epsilon_0$ elsewhere. This solution is correct because it meets all the conditions, as a moments reflection confirms. Note that the charge density on the conductors is unaltered, being equal to

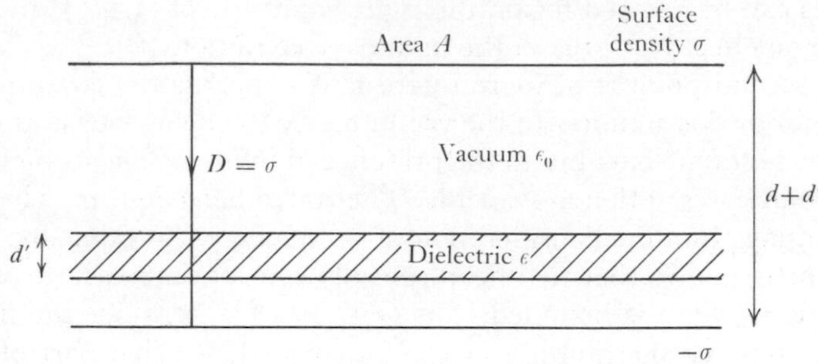

Figure 3.23

the outward normal component of **D**; but the potential difference between the conductors changes.

As a simple example consider the parallel plate capacitor shown in figure 3.23, in which a dielectric slab of thickness d' and permittivity $\epsilon = \epsilon_0 \kappa$ is placed between the plates whose separation is $d+d'$. Then between the plates **D** is everywhere normal to them and of uniform magnitude σ, where σ is the surface charge density on one plate. Hence **E** is normal to the plates, of magnitude σ/ϵ in the dielectric slab and σ/ϵ_0 elsewhere. The potential difference between the plates is therefore

$$V = \frac{\sigma}{\epsilon_0} d + \frac{\sigma}{\epsilon} d' = \frac{\sigma}{\epsilon_0}(d + d'/\kappa), \tag{3.120}$$

giving capacity
$$C = \sigma A/V = \epsilon_0 A/(d+d'/\kappa), \tag{3.121}$$

where A is the plate area. This result shows the expected agreement with capacities $\epsilon_0 A/d$ and $\epsilon_0 \kappa A/d'$ in series.

A further evident generalization is to fill regions between different pairs of equipotentials with different homogeneous dielectrics, leaving **D** unaltered. In the limit in which adjacent equipotentials approach one another it appears that the solution to any problem with conductors in a vacuum can be generalized to the case where the vacuum is replaced by any dielectric for which the permittivity is constant over each equipotential surface: **D** is unaltered, and **E** is given by $\mathbf{D}/\epsilon$.

The correctness of this result is confirmed mathematically by noting that in the proposed solution $\operatorname{div} \mathbf{D} = \rho$, as required, and $\operatorname{curl} \mathbf{D} = 0$ so that

$$\operatorname{curl} \mathbf{E} = \operatorname{curl}(\mathbf{D}/\epsilon) = -\mathbf{D} \wedge \operatorname{grad}(1/\epsilon),$$

which vanishes by virtue of the hypothesis that $\operatorname{grad} \epsilon$ is parallel to **D**.

As an example, suppose the spherical capacitor of figure 3.15 is filled with a dielectric of permittivity $\epsilon(r)$. Then **D** is radial, with $D = q/(4\pi r^2)$. Hence **E** is radial, with $E = q/(4\pi\epsilon r^2)$, the voltage is

$$\int_a^b E \, \mathrm{d}r,$$

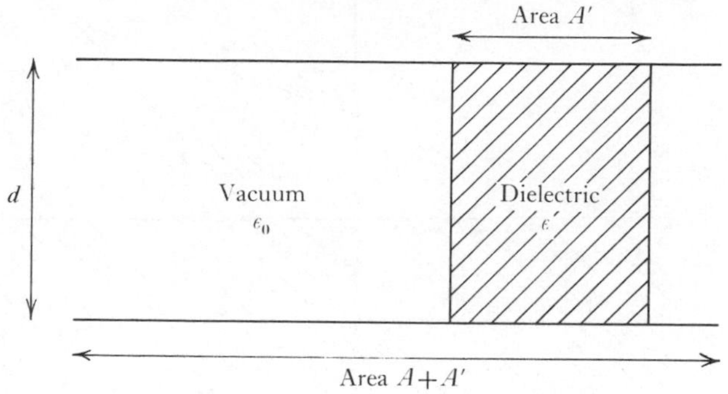

Figure 3.24

and the capacity therefore

$$C = 4\pi \Big/ \int_a^b \frac{dr}{\epsilon r^2}. \tag{3.122}$$

Another way of generalizing a vacuum problem is by filling an entire field tube with a homogeneous dielectric. In this case, if each conductor is maintained at its original potential, **E**, with ϕ, is unaffected; and **D** is ϵ**E** in the dielectric, ϵ_0**E** elsewhere. Again it is clear that all conditions are satisfied. Note that the charge density on those parts of conducting surfaces in contact with dielectric increases by the factor $\kappa = \epsilon/\epsilon_0$, whereas that on the remaining parts is unaltered.

Figure 3.24 shows a parallel plate capacitor with a dielectric cylinder normal to and terminating on the plates. Here **E**, between the plates, is everywhere normal to them and of uniform magnitude V/d, where V is the voltage and d the plate separation. Hence **D** is normal to the plates, of magnitude $\epsilon V/d$ in the dielectric and $\epsilon_0 V/d$ elsewhere. The capacity is therefore

$$C = \frac{\epsilon_0}{d}A + \frac{\epsilon}{d}A' = \frac{\epsilon_0}{d}(A + \kappa A'), \tag{3.123}$$

where A' is the area of the cross-section of the dielectric cylinder and $A+A'$ that of the plate. This is, of course, equivalent to capacities $\epsilon_0 A/d$ and $\epsilon A'/d$ in parallel.

The further implication now is that the solution to any problem with conductors in a vacuum can be generalized to the case where the vacuum is replaced by any dielectric for which the permittivity is constant along each field line; the potential ϕ is then unaltered, and **D** is ϵ**E**. Confirmation of this result comes from noting that the equation for ϕ, (3.117) with $\rho = 0$, reduces to $\nabla^2\phi = 0$ by virtue of the hypothesis that grad ϵ is orthogonal to grad ϕ.

If the cylindrical capacitor discussed in § 3.1.5 were filled with a dielectric of permittivity $\epsilon(\theta)$, and had potential drop $V_0 \log (b/a)$, then the field would

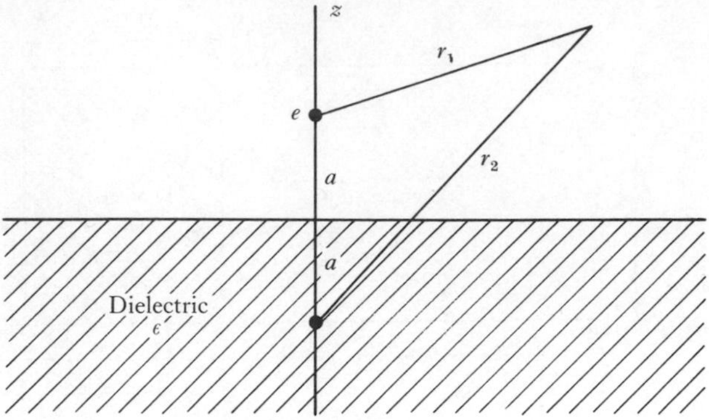

Figure 3.25

be radial with $E = V_0/r$. Thus $D = V_0\epsilon(\theta)/r$, giving the total charge on unit length of one cylinder

$$V_0\int_0^{2\pi}\epsilon(\theta)\,\mathrm{d}\theta,$$

so that the capacity per unit length is

$$\int_0^{2\pi}\epsilon(\theta)\,\mathrm{d}\theta/\log(b/a). \tag{3.124}$$

3.4.5 *Examples: images*

A few dielectric problems can be solved by methods akin to those discussed in § 3.3.3.

Imagine a point charge e at $(0, 0, a)$ in the presence of a homogeneous dielectric of permittivity $\epsilon = \epsilon_0\kappa$ that occupies the half-space $z < 0$. Try the potential function

$$\phi = \begin{cases} \dfrac{e}{4\pi\epsilon_0 r_1}+\dfrac{e'}{4\pi\epsilon_0 r_2}, & z > 0, \\ \dfrac{e''}{4\pi\epsilon_0 r_1}, & z < 0, \end{cases} \tag{3.125}$$

where r_1 and r_2 are distances from $(0, 0, a)$ and $(0, 0, -a)$ respectively (see figure 3.25). This potential goes to zero appropriately as $r_1 \to \infty$, and has the correct singularity at $r_1 = 0$. It only remains to find e' and e'' so that the boundary conditions at $z = 0$ are satisfied.

Continuity of ϕ across $z = 0$ requires

$$e+e' = e'', \tag{3.126}$$

and continuity of the normal component of $\mathbf{D}$ requires

$$e-e' = \kappa e''. \tag{3.127}$$

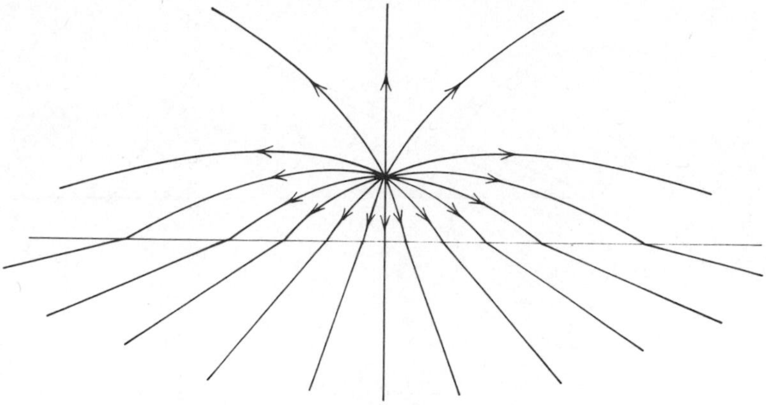

Figure 3.26

Hence
$$e' = -\frac{\kappa - 1}{\kappa + 1} e, \quad e'' = \frac{2}{\kappa + 1} e, \tag{3.128}$$
and the solution is obtained.

The field in $z > 0$ is identical to the vacuum field of two point charges, the given charge and the 'image' charge e' at $(0, 0, -a)$. The field in $z < 0$ is identical to the vacuum field of a point charge e'' at $(0, 0, a)$. Illustrative field lines are sketched in figure 3.26.

The effect of the polarization of the dielectric is represented in $z > 0$ by the image charge e'. This is surprising in that the dielectric as a whole is uncharged, and image charges would therefore be expected to have zero total charge. The paradox arises because of the assumed infinite extent of the dielectric, as is made clear a little later.

It is obvious that the problem of a uniform line charge along $y = 0$, $z = a$ in the presence of the dielectric half-space has a similar solution in terms of images. Specifically, if e is the charge per unit length,
$$\phi = \begin{cases} -\dfrac{e}{2\pi\epsilon_0} \log r_1 - \dfrac{e'}{2\pi\epsilon_0} \log r_2, & z > 0, \\ -\dfrac{e''}{2\pi\epsilon_0} \log r_1, & z < 0, \end{cases} \tag{3.129}$$
where e' and e'' are given by (3.128), and r_1 and r_2 are distances from the line charge and its image respectively.

It may be asked whether problems involving dielectric cylinders or spheres can be solved by images, on the lines of the treatment of conductors with these geometries in §3.3.3. It happens that a uniform line charge parallel to a circular cylindrical, homogeneous dielectric is soluble in terms of a finite number of image line charges, though a point charge in the presence of a dielectric sphere is not.

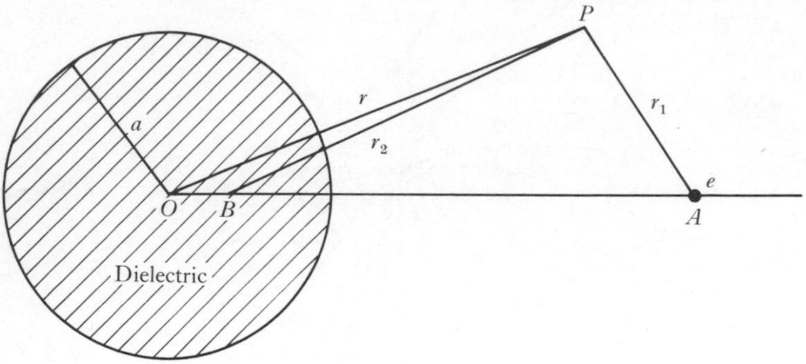

Figure 3.27

Figure 3.27 represents a uniform line charge e at A parallel to and distance b from the axis O of a dielectric cylinder $r < a$. B is inverse to A, that is, $OB = a^2/b$; and r_1 and r_2 are distances from A and B respectively.

If image line charges are to succeed in representing the field, presumably, for the region $r > a$, one will be located at B. If this is e' there must, somewhere, be another line charge $-e'$, because the dielectric has no net charge; and the only conceivable remaining location is at O. Try, therefore,

$$\phi = \begin{cases} -\dfrac{1}{2\pi\epsilon_0}(e\log r_1 + e'\log r_2 - e'\log r), & r > a, \\ -\dfrac{e''}{2\pi\epsilon_0}\log r_1 + \phi_0, & r < a. \end{cases} \tag{3.130}$$

It remains to be seen if the boundary conditions at $r = a$ can indeed be satisfied with suitable choice of the constants e', e'' and ϕ_0.

When the field point P is on $r = a$ the triangles OBP and OPA are similar (cf. figure 3.19, and the discussion in § 3.3.3). With $r = a$, $r_2 = ar_1/b$, it is seen that the continuity of ϕ requires

$$e\log r_1 + e'\log(ar_1/b) - e'\log a = e''\log r_1 - 2\pi\epsilon_0\phi_0,$$

which implies $2\pi\epsilon_0\phi_0 = e'\log b$ and

$$e + e' = e''. \tag{3.131}$$

Likewise the continuity of the normal component of $\mathbf{D}$ requires

$$\frac{e}{r_1}\cos\beta - \frac{be'}{ar_1}\cos\alpha + \frac{e'}{a} = \kappa\frac{e''}{r_1}\cos\beta,$$

where α and β are the angles shown in figure 3.19; and since the coefficient of e' is

$$-\frac{b\cos\alpha - r_1}{ar_1} = -\frac{\cos\beta}{r_1}$$

the relation is simply

$$e - e' = \kappa e''. \tag{3.132}$$

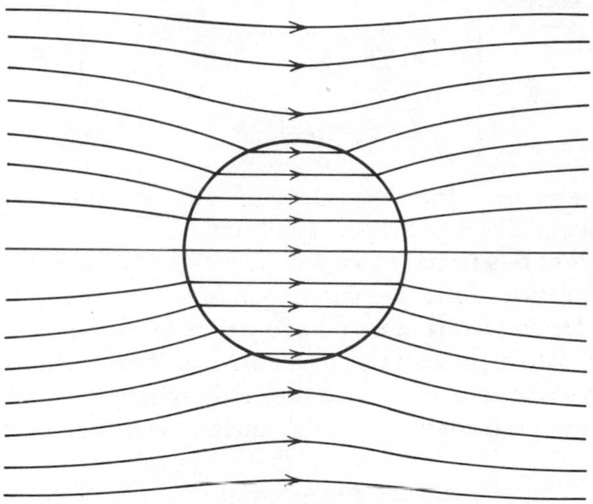

Figure 3.28

The method therefore succeeds. Furthermore (3.131) and (3.132) are identical with (3.126) and (3.127) respectively. Hence e' and e'' are given by (3.128), and there is evidently a very close parallel with the half-space problem. In fact, if both a and b tend to infinity with $b-a$ remaining fixed, the half-space solution is clearly recovered, with the additional information that there is an image charge $-e'$ which recedes to infinity. This explains the paradox of the apparently non-zero net image charge in the half-space solution.

There is another limiting case, which is of considerable interest. If both e and b tend to infinity with $e/b = -2\pi\epsilon_0 E_0$ remaining finite the problem becomes that of a dielectric cylinder in a given uniform field $\mathbf{E}_0$. The image system for the vacuum region is then a line dipole at O, directed along $\mathbf{E}_0$, of moment

$$\frac{a^2 e'}{b} = -\frac{\kappa-1}{\kappa+1}\frac{a^2 e}{b} = 2\pi\epsilon_0 \frac{\kappa-1}{\kappa+1} a^2 E_0. \tag{3.133}$$

The field in the dielectric is uniform, parallel to $\mathbf{E}_0$, and of magnitude

$$-\frac{e''}{2\pi\epsilon_0 b} = \frac{2}{\kappa+1} E_0. \tag{3.134}$$

The field lines are sketched in figure 3.28. For the spacing of the lines to be an inverse measure of the field strength they must be identified as **D** lines, since it is the flux of **D**, not **E**, that remains constant across the surface of the dielectric. Inside the dielectric, though the magnitude of **E** is less than E_0, that of **D** exceeds $\epsilon_0 E_0$.

In analytic form

$$\phi = \begin{cases} -E_0 r\cos\theta + \dfrac{\kappa-1}{\kappa+1}E_0\, a^2 \dfrac{\cos\theta}{r}, & r > a, \\ -\dfrac{2}{\kappa+1}E_0 r\cos\theta, & r < a, \end{cases} \qquad (3.135)$$

where θ measures angle from the direction of $\mathbf{E}_0$, and the result of problem 2.8 for the potential of a line dipole has been used.

The success of this form of solution stems simply from the fact that the angular dependence of the line dipole potential matches that of a uniform field. The analogous feature of the point dipole potential was noted previously in connection with a conducting sphere in a uniform field (see (3.88) and (3.89)), and also covers the case of a dielectric sphere in a uniform field.

For this latter problem the potential function is in fact

$$\phi = \begin{cases} -E_0 r\cos\theta + \dfrac{\kappa-1}{\kappa+2}a^3 E_0 \dfrac{\cos\theta}{r^2}, & r > a, \\ -\dfrac{3}{\kappa+2}E_0 r\cos\theta, & r < a, \end{cases} \qquad (3.136)$$

where r, θ are now *spherical* polar coordinates (there being axial symmetry about $\theta = 0$). By inspection (3.136) gives continuity across $r = a$ of both ϕ and the normal component of $\mathbf{D}$. The field inside the sphere is uniform.

Problems 3

3.1 For a set of collinear point charges there are field tubes with axial symmetry about the line of charges. Establish equation (3.7) for the field lines by applying Gauss' theorem to the closed surface formed by such a tube and two adjacent cross-sections normal to the axis.

3.2 (*a*) Show that, for equal point charges at A and B, a field line that is orthogonal to AB at B ultimately makes the angle $\pi/3$ with AB.

(*b*) Equal charges are situated at points A, B and C, where $AB = BC$. Locate the neutral points, and sketch the field lines through them by finding the directions of these field lines both at the neutral points and at infinity. Sketch some other representative field lines.

3.3 Line charges $e_1, e_2, \ldots, e_n$ per unit length are parallel to the z axis, and intersect the plane $z = 0$ in the points $E_1, E_2, \ldots, E_n$, respectively. Show that each field line in the plane $z = 0$ is the locus of points P for which

$$\sum_{i=1}^{n} e_i\theta_i = \text{constant},$$

where θ_i is the angle between E_iP and any fixed line in $z = 0$.

Deduce that the field lines of parallel line charges e and $-e$ per unit length are circles.

3.4 A circular disc of radius a has uniform surface charge density σ. Show that the potential at the rim of the disc is $\sigma a/(\pi\epsilon_0)$.

3.5 Two infinite parallel conducting planes p_1 and p_2, a distance $2d$ apart, are kept at zero potential. A third infinite plane, p, carrying a uniform surface charge density σ, lies between them at distance $d-x$ from p_1. Find the charge densities on p_1 and p_2, and the potential of p.

Find also the electric force on p, and show that if the plane be subject additionally to an oppositely directed mechanical force λx per unit area, then the equilibrium position midway between p_1 and p_2 is stable if and only if $\lambda > \sigma^2/(2\epsilon_0 d)$.

3.6 A conducting film of soap solution of surface tension T forms an uncharged spherical bubble of radius a. The bubble then acquires a charge, so that it expands isothermally to radius b and has potential V. Show that

$$p_0(b^3-a^3)+4T(b^2-a^2) = \tfrac{1}{2}\epsilon_0 V^2 b,$$

where p_0 is the external pressure.

3.7 From the results of problem 2.8 for the potential and field of a line dipole show that the equipotentials are circular cylinders and the field lines circles, and verify their orthogonality.

Verify also that the equipotentials (3.9) and the field lines (3.3) of a point dipole are orthogonal.

3.8 Two electric dipoles, each of moment p, are distance a apart, one being oriented along the line joining them, and the other perpendicular to that line. Find the force and couple on each dipole.

3.9 An electric dipole of moment $\mathbf{p}$, which is fixed in magnitude but variable in direction, moves with velocity $\mathbf{v}$ in a magnetic field $\mathbf{B}$. Show that the force on the dipole is

$$\mathbf{v}\wedge(\mathbf{p}.\mathrm{grad})\,\mathbf{B}+\frac{\mathrm{d}\mathbf{p}}{\mathrm{d}t}\wedge\mathbf{B}.$$

3.10 Charge q is uniformly distributed over the surface of a sphere of radius a. Write down the energy from (3.57), and verify that the same result is given by direct evaluation of (3.66).

Calculate the energy if q is the electron charge and a the classical electron radius.

3.11 Charge q is uniformly distributed throughout the sphere $r < a$. Verify, by direct evaluation of the integrals, that (3.55) and (3.66) give the same result for the energy, namely $3q^2/(20\pi\epsilon_0 a)$. What fraction of the latter integral is contributed by the region $r < a$?

3.12 Charge is distributed with density $\rho = qk^2\mathrm{e}^{-kr}/(4\pi r)$, where r is distance from the origin and q and k (>0) are constants. Show that the total charge is q. Use Gauss' theorem to find the field, and deduce the potential. Show that the energy is $q^2k/(16\pi\epsilon_0)$.

What charge distribution gives the potential $\phi = q\mathrm{e}^{-kr}/(4\pi\epsilon_0 r)$?

3.13 Electrostatic fields $\mathbf{E}_1$ and $\mathbf{E}_2$ are superposed. Show that the interaction energy (that additional to the sum of the energies of the individual fields in isolation) is

$$\epsilon_0\int_{\text{all space}}\mathbf{E}_1.\mathbf{E}_2\,\mathrm{d}\tau.$$

Consider the case in which $\mathbf{E}_1$ is due to a point charge e_1 at the origin, and $\mathbf{E}_2$ to a point charge e_2 at (0, 0, a). Show by introducing spherical polar coordinates that the above integral for the interaction energy can be written

$$\frac{e_1e_2}{8\pi\epsilon_0}\int_0^\infty\int_0^\pi\frac{(r-a\cos\theta)\sin\theta}{(r^2+a^2-2ar\cos\theta)^{\frac{3}{2}}}\,\mathrm{d}r\,\mathrm{d}\theta.$$

Evaluate this by first doing the r integration (change the variable to $s = r - a\cos\theta$). The result $e_1 e_2/(4\pi\epsilon_0 a)$ is expected: why?

3.14 A capacitor consists of a conducting plate in the form of two opposite quadrants of a circle of radius a (joined at the centre) suspended in a horizontal plane above an identical parallel plate. The plates are distance b ($\ll a$) apart, with their centres in the same vertical line, and the upper is rotated through an acute angle θ relative to the lower. To the approximation in which edge effects are neglected, find the capacity, and also the energy associated with a potential difference V between the plates.

Derive from the energy both the increase in the tension of the suspension due to the attraction between the plates, and also the torque that must be exerted by the suspension to maintain equilibrium.

3.15 Rectangular conducting plates occupy the portions $0 < x < a$, $0 < y < b$ of the planes $z = 0$, $z = 4d$, and carry charges $-q$, q, respectively. Others occupy the portions $\xi < x < a+\xi$, $0 < y < b$ of the planes $z = 3d/2$, $z = 5d/2$, and carry charges $2q$ and $-2q$, respectively. Show that, if d is much less than a, b and ξ (so that edge effects may be neglected), the energy is $2q^2d/[\epsilon_0 b(a-\frac{1}{2}\xi)]$.

Deduce that there is a force on each plate tending to reduce ξ, and find its value.

3.16 Of three concentric, spherical, conducting shells the inner and outer, which have radii a and c respectively, are earthed, whilst the intermediate one, of radius b, carries a charge q. Show that the energy is

$$\frac{1}{8\pi\epsilon_0}\frac{(c-b)(b-a)}{b^2(c-a)}q^2.$$

Show that there is no net electrical force tending to expand or contract the intermediate shell if $b = 2ac/(a+c)$.

3.17 Show that, if r, θ ($0 \leqslant \theta \leqslant 2\pi$) are cylindrical polar coordinates, and A is a constant, then

$$\phi = Ar^{\frac{1}{2}}\sin(\tfrac{1}{2}\theta)$$

is the potential due to a semi-infinite, charged conducting plane. Find the electric field everywhere, and the charge density on the plane.

3.18 Confirm that the cylindrical polar coordinate θ is a solution of Laplace's equation.

Consider (a) a pair of conducting plates at potentials $\pm V$, which together form a complete plane, save for an infinitesimal gap between the straight edges of the plates; (b) a pair of identical conducting sheets at potentials $\pm V$, which together form an infinitely long, circular cylindrical surface, complete save for infinitesimal gaps between the straight edges of the sheets. Obtain in each case the potential, the field and the charge density on the conductors.

3.19 Indicate how the solutions of problem 3.18 can be generalized to the cases in which (a) the conducting plates together form a wedge of arbitrary angle rather than a plane; (b) the identical conducting sheets have cross-sections that are arbitrary arcs of circles rather than semi-circles, and together form a cylinder whose cross-section is an ogive rather than a circle.

3.20 Suppose in problem 3.5 that instead of the plane p a point charge e is situated between p_1 and p_2 at distance $d-x$ from p_1. What charges are induced on p_1 and p_2?

Use images to show that the equilibrium position $x = 0$ is unstable, and that for a particle of mass m a small displacement from this position grows initially as $e^{t/\tau}$, where

$$\tau^2 = 32\pi\epsilon_0 md^3 \bigg/ \left(7e^2 \sum_{r=1}^{\infty} 1/r^3\right).$$

3.21 A point charge e is distance d from the centre of a conducting sphere of radius $a(< d)$. What charge must the sphere carry for there to be no force on the point charge?

3.22 The surface of a charged conductor is defined by two spherical surfaces that intersect orthogonally. Show that the field can be represented by three image point charges.

Sketch the field lines for the case when the spheres have equal radii.

3.23 A point charge e is distance x from the centre of a conducting sphere of radius a ($< x$). Find the energy W in the two cases (*a*) sphere earthed, (*b*) sphere uncharged. Explain carefully why in each case the attractive force on the point charge is $\partial W/\partial x$, and verify that this is so by direct application of the inverse square law to the charge and the image charges.

3.24 A dipole of moment p is distance d from an infinite, plane, earthed conductor and makes an angle α with the normal to the plane. Find the force and couple on the dipole, and also the surface charge density on the plane.

3.25 A dipole of moment p is distance d from the centre of an earthed conducting sphere of radius a ($< d$). Find the image system (*a*) when the dipole lies along the line joining it to the centre of the sphere, (*b*) when it is perpendicular to that line. What is the total charge on the sphere in each case?

3.26 A wire lies along $x = a$, $y = b$ and conducting sheets occupy $y = 0$, $x \geqslant 0$ and $x = 0$, $y \geqslant 0$. Find an approximation to the capacity (per unit length in the z direction) in the case when the radius r of the cross-section of the wire is much less than a and b.

3.27 A capacitor formed by coaxial, circular cylindrical, conducting shells of radii a and b ($> a$) has capacity $2\pi\epsilon_0/\log(b/a)$ per unit length. Find approximately the change in the capacity if the cylinders are slightly displaced so that the axes are separated by a distance d ($\ll a, b$).

3.28 Four identical conducting spheres are centred at the vertices of a regular tetrahedron. When two spheres each have a charge q and the other two each have a charge $-q$ the first pair are at potential V. If, instead, one sphere has charge Q and the others are uncharged, what is the potential difference between them?

3.29 What is the maximum energy per cubic metre that can be stored in a capacitor filled with a homogeneous medium of dielectric constant 3 which can withstand without breakdown a field up to, but not beyond, 10^7 volt m^{-1}?

3.30 The region between two concentric spherical conducting shells is filled on one side of a diametral plane with a homogeneous medium of dielectric constant κ_1, and on the other side with a homogeneous medium of dielectric constant κ_2. The shells are insulated; the inner, of radius a, carries charge q_1, and the outer, of radius b, charge q_2. Find the electric field at all points.

3.31 A conducting spherical shell of mass M and radius a floats on oil of dielectric constant κ. When it is uncharged a fraction $1/\lambda$ ($< \frac{1}{2}$) of the volume of the sphere is submerged. Show that the aquisition of a charge q causes the sphere to be half submerged if

$$q^2 = 4\pi\epsilon_0(\lambda-2)\frac{(\kappa+1)^2}{\kappa-1}Mga^2.$$

3.32 Calculate the force on a dipole of moment p that is in vacuum at distance d from, and perpendicular to, the plane face of a semi-infinite homogeneous medium of dielectric constant κ.

3.33 A conducting sphere of radius a, covered by a concentric spherical shell of uniform dielectric constant κ, of internal and external radii a and b, is placed in a previously uniform electric field $\mathbf{E}_0$. Find the potential everywhere, and confirm that it agrees with the results given in the text (*a*) when $b = a$, (*b*) when $a = 0$.

3.34 An unbounded, homogeneous medium of dielectric constant κ is permeated by a uniform electric field $\mathbf{E}_0$. A sphere of radius a is then excised from the dielectric, and a dipole of moment p placed at the centre of the cavity and oriented in the direction opposite to $\mathbf{E}_0$. Find the potential everywhere, and sketch the lines of force outside the cavity in the case $\kappa = 2$, $p = 19a^3E_0/(12\pi\epsilon_0)$.

3.35 A dipole of moment p is at the centre of a homogeneous spherical shell, of internal and external radii a and b, and dielectric constant κ. Show that the external field is the same as if the shell were absent and the moment of the dipole were λp, where

$$1/\lambda = 1+\frac{2}{9}\frac{(\kappa-1)^2}{\kappa}\left(1-\frac{a^3}{b^3}\right).$$

3.36 Show that if there is a small increase $\delta\epsilon$ (possibly a function of position) in the permittivity of the medium between charged conductors, then the energy increases by

$$\frac{1}{2}\int_{\text{all space}} E^2\delta\epsilon\,\mathrm{d}\tau$$

in the case when the potential of each conductor is kept constant, and decreases by the same amount when the charge on each conductor is kept constant.

Deduce that an increase in the permittivity anywhere in the dielectric between the plates of a capacitor increases the capacity.

4

THE MAGNETIC FIELD OF STEADY AND SLOWLY VARYING CURRENTS

4.1 Steady currents in a conductor

4.1.1 *Conductivity*

This chapter gives, in the main, an investigation of the time-independent magnetic field. It also includes some discussion of magnetic fields that vary sufficiently slowly for equation (2.52), namely

$$\frac{1}{\mu_0}\operatorname{curl}\mathbf{B} = \mathbf{J}, \tag{4.1}$$

to remain valid. The condition that the displacement current term $\epsilon_0 \dot{\mathbf{E}}$, which appears additionally on the right hand side in the exact Maxwell equation (2.106), be negligible is roughly that the length scale L and the time scale T under consideration satisfy the inequality

$$L/T \ll c, \tag{4.2}$$

as can be seen by an order of magnitude argument invoking the other Maxwell equation (2.104).

The fundamental equations of magnetostatics were introduced in § 2.2, and the concept of field lines and tubes, explained with reference to the electrostatic field in § 3.1.1, can likewise be applied to the magnetic field. At each point on a line the tangent is in the direction of the vector $\mathbf{B}$; and since $\operatorname{div}\mathbf{B} = 0$ everywhere, the area of normal cross-section of an infinitesimal field tube is inversely proportional to the field strength, and field tubes are closed (cf. the discussion of steady current flow in § 2.2.1).

The description of the magnetostatic field is somewhat more complicated than that of the electrostatic field. The source is a vector quantity, current density $\mathbf{J}$, rather than a scalar quantity, charge density ρ; and the generally valid potential representation is in terms of the curl of a vector $\mathbf{A}$ rather than the gradient of a scalar ϕ. There are, it is true, problems advantageously treated in terms of the magnetic scalar potential in a manner entirely analogous to electrostatics, but these only form a not fully representative part of the whole.

In electrostatics the common situation is that in which charges on the surfaces of conductors are in equilibrium under the combined action of their own electric field and the forces confining them to the conductors. In magnetostatics there can be an analogous situation, which will be considered later. The more familiar case, however, is when current is made to flow in a conductor by the maintenance of an electric field through an external agency, such as a battery; then the pattern of current flow simply follows that of the electric field lines. A magnetic field is, of course, generated by the current, but the effect of its own magnetic force on the distribution of the current is ordinarily negligible.

The conduction electrons kept in motion by a steady electric field in a conducting medium experience some resistance. They might thus be expected to move, on average, with a velocity proportional at each point to the electric field, analogous to the terminal velocity of a body sliding down an inclined plane. This would imply current density proportional to electric field, a conclusion which is confirmed experimentally, being no more than a version of Ohm's law.

If the result is written

$$\mathbf{J} = \sigma \mathbf{E}, \tag{4.3}$$

σ is called the *conductivity*. The dimensions of σ are mho m^{-1}, where mho signifies amp $volt^{-1}$ (reciprocal ohm). Some values are given in the table below.

Substance	Conductivity (mho m^{-1})
silver	6.8×10^7
copper	5.8×10^7
lead	4.5×10^4
sea water	4
fresh water	10^{-3}

The conductivity of different substances ranges over many orders of magnitude. If a very good conductor $\mathscr{C}$ is embedded in a medium of appreciably lower conductivity it may be an adequate approximation to treat $\mathscr{C}$ as a *perfect* conductor, that is, one for which the conductivity is infinite. To this approximation the implication of (4.3) is that the electric field inside $\mathscr{C}$ is zero.

4.1.2 *Description of flow pattern*

The calculation of the current distribution in a conductor is mathematically equivalent to finding the electrostatic field in a dielectric, as is now shown.

Consider two perfectly conducting electrodes embedded in a medium of conductivity σ and maintained by a battery at a potential difference V. The steady flow of current in the medium from one electrode to the other is governed by (4.3), together with

$$\operatorname{div} \mathbf{J} = 0 \tag{4.4}$$

and

$$\mathbf{E} = -\operatorname{grad} \phi, \tag{4.5}$$

where ϕ is constant on each electrode, the values differing by V. The key observation is that, if in these equations $\mathbf{J}$ is replaced by $\mathbf{D}$, and σ, which may be a function of position, by ϵ, then the equations are identical with those governing the electrostatic field of conductors in a dielectric.

Moreover, the total current flowing between the electrodes through the medium, and returning through the battery via connecting wires, is given by

$$I = \int \mathbf{J} \, . \, d\mathbf{S}, \tag{4.6}$$

where the integral is taken over the surface of the positive electrode: this current is, of course, proportional to V, and the ratio

$$R = V/I \tag{4.7}$$

defines the *resistance* between the electrodes. Analogously, in the dielectric case, the total charge on the conductor at higher potential is

$$q = \int \mathbf{D} \, . \, d\mathbf{S},$$

that on the other conductor being $-q$, and the ratio

$$C = q/V$$

is the capacity of the capacitor formed by the two conductors. Thus the capacity in the one case is analogous to reciprocal resistance, called *conductance*, in the other case.

It should perhaps be emphasized that all the current leaving one electrode enters the other, and returns through the battery and the connections to it; there is no flow to infinity, because there is no return route from infinity. The dielectric analogy is therefore solely with the case in which the conductors in the dielectric carry equal and opposite charges, so that all the field lines proceed from one conductor to the other. Further, the analogous problem may not be realizable: in particular, if the conducting medium is bounded by a non-conducting medium, a vacuum, for example, the lines of current flow must be tangential to the bounding surface; whereas there is no comparable boundary condition in any practical dielectric problem since no dielectric has zero permittivity.

Another point to note is that, in the current flow problem, there will in general be a charge density ρ. This is determined, when the appropriate solution of equations (4.3), (4.4) and (4.5) has been found, from the relations $\rho = \operatorname{div} \mathbf{D}$, $\mathbf{D} = \epsilon\mathbf{E}$. Here ϵ is the actual permittivity of the conducting medium and $\mathbf{D}$ the actual displacement vector (there should be no confusion with the quantities analogous to σ and $\mathbf{J}$; $\mathbf{E}$ alone designates the same quantity in the actual and in the analogous problem). For example, at an interface between two media of different conductivities the normal component of $\mathbf{J}$ is continuous, as implied by (4.4), so in general the normal component of **D** is discontinuous and there is correspondingly a surface charge density.

Evidently, with due attention to detail, many of the results of § 3.4 can be applied to steady current flow. Two such are now mentioned.

4.1.3 *Energy and reciprocal theorems*

Consider energy. The rate at which the electric field does work on the conduction electrons is

$$\mathbf{J}.\mathbf{E} \tag{4.8}$$

per unit volume, and this represents the rate of dissipation of energy in the medium, known as Joule heating. Since (4.8) can be written $-\operatorname{div}(\phi\mathbf{J})$, the volume integral

$$\int \mathbf{J}.\mathbf{E}\,d\tau \tag{4.9}$$

taken over the region exterior to the electrodes transforms to

$$\int \phi\mathbf{J}.d\mathbf{S}$$

taken over the surfaces of both electrodes; which confirms that (4.9) is equal to

$$IV, \tag{4.10}$$

the rate at which the battery works to maintain the current.

Now the replacement of **J** by **D** in (4.8) gives **E**.**D**, which is twice the field energy density in the dielectric problem (see (3.119)). The analogue of the minimum energy theorem of § 3.4.3 can therefore be stated as follows: for a system of perfectly conducting electrodes embedded in a conducting medium the rate of dissipation of energy for the actual steady current flow is less than that for any different (hypothetical) current flow in which the total current leaving or entering each electrode is unaltered.

As another example, consider Green's reciprocal theorem. The analogue is that, for a system of n electrodes,

$$\sum_{i=1}^{n} I_i V_i' = \sum_{i=1}^{n} I_i' V_i, \tag{4.11}$$

where currents I_i leave the electrodes when their potentials are V_i, and currents I_i' when their potentials are V_i'.

4.2 Currents and fields

4.2.1 *Uni-directional axially symmetric current*

In the common case when the conducting medium in which the current flows is a thin wire the situation is relatively simple and the magnetic field outside the wire can be calculated on the assumption that the current flows along a line.

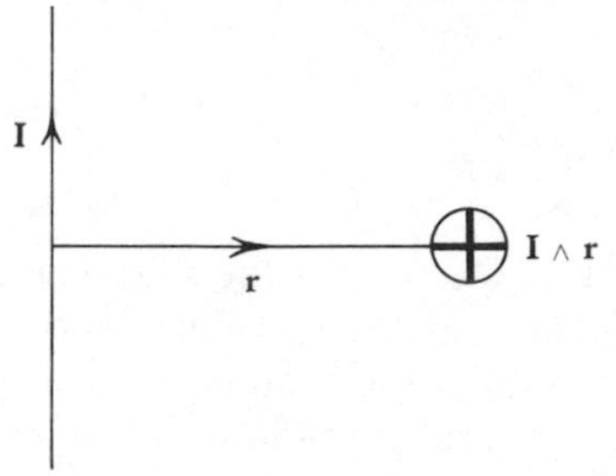

Figure 4.1

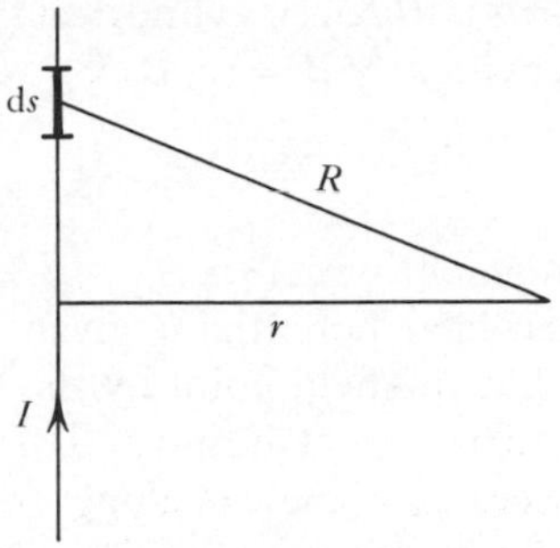

Figure 4.2

Suppose that steady current I flows along an infinite straight line. The magnetostatic field can be obtained from the Biot–Savart law (2.55), as was in effect done in § 1.3.1. The magnetic field lines are circles centred on the line current, in planes normal to it. The direction of the field is in the sense of a right-handed screw relative to the sense of current flow, and the magnitude is given by (1.15), namely

$$B = \frac{\mu_0 I}{2\pi r}, \tag{4.12}$$

where r is the distance from the line current. Vectorially,

$$\mathbf{B} = \frac{\mu_0}{2\pi}\frac{\mathbf{I}\wedge\mathbf{r}}{r^2}, \tag{4.13}$$

as indicated in figure 4.1.

Steady current has closed lines of flow, and in principle it is to be imagined in the present case that there is a return circuit at infinity. Mere inspection of the Biot–Savart law shows that any such return circuit makes no contribution to the field.

It is also instructive to note that once it is recognized from the Biot–Savart law that the geometry of the field lines is an obvious consequence of symmetry, then the magnitude of the field, (4.12), is immediately obtained by applying the integral form of Ampère's circuital law, (2.51).

This simple problem may also be used to illustrate the application of the vector and scalar potentials. For the former, (2.63) gives

$$\mathbf{A} = \frac{\mu_0}{4\pi}\mathbf{I}\int\frac{\mathrm{d}s}{R}, \tag{4.14}$$

where R is distance from the line element $\mathrm{d}s$ (see figure 4.2); and although (4.14) is only a formal statement, since the integral along the infinite line diverges, the mathematics is analogous to the calculation of the electrostatic potential of an infinite uniform line charge, and consequently (cf. (3.32))

$$\mathbf{A} = -\frac{\mu_0\mathbf{I}}{2\pi}\log r. \tag{4.15}$$

To form curl **A** use cylindrical polar coordinates (r, θ, z), with $\mathbf{I} = (0, 0, I)$. Then evidently $\mathbf{B} = (0, B_\theta, 0)$, where

$$B_\theta = -\frac{\partial A_z}{\partial r} = \frac{\mu_0 I}{2\pi r},$$

in agreement with (4.12).

The scalar potential is given by (2.49), where Ω is the solid angle subtended at the field point by the loop formed by the rectilinear current and the return line at infinity. To find this solid angle note that two planes intersecting, at angle θ, along a diameter of a sphere of unit radius intercept area 2θ on the surface of the sphere. Hence

$$\phi = -\frac{I\theta}{2\pi}, \tag{4.16}$$

and so $\mathbf{B} = -\mu_0 \operatorname{grad} \phi = (0, B_\theta, 0)$, where

$$B_\theta = -\frac{\mu_0}{r}\frac{\partial \phi}{\partial \theta} = \frac{\mu_0 I}{2\pi r}.$$

Suppose now that the simple line current is replaced by some unidirectional, axially symmetric current distribution; that is, by a current density **J** with cartesian components $(0, 0, J)$, where J is a function only of distance r from the z axis. Then clearly symmetry still dictates that the field lines are circles $r =$ constant, so the field strength can be obtained from the circuital law (2.51). It is

$$B = \frac{\mu_0}{r}\int_0^r r' J(r')\,\mathrm{d}r'. \tag{4.17}$$

If, for example, the current flow were confined to $r < a$, being uniform within the cylinder and giving total current I, then

$$J = \begin{cases} I/(\pi a^2), & r < a, \\ 0, & r > a, \end{cases} \tag{4.18}$$

and

$$B = \begin{cases} \mu_0 I r/(2\pi a^2), & r < a, \\ \mu_0 I/(2\pi r), & r > a. \end{cases} \tag{4.19}$$

Or if the current were confined to the *surface* of the cylinder $r = a$, then

$$B = \begin{cases} 0, & r < a, \\ \mu_0 I/(2\pi r), & r > a. \end{cases} \tag{4.20}$$

It should perhaps be mentioned that here, and elsewhere until stated to the contrary, the media in which current flows are assumed to have no significant intrinsic magnetic property. Magnetic media are not considered until §4.5.

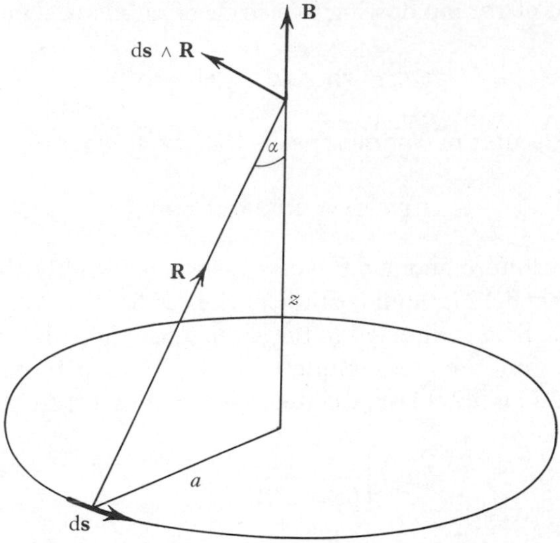

Figure 4.3

4.2.2 *Circular current loop*

Suppose current I flows in a circular loop of radius a. Then the field at a point P on the axis of the loop is easily calculated, either from the Biot–Savart law or by using the scalar potential.

The field at P is evidently directed along the axis. Now the component of $\mathbf{ds} \wedge \mathbf{R}$ in this direction is $R\,ds \sin\alpha$, where $\mathbf{R}$ is the vector from element $\mathbf{ds}$ of the loop to P, and α is the angle R makes with the axis (see figure 4.3). Therefore (2.55) gives field strength

$$B = \frac{\mu_0 I}{4\pi}\frac{2\pi a}{R^2}\sin\alpha = \tfrac{1}{2}\mu_0 I a^2/(a^2+z^2)^{\frac{3}{2}}, \tag{4.21}$$

where z is the distance of P from the centre of the loop.

Alternatively, the scalar potential at P, given by (2.49), is

$$\phi = \frac{I}{4\pi}2\pi(1-\cos\alpha) = \tfrac{1}{2}I(1-z/R). \tag{4.22}$$

Hence, from (2.46),
$$B = \tfrac{1}{2}\mu_0 I\frac{\partial}{\partial z}\left[\frac{z}{(a^2+z^2)^{\frac{1}{2}}}\right], \tag{4.23}$$

which checks with (4.21).

Incidentally, the vector potential $\mathbf{A}$ is obviously zero on the axis; its derivatives transverse to the axis are required to obtain $\mathbf{B}$ on the axis.

At the centre of the loop the field strength is

$$\mu_0 I/(2a). \tag{4.24}$$

Thus a current of 10 amp flowing in a circle of radius 10^{-1} m produces at its centre a field

$$50\mu_0 = 2\pi \times 10^{-5}\,\text{weber m}^{-2}.$$

A magnetic field unit in common use is the gauss, where

$$1\ \text{gauss} = 10^{-4}\,\text{weber m}^{-2}.$$

The field is therefore about 0.6 gauss, which is roughly the value of the earth's magnetic field in high latitude at the earth's surface.

Note that the field is directed in the same sense right along the axis of the loop. By imagining the axis completed by a circuit at infinity it is easy to verify Ampère's circuital law; for the integral of (4.23) along the axis gives

$$\tfrac{1}{2}\mu_0 I\left[\frac{z}{(a^2+z^2)^{\frac{1}{2}}}\right]_{-\infty}^{\infty} = \mu_0 I. \tag{4.25}$$

It is a more complicated matter to obtain the field at points off the axis of the loop. However, certain features can be comparatively easily deduced.

If $\mathbf{r}$ be the radius vector from the centre of the loop, then, apart from the factor $\mu_0 I/a$, the field depends only on $\mathbf{r}/a$.

For $r/a \gg 1$ the field is approximately that of a magnetic dipole of moment $\pi a^2 I$ located at the centre of the loop and directed along the axis: this being nothing other than a statement of the law introduced in § 2.2.2 (compare (4.21) with (2.45)). For $r/a \ll 1$ the field is approximately that at the centre of the loop.

More detail can be obtained fairly easily for points near the axis. For if ρ denotes distance from the axis,

$$B_\rho = f(z)\rho + O(\rho^2);$$

and $f(z)$ can be found from $\operatorname{div}\mathbf{B} = 0$, which is here

$$\frac{1}{\rho}\frac{\partial}{\partial\rho}(\rho B_\rho) + \frac{\partial B_z}{\partial z} = 0,$$

giving

$$f(z) = -\frac{1}{2}\left(\frac{\partial B_z}{\partial z}\right)_{\rho=0}.$$

Thus, from (4.21),

$$B_\rho = \tfrac{3}{4}\mu_0 I\frac{a^2 z\rho}{(a^2+z^2)^{\frac{5}{2}}} + O(\rho^2). \tag{4.26}$$

Furthermore the θ component of $\operatorname{curl}\mathbf{B} = 0$ is

$$\frac{\partial B_z}{\partial\rho} = \frac{\partial B_\rho}{\partial z};$$

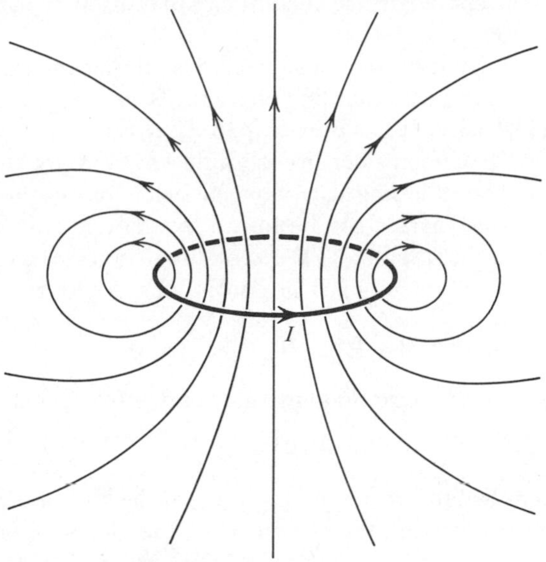

Figure 4.4

and the substitution of (4.26) in the right hand side, followed by integration with respect to ρ, gives

$$B_z = \tfrac{1}{2}\mu_0 I a^2 \frac{1}{(a^2+z^2)^{\frac{3}{2}}}\left[1+\frac{3}{4}\frac{a^2-4z^2}{(a^2+z^2)^2}\rho^2+O(\rho^3)\right]. \tag{4.27}$$

Another important approximation is available when the field point is close to the loop; that is, $\rho/a \simeq 1$, $z/a \ll 1$. In this case the field may be calculated on the assumption that the current loop is replaced by an infinite straight line, in the plane of the loop and tangent to it at the point nearest the field point. The field lines in the vicinity of the loop are thus almost circular. This idea is not restricted to loops of any particular shape. Roughly speaking it provides a working approximation to the field of any current loop for points at distances from the loop that are small compared with the local radius of curvature of the loop. Moreover, it can also be used to calculate the field *inside* a wire carrying a current, if the distribution of current across the wire is known.

The features that have been discussed give, in combination, a fair idea of the geometry of the field lines. These are indicated in figure 4.4.

4.2.3 *Surface currents and superconductors*

The concept of surface charge is helpful in indicating how the electrostatic field can be pictured as partitioned into separate regions where the fields are independently supported by appropriate charge densities on equipotential surfaces. The concept of surface current can play a somewhat similar role in magnetostatics.

Suppose that current flows in a surface S, and that the rate of passage of charge across each infinitesimal line element $d\mathbf{s}$ in S is $\mathbf{j}.d\mathbf{s}$. Then $\mathbf{j}$ is the surface current density. It has unit amp m^{-1}.

A finite surface current density signifies an infinite volume current density. The effect of this singularity on the behaviour of the magnetic field in the vicinity of the surface can be found by analysis quite similar to that in § 3.1.3. The fact that the flux of $\mathbf{B}$ out of any closed surface is zero implies that the normal component of $\mathbf{B}$ is continuous across the surface; or in symbols (cf. (3.10))

$$[\mathbf{n}.\mathbf{B}] \equiv B_{n2} - B_{n1} = 0. \tag{4.28}$$

Furthermore, Ampère's circuital law (2.51) gives (cf. (3.11))

$$[\mathbf{n} \wedge \mathbf{B}] = \mu_0 \mathbf{j}; \tag{4.29}$$

for if (2.51) is applied to a rectangular circuit of the kind shown in figure 3.7, there is current $j\,ds$ through the circuit when the long sides of the rectangle (of length ds) are perpendicular to $\mathbf{j}$, and none when they are parallel to $\mathbf{j}$.

The combination of (4.28) and (4.29) can be expressed in the single statement (cf. (3.12))

$$[\mathbf{B}] = \mu_0 \mathbf{j} \wedge \mathbf{n}. \tag{4.30}$$

That the corresponding values $\mathbf{B}_1$ and $\mathbf{B}_2$ of $\mathbf{B}$ on either side of a surface current are different raises the question of what magnetic force is exerted on the current. An argument analogous to that which lead to (3.17) shows the force to be

$$\tfrac{1}{2}\mathbf{j} \wedge (\mathbf{B}_1 + \mathbf{B}_2) \tag{4.31}$$

per unit area.

In practical electrostatics the generation of surface charge is associated with the presence of conductors in a way that has been fully explained. It has been stressed that the equilibrium state is reached very quickly, and a quantitative assessment of the time scale involved follows from (4.3), if it is provisionally accepted that this relation holds also for time varying fields. For substitution into $\operatorname{div}\mathbf{E} = \rho/\epsilon_0$ gives, for a homogeneous conductor,

$$\operatorname{div}\mathbf{J} = \frac{\sigma}{\epsilon_0}\rho,$$

which in turn implies, from the charge conservation relation (1.10), that

$$\dot{\rho} + \frac{\sigma}{\epsilon_0}\rho = 0. \tag{4.32}$$

Thus
$$\rho(\mathbf{r}, t) = \rho(\mathbf{r}, 0)\, e^{-\sigma t/\epsilon_0}. \tag{4.33}$$

The volume charge density at any interior point of a conductor is therefore predicted to decay exponentially, diminishing by the factor $1/e$ during every time interval ϵ_0/σ. If $\sigma = 5\times 10^7$ this interval is about 0.2×10^{-18} sec, indicating how rapidly any charge initially inside a good conductor would be replaced by charge on the surface.

In practical magnetostatics the ease with which current can be controlled by being constrained to flow along thin insulated wire makes simple the deliberate manufacture of what is in effect surface current; as, for example, in a *solenoid*, where the wire is wrapped closely round a tubular surface. However, there is also a situation somewhat analogous to the exclusion, in the steady state, of electric field, and consequently charge density, from the interior of a conductor; namely the exclusion of magnetic field and current density from the interior of a so-called *superconductor*. Just as the conductor naturally supports surface charge density, so the superconductor naturally supports surface current density, in a way which is now briefly described.

In 1911, soon after the liquefaction of helium was achieved, it was discovered that below a critical temperature within the range of the extremely low temperatures involved (~ 1 °K) certain substances lose any measurable resistance to current flow. The phenomenon was called superconductivity, and its first interpretation was in terms of the relation (4.3), to the effect that for all practical purposes the conductivity σ was infinite, being at least orders of magnitude greater than the highest values recorded in the table in §4.1. With such a superconducting substance $\mathbf{E}$ must therefore be zero, whence it follows from Maxwell's equation (2.104), which is universally valid, that $\mathbf{B}$ is time independent.

To illustrate the consequences of this reasoning suppose that an uncharged superconductor is placed in a given magnetostatic field $\mathbf{B}_0$ whose sources are not subject to variation. If there were originally no magnetic field inside the superconductor, then its interior remains field free. The boundary condition (4.28) must be satisfied, and the onus is therefore on the exterior field to adjust itself so that the normal component at the surface, being continuous, is zero. The total field $\mathbf{B}$ therefore consists of the invariable field $\mathbf{B}_0$, together with a field, $\mathbf{B}_1$ say, which is such that $\mathbf{B} = \mathbf{B}_0 + \mathbf{B}_1$ is zero in the interior of the superconductor and has zero normal component just outside its surface.

The origin of $\mathbf{B}_1$ can only be a current distribution on the surface of the superconductor, since (4.1) precludes the possibility of any volume current density. If the surface current density is $\mathbf{j}$ its relation to the magnetic field just outside the surface is determined by (4.29), and may be written

$$\mathbf{j} = \frac{1}{\mu_0}\mathbf{n}\wedge\mathbf{B}. \tag{4.34}$$

Thus, in the steady state, currents in the surface of the superconductor screen the interior from magnetic field somewhat as charges on the surface of any conductor screen the interior from electric field.

The result constrasts with what would happen if an ordinary good conductor, say copper at room temperature, were placed in the field $\mathbf{B}_0$. In this case **E**, and hence $\dot{\mathbf{B}}$, are not undetectably small inside the conductor, and the ultimate equilibrium state is simply that in which the field is $\mathbf{B}_0$ everywhere. The rate of attainment of equilibrium depends on the quality and geometry of the conductor. An order of magnitude argument from Maxwell's equations gives the time scale for the rate of progress towards equilibrium as

$$T \sim \mu_0 \sigma L^2, \tag{4.35}$$

where L is a characteristic length of the conductor. Thus, if L is 1 m, T for copper is about 60 sec.

Whilst the specific deduction just made about the behaviour of a superconductor placed in a magnetic field is correct, in fact the relation (4.3) with σ infinite is inadequate to describe true superconductivity, even in the crudest phenomenological fashion. For, with regard to the magnetic field, σ infinite only implies $\dot{\mathbf{B}} = 0$; from which it might be presumed, for example, that if a conductor initially pervaded by a magnetic field were cooled below the critical temperature, then the field would be locked in its interior. For some time after the discovery of superconductivity this was indeed thought to be the case; but experimental evidence eventually pointed to the conclusion that the magnetic field was expelled from a *pure* superconductor (the Meissner effect, 1933). It is now recognized that, regardless of the way in which the superconducting state is reached, the magnetic field vanishes inside a superconductor except for a transition layer at the surface. Moreover, the field penetrates only to a depth of the order of 10^{-7} m, so that it is acceptable to treat the superconductor in magnetostatics simply as a region, bounded by surface current, throughout which $\mathbf{B} = 0$.

The application of (4.31), with $\mathbf{B}_1 = 0$ and, from (4.34), $\mathbf{B}_2 = -\mu_0 \mathbf{n} \wedge \mathbf{j}$, gives the force per unit area on a superconductor as

$$\tfrac{1}{2}\mu_0 j^2 = \tfrac{1}{2}B^2/\mu_0 \tag{4.36}$$

along the *inward* normal (cf. (3.24)).

4.2.4 *Partitioning the field*

Much as, in electrostatics, complete equipotential surfaces can be replaced by conducting sheets without disturbing the field, so in magnetostatics closed surfaces that are everywhere tangential to the field lines can be replaced by superconducting sheets without disturbing the field.

To describe the situation in more detail suppose a given steady current density **J** produces a magnetostatic field **B**. Let S be any closed surface such that, at each point on S, **B** is tangential to S. Let $\mathbf{J}'$, $\mathbf{B}'$ be zero everywhere outside S, and equal to **J**, **B**, respectively, everywhere inside S. Then it may be observed that surface current density

$$\mathbf{j} = -\mu_0 \mathbf{n} \wedge \mathbf{B} \tag{4.37}$$

on S, where **n** is the unit vector along the outward normal, together with $\mathbf{J}'$ gives rise to the field $\mathbf{B}'$.

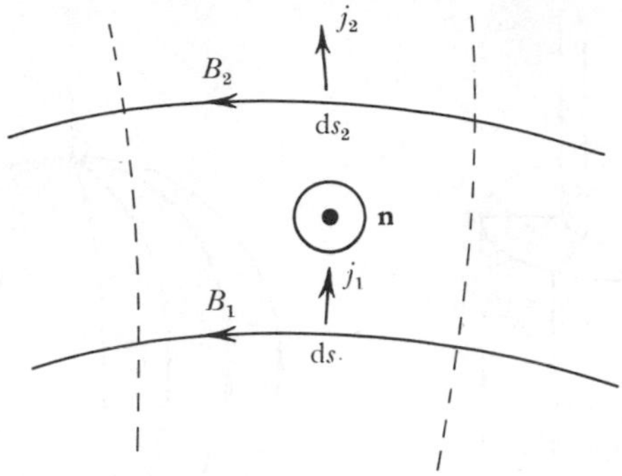

Figure 4.5

The observation rests on the fact that the field equations are satisfied everywhere, the surface density **j** being necessary and sufficient to allow for the discontinuity in **B**′ across S. Direct confirmation that the combination of **J**′ and **j** is solenoidal can be obtained by referring to figure 4.5, which shows adjacent field lines and their orthogonal trajectories on an element dS of S; for the application first of Ampère's circuital law and then of (4.37) gives

$$\mathbf{n}.\mathbf{J}'\,\mathrm{d}S = \frac{1}{\mu_0}(B_2\,\mathrm{d}s_2 - B_1\,\mathrm{d}s_1) = j_2\,\mathrm{d}s_2 - j_1\,\mathrm{d}s_1. \qquad (4.38)$$

It is, of course, also the case that surface current $-\mathbf{j}$ on S, together with $\mathbf{J}-\mathbf{J}'$, gives rise to the field $\mathbf{B}-\mathbf{B}'$, which is zero inside S and equal to **B** outside S.

The electrostatic analogy is when the fields on either side of an equipotential surface are independently maintained with the help of appropriate surface charge densities. As described in § 3.1.4, a physical picture of this process is obtained by conceiving the equipotential surface to be replaced by a conducting sheet. Likewise in the present case it may be imagined that S is replaced by a superconducting sheet, the inner and outer surfaces of which carry the surface current densities **j** and $-\mathbf{j}$ respectively.

The result at the end of § 4.1.2 furnishes a simple example. Suppose the cylindrical surface $r = a$ is superconducting, with current, of uniform density $I/(2\pi a)$, flowing parallel to the z axis on its outer surface and in the opposite direction on its inner surface. Then the current on the outer surface, by itself, is responsible for the field (4.20); that on the inner surface, together with the current I along the z axis, generates the field

$$\mathbf{B} = \begin{cases} I/(2\pi r), & r < a, \\ 0\quad, & r > a. \end{cases} \qquad (4.39)$$

Since the argument requires only that the superconducting sheet be everywhere tangential to the field lines, the fact that these are circles r = constant

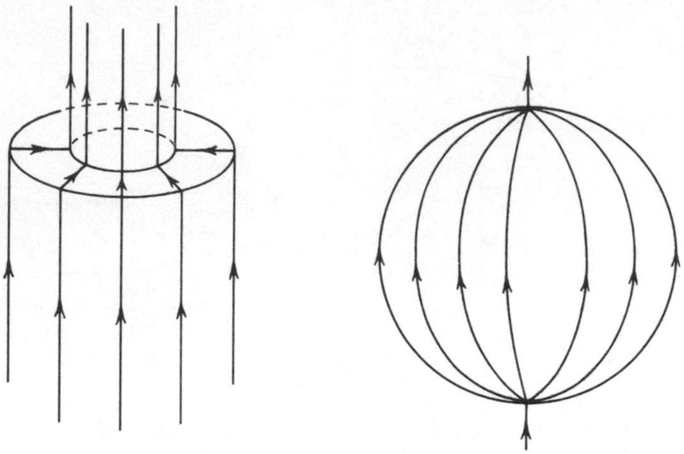

(a) (b)

Figure 4.6

means that any surface of revolution about the z axis will serve. Examples are suggested in figure 4.6; in each case axially symmetric current flowing as indicated in the surface of revolution generates a field that interior to the surface is zero, and exterior to the surface is precisely as though the whole current were concentrated along the axis.

Revert now to the case of some arbitrary static field $\mathbf{B}$, and select a field tube within which there is no current, as depicted in figure 4.7. Then the observation is that if current density (4.37) were maintained on the surface of the tube it would generate the field that is identical to $\mathbf{B}$ everywhere within the tube and is zero everywhere outside the tube. Moreover, $\mathbf{j}$ itself is solenoidal, $\mathbf{J}'$ being zero in (4.38), and could therefore be supported on the surface of a superconductor. An example is again provided by the field of current I flowing along the z axis, with a field tube swept out by rotating about the z axis any closed curve, coplanar with the axis, that does not intersect it. Another particularly simple example is afforded by a *uniform* field, $\mathbf{B}_0$ say: a field tube is then an infinitely long cylinder, of arbitrary cross-section, whose generators are parallel to $\mathbf{B}_0$; and current of uniform density B_0/μ_0 flowing round such a cylinder, normal to its generators, produces the field $\mathbf{B}_0$ inside and zero field outside.

4.2.5 *Inductors: self and mutual inductance*

The concept of *inductance* plays a role in magnetostatics somewhat akin to that of capacitance in electrostatics.

An electrostatic field can be confined, by suitable surface charge densities, to the region between the 'plates' of a capacitor when these plates are closed conducting surfaces, one of which lies entirely within the other. Likewise,

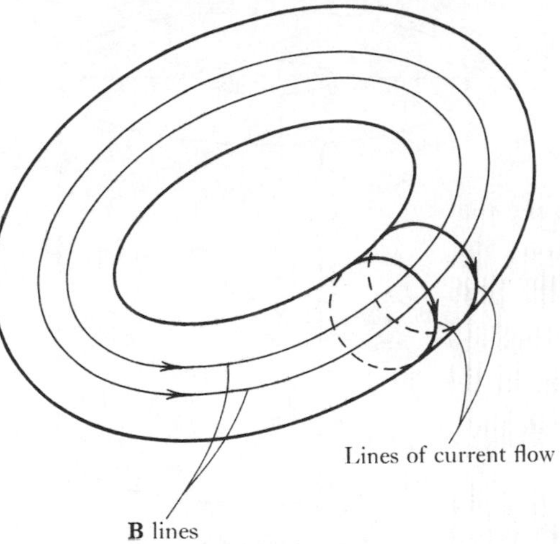

Figure 4.7

as has just been demonstrated, a magnetostatic field can be confined, by suitable surface current densities, to the region within a closed superconducting surface. The 'ring-like' surface performing this function is conveniently called an ideal *inductor*. The field lines of **B** form loops round the entire length of the ring, and the lines of flow of the surface current that supports the field are normal to the **B** lines at the surface (see figure 4.7).

For the capacitor, the voltage V across the plates is proportional to the magnitude q of the charge on each, and the capacitance C is q/V. For the inductor the flux Φ of **B** is proportional to the current I, and the inductance is defined as

$$L = \Phi/I. \tag{4.40}$$

Φ is given (uniquely) by the flux of **B** across any surface intersected just once by each field line; and I by the current that crosses any line intersected just once by each line of current flow.

The unit of inductance is called a henry, 1 henry being 1 weber amp^{-1}. The dimensions of μ_0 are often stated as henry m^{-1} (cf. (1.17)).

To take a specific example, consider first the field due to a current I flowing down the z axis. Now isolate that part of the field tube bounded by the planes $z = 0$, $z = d$ and the cylinders $r = a$, $r = b$, where r denotes distance from the z axis. These surfaces form the inductor shown in figure 4.8; the sense of current flow is indicated by the arrows, and the total current, flowing up the outer cylinder and down the inner cylinder, is, of course, I. The magnetic field at any point inside the inductor has magnitude

$$B = \mu_0 I/(2\pi r) \tag{4.41}$$

so that
$$\Phi = \frac{\mu_0 Id}{2\pi}\int_a^b \frac{\mathrm{d}r}{r} = \frac{\mu_0 Id}{2\pi}\log(b/a).$$

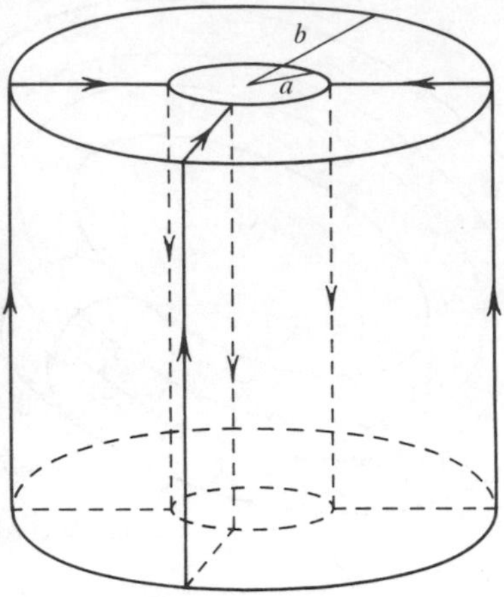

Figure 4.8

Hence the inductance is
$$L = \frac{\mu_0 d}{2\pi} \log (b/a). \tag{4.42}$$

For an inductor that isolates a tube of the same field, but has a cross-section of arbitrary shape, a simple approximate result can be obtained for the case when the tube is narrow in the sense that the ratio of the longest chord of the normal cross-section to the shortest field line is much less than unity. For then, in the expression (4.41) for B, r may be replaced by a constant radius a, and
$$L = \mu_0 A/(2\pi a), \tag{4.43}$$
where A is the area of normal cross-section.

The ideal inductor so far described imprisons a magnetic field, and may be compared with the capacitor that imprisons an electric field. The definition of capacity, however, applies equally to a pair of conductors exterior to one another, with a field that spreads out to infinity. Likewise the definition of inductance applies also to the case when a surface 'ring' current on a superconductor generates a magnetostatic field that pervades the entire *exterior* region.

For example, consider first the field generated by a current I flowing in a circular loop of radius a, the topology of which is sketched in figure 4.4. Now envisage a ring-like surface, enclosing the loop, to which at each point the field line through that point is tangent. Then the appropriate current flowing in the surface maintains the field exterior to the surface and gives zero field throughout the interior, as indicated in figure 4.9.

I is now the total current flowing right round the ring; and Φ is the flux of **B** through the ring. The inductance, for a ring of uniform cross-section, is
$$L = 2\pi \int_0^{a'} B(r)\, \mathrm{d}r, \tag{4.44}$$

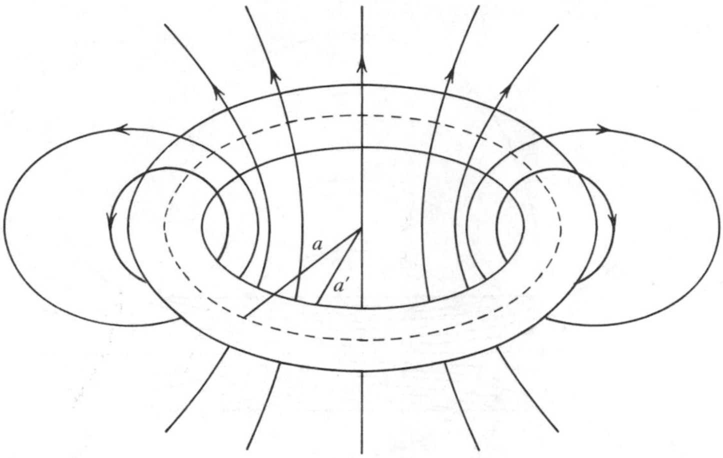

Figure 4.9

where $B(r)$ is the magnetic field in the plane of a circular loop of radius a carrying unit current, r being distance from the centre of the loop, and a' ($< a$) is the inner radius of the ring. For the case $a'/a \simeq 1$ L can be evaluated approximately (see § 4.4.1), with the result

$$L = \mu_0 a[\log(8a/b) - 2], \tag{4.45}$$

where $b = a - a'(\ll a)$.

There is a further important concept applicable to inductors which is now considered, namely *mutual* inductance.

The mutual inductance of any two inductors in specified relative positions is defined as the flux of **B** through one due, solely, to unit current flowing round the other. It is shown later (§ 4.3.4) that it is immaterial to the definition which inductor carries the unit current.

More generally, if there are n inductors in specified relative positions, the flux through the ith can be written

$$\Phi_i = \sum_{j=1}^{n} L_{ij} I_j, \tag{4.46}$$

where I_j is the current flowing round the jth. Then L_{ii} is the self-inductance of the ith inductor, and L_{ij} $(i \neq j)$ is the mutual inductance of the ith and jth inductors, having the same value as L_{ji}. The values of each inductance depend, in general, on the shape and disposition of all the inductors. The relation (4.46) may be compared with (3.106).

A word should be said on the matter of sign. A convenient convention is to adopt a direction for positive current in each inductor, and then to evaluate the relevant magnetic flux in the associated right-handed screw direction, so that it may be positive or negative. With this convention, therefore,

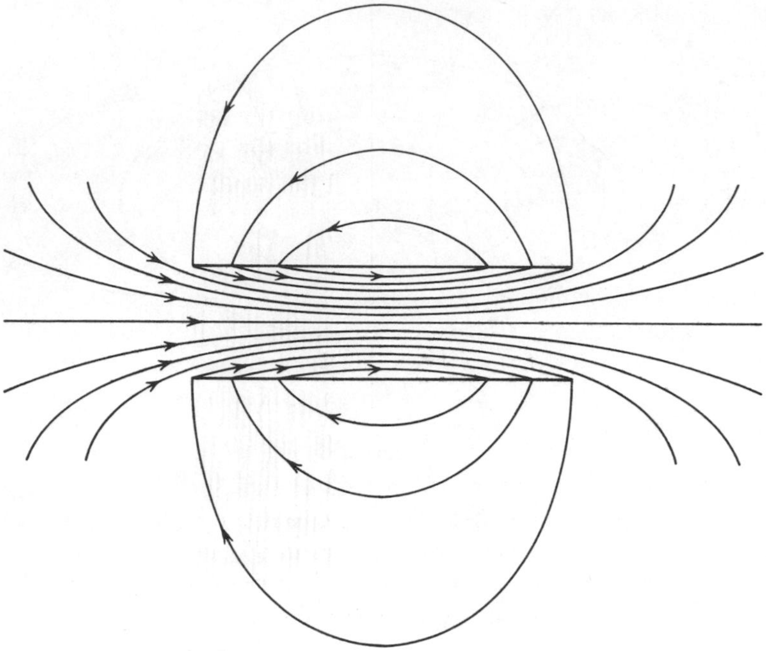

Figure 4.10

mutual inductance can be of either sign, whereas self-inductance is positive.

When in isolation the ideal inductor described has the feature that every magnetic field line links every line of current flow; there is no ambiguity, if some jargon, in saying that flux Φ threads current I. In practice, however, this state of affairs is achieved at best only approximately. The pattern of current flow on the surface of an ideal inductor in isolation is unique, as shown in §4.2.6, and that in a common conductor will depart from the ideal. Strictly, then, in the latter case, the definition of inductance (4.40) is not appropriate, and an important relation between inductance and energy, introduced in §4.3.1, must be invoked to provide an alternative definition that remains applicable in practical cases.

To illustrate the point, consider a conducting surface in the shape of a circular cylinder of radius a, that carries current, flowing normal to the generators, of uniform surface density j. If the cylinder is infinitely long this represents an ideal inductor; the resulting field is uniform inside, of magnitude $B_0 = \mu_0 j$, and zero outside. The inductance, though, is zero, because, for finite field, the total current is infinite. If, on the other hand, the cylinder has finite length l, there is some 'leakage' of flux in the manner indicated in figure 4.10. For $l \gg a$ this is comparatively slight, and the field is approximately B_0 inside the cylinder, except within a distance from the ends of order a. The inductance is then approximately

$$L = \mu_0 \pi a^2/l,$$

which may be compared with (4.43).

If the current remains uniform the leakage becomes more pronounced as l/a decreases, and the definition of inductance (4.40) is not applicable. In particular, if $l/a \ll 1$ the conductor is more akin to a wire loop, and certainly much then depends on the detail of the field close to it, as is indicated by the formula (4.45) for small values of b.

Were the finite cylinder a superconductor, it would be a ring-like ideal inductor of the kind previously described. This implies that there is *some* distribution of surface current on the cylinder that results in a magnetic field every line of which threads every line of current flow; but this distribution is not what would ordinarily be met in practice.

4.2.6 *Uniqueness theorem: superconducting sphere in uniform field*

Implicit in the previous discussion is the requirement that the topology of an inductor be ring-like; more specifically, that the regions bounded by its surface be *doubly* connected. A superconductor whose surface is the boundary between simply connected regions cannot support a surface current in the absence of excitation by an external field. That this must be so can be seen by the following argument.

Consider a closed superconducting surface S that bounds a simply connected region. Suppose it were possible for current to flow in S in the absence of any source of field other than the current itself, and let the field of the current be $\mathbf{B}$. Then throughout the interior of S, $\mathbf{B}$ must be expressible as the gradient of a single-valued scalar, ϕ, that satisfies $\nabla^2\phi = 0$, as noted in the final paragraph of §2.2.3. But the normal component of $\mathbf{B}$ vanishes on S, since S is superconducting; so therefore does the normal derivative of ϕ. It follows from §3.3.2 that ϕ is constant, and hence $\mathbf{B}$ is identically zero within S. It is likewise identically zero outside S. There is thus no field, and there can be no current.

It was shown in §4.2.5 that examples of self-consistent current and field for ring-like superconductors are in principle easy to recognize. Proof of their existence for any ring-like superconductor cannot be given here, but uniqueness is easily established.

Consider, for the sake of definiteness, the space exterior to a solid superconducting ring round the length of which flows total current I. Throughout this space

$$\mathbf{B} = -\mu_0 \operatorname{grad} \phi, \tag{4.47}$$

where ϕ is not single-valued, but increases by I as the field point travels round a loop that links the ring. Suppose now that ϕ_1 and ϕ_2 each satisfy the conditions, for the same current I. Then $\psi = \phi_1 - \phi_2$ *is* single-valued, and since its normal derivative vanishes on the superconducting boundary ψ must be a constant. Thus there cannot be more than one field **B**, and hence one current distribution, for a given total current I.

It is a corollary to what has been said that the boundary value problem specified by placing a superconductor in a given externally maintained magnetostatic field $\mathbf{B}_0$, as described in § 4.2.3, has only one solution provided the superconducting region is simply connected. If it is doubly connected, so forming an inductor, there is an infinity of solutions, one associated with each value I assigned to the total current flowing round the inductor.

A case easily solved is that in which the field $\mathbf{B}_0$ is uniform and the superconductor is a sphere. The problem is mathematically equivalent to the hydrodynamic one of a rigid sphere in a uniform stream of ideal fluid, the magnetic field lines being the same as the streamlines. It is readily solved in terms of the potential functions (3.88) and (3.89) used in the electrostatic problem of a sphere in a uniform electric field. For if the total field outside the sphere $r = a$ is written as (4.47), the solution is evidently given by

$$\mu_0 \phi = -B_0\left(r+\frac{a^3}{2r^2}\right)\cos\theta. \tag{4.48}$$

The first term in (4.48) represents the given uniform field, B_0 parallel to $\theta = 0$; and the second term is the additional field due to the induced surface currents in the sphere, its coefficient being determined by the boundary condition $\partial\phi/\partial r = 0$ on $r = a$.

The magnitude of the surface current density is

$$\frac{1}{r}\frac{\partial\phi}{\partial\theta}$$

evaluated on $r = a$; that is

$$\frac{3}{2\mu_0}B_0\sin\theta. \tag{4.49}$$

The lines of current flow are circles in planes normal to $\theta = 0$, the sense of flow being a left-handed screw about $\theta = 0$, which is clearly necessary to achieve cancellation of the given field B_0 throughout $r < a$.

4.3 Energy

4.3.1 *Definition: energy of inductors*

A steady current distribution has potential energy represented by the work done in building up the distribution against the reaction of the induction forces. The presence of induction forces is, of course, dictated by Faraday's law, because any change in the current distribution is accompanied by a corresponding change in the magnetic field.

Consider, for example, a single isolated ideal inductor, which in its final state maintains total current I threaded by flux Φ. Then if at some stage of the build up the current is λI, the corresponding flux is $\lambda\Phi$ and the rate of change of flux is $\Phi\,d\lambda/dt$. The rate at which work is done against the induction forces is therefore

$$\lambda I\Phi\,d\lambda/dt, \tag{4.50}$$

so that the total work done in taking λ from zero to unity is

$$W = \tfrac{1}{2}I\Phi. \tag{4.51}$$

This can also be expressed, using (4.40), in either of the alternative forms

$$W = \tfrac{1}{2}LI^2 \tag{4.52}$$

and

$$W = \tfrac{1}{2}\Phi^2/L. \tag{4.53}$$

For a system consisting of n ideal inductors ultimately carrying currents $I_1, I_2, \ldots, I_n$ threaded by fluxes $\Phi_1, \Phi_2, \ldots, \Phi_n$, respectively, the same argument leads to the result

$$W = \frac{1}{2}\sum_{i=1}^{n} I_i\Phi_i. \tag{4.54}$$

Here the self-energy of each inductor is, of course, included, in that a contribution to the flux Φ_i through the ith inductor comes from the field generated by I_i.

Evidently (4.54) can be expressed in terms of the currents alone, in the form corresponding to (4.52), by introducing (4.46). The result is

$$W = \frac{1}{2}\sum_{i,j=1}^{n} L_{ij}I_iI_j. \tag{4.55}$$

4.3.2 *Energy of distributed current*

In the case when there is a general flow of steady current, specified by a volume current density $\mathbf{J}$, essentially the same reasoning that led to the formula (4.51) can be applied by resolving the distribution into tubes of flow of infinitesimal cross-section that take the place of the inductors. The contribution to W associated with each tube is

$$\tfrac{1}{2}J\,dS\,\Phi \tag{4.56}$$

where dS is the normal cross-section of the tube ($J\,dS$ being, of course, constant along the tube), and Φ is the flux of $\mathbf{B}$ through the loop formed by the tube.

It now turns out that W can be neatly expressed as a volume integral by introducing the vector potential $\mathbf{A}$ such that $\mathbf{B} = \operatorname{curl}\mathbf{A}$ (see §2.2.5). For application of Stokes' theorem then gives

$$\Phi = \oint \mathbf{A}\,.\,d\mathbf{s}, \tag{4.57}$$

where the line integral is round the loop formed by the tube; and since $\mathcal{J}\mathrm{d}\mathbf{s} = \mathbf{J}\,\mathrm{d}s$ substitution of (4.57) into (4.56) evidently yields, on summing over all tubes,

$$W = \frac{1}{2}\int \mathbf{J}\,.\,\mathbf{A}\,\mathrm{d}\tau, \tag{4.58}$$

where the integration is over the entire current distribution. This formula can be regarded as the analogue of formula (3.55) for electrostatic energy.

4.3.3 *Field energy*

The energy of a steady current distribution, (4.58), can be expressed in terms of the field **B** alone.

Since $\mu_0 \mathbf{J} = \operatorname{curl} \mathbf{B}$ the integrand of (4.58) can be written (see (A. 15)) as $1/\mu_0$ times

$$\operatorname{div}(\mathbf{B} \wedge \mathbf{A}) + \mathbf{B}\,.\operatorname{curl}\mathbf{A}.$$

The volume integral of the divergence can be transformed into the surface integral

$$\int (\mathbf{B} \wedge \mathbf{A})\,.\,\mathrm{d}\mathbf{S}, \tag{4.59}$$

which vanishes if taken over the sphere at infinity because **A** and **B** are $O(1/r^2)$ and $O(1/r^3)$ respectively as $r \to \infty$. Hence

$$W = \frac{1}{2\mu_0}\int \mathbf{B}^2\,\mathrm{d}\tau, \tag{4.60}$$

where the volume integral is over all space. This form is analogous to (3.66).

As an example, take the field associated with the inductor shown in figure 4.8. It is confined to the region $a < r < b$, $0 < z < d$, and its magnitude is given by (4.41). Hence (4.60) is

$$W = \frac{\mu_0 I^2 d}{4\pi}\int_a^b \frac{\mathrm{d}r}{r} = \frac{\mu_0 I^2 d}{4\pi}\log(b/a), \tag{4.61}$$

in agreement with (4.52), since L is given by (4.42).

The origin of the field expression for the energy is shown particularly clearly by considering the cylindrical inductor described at the end of §4.2.5. If the cylinder, of length l and normal cross-sectional area A, is so narrow that leakage can be neglected, the current and flux are related to the field through the equations

$$I/l = B/\mu_0, \quad \Phi = AB; \tag{4.62}$$

the cross-section need not be circular, and was only taken so in §4.2.5 for convenience of description. It follows from (4.62) that

$$\tfrac{1}{2}I\Phi = \frac{1}{2\mu_0}B^2 Al, \tag{4.63}$$

in which the left hand side is the form first obtained for the energy of an inductor, and the right hand side is $B^2/(2\mu_0)$ times the volume of the region inside the cylinder.

If the general magnetostatic field is conceived as an aggregate of elements of infinitesimal field tubes, in which each element is isolated within a cylindrical inductor, then it is natural to regard magnetic energy as localized in the field with volume density

$$\frac{1}{2\mu_0}\mathbf{B}^2. \tag{4.64}$$

4.3.4 *Reciprocal theorem*

From a comparison of the energy expressions (4.58) and (3.55) it might be expected that Green's reciprocal theorem (3.103) has the analogue

$$\int \mathbf{J}\,.\,\mathbf{A}'\,d\tau = \int \mathbf{J}'\,.\,\mathbf{A}\,d\tau, \tag{4.65}$$

where current densities $\mathbf{J}$ and $\mathbf{J}'$ separately give rise to respective vector potentials $\mathbf{A}$ and $\mathbf{A}'$, and the integrals are over all space. The truth of (4.65) is easily confirmed; for μ_0 times the left hand side is

$$\int \mathbf{A}'\,.\,\mathrm{curl}\,\mathbf{B}\,d\tau = \int [\mathrm{div}\,(\mathbf{B}\wedge\mathbf{A}') + \mathbf{B}\,.\,\mathrm{curl}\,\mathbf{A}']\,d\tau$$

$$= \int \mathbf{B}\,.\,\mathbf{B}'\,d\tau,$$

which by symmetry is also μ_0 times the right hand side.

A special case is when the current is carried solely on the surface of n fixed ideal inductors. Then

$$\sum_{i=1}^{n} I_i\Phi_i' = \sum_{i=1}^{n} I_i'\Phi_i, \tag{4.66}$$

where total currents I_i flowing round the inductors correspond to respective fluxes Φ_i through them, and current I_i' to fluxes Φ_i'.

It is an immediate corollary that $L_{ij} = L_{ji}$. For the introduction of (4.46) into (4.66) gives

$$\sum_{i,j=1}^{n} L_{ij} I_i I_j' = \sum_{i,j=1}^{n} L_{ij} I_i' I_j,$$

from which the result follows by taking just one of the I_i and one of the I_i' to be non-zero.

4.4 Circuits

4.4.1 *Inductance of circuits*

When current flows round a wire circuit it is for most purposes permissible to neglect the thickness of the wire. An exception, though, is the calculation of the self-inductance of a wire loop.

For $b \ll a$ the expression (4.45) gives the inductance of a circular loop of superconducting wire, where a is the radius of the loop and b the radius of the wire (the cross-section of the wire is effectively circular since the field lines of a line current approximate to circles in the vicinity of the current). In practice the wire of a circuit is unlikely to be superconducting; steady or 'slowly varying' current will be distributed more or less uniformly over the cross-section rather than on the surface, and the associated magnetic field will penetrate into the wire. In such a case, as explained at the end of § 4.2.5, the definition (4.40) of inductance is not strictly applicable. However, an effective alternative definition is suggested by the formula (4.52) for the energy of an isolated ideal inductor carrying current I.

For an ordinary wire circuit carrying current I the energy is still proportional to I^2, as is obvious from either of the expressions (4.58) and (4.60) from which the energy could be calculated. It can therefore be written

$$W' = \tfrac{1}{2}L'I^2, \tag{4.67}$$

and this formula may be used to *define* the inductance L'.

Dashes are used in (4.67) to emphasize that, even were the respective wires of identical size and shape, W' and L' in (4.67) differ from W and L in (4.52), and indeed depend on the precise current distribution: however, the difference is small when the thickness of the wire is much less than its minimum radius of curvature; the logarithmic term in (4.45), for example, is also the dominant term in L' for $b/a \ll 1$. For such thin wires the contribution to the integral (4.60) from the region inside the wire may be estimated by assuming that the field at any interior point is essentially that of an infinite straight wire coincident with the actual wire in the vicinity of the point (cf. the remarks at the end of § 4.2.2). For an infinite straight wire of radius b, in which total current I flows uniformly, the field is given by (4.19) with a replaced by b; and the contribution to (4.60) from the region $r < b$, is, per unit length,

$$\frac{\mu_0 I^2}{4\pi b^4}\int_0^b r^3\,\mathrm{d}r = \frac{\mu_0 I^2}{16\pi}. \tag{4.68}$$

The corresponding contribution to the self-inductance of a wire circuit of length l is therefore

$$\frac{\mu_0 l}{8\pi}. \tag{4.69}$$

This is sometimes called the *internal* self-inductance. For a wire of radius b in the form of a circle of radius a it is $\frac{1}{4}\mu_0 a$, and would be added to (4.45) to give the total self-inductance.

Yet another definition of inductance is available from the statement of Faraday's law. Again referring first to the superconducting wire loop, if the current along it were varied the applied voltage necessary to overcome the self-induced voltage would be

$$V = \frac{\mathrm{d}\Phi}{\mathrm{d}t} = L\frac{\mathrm{d}I}{\mathrm{d}t}, \tag{4.70}$$

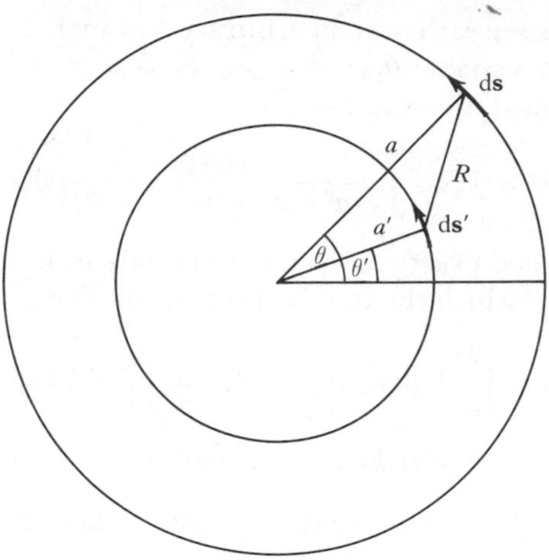

Figure 4.11

which of course agrees with (4·52) through $\mathrm{d}W/\mathrm{d}t = IV$. Now a relation of the form (4·70) is also applicable to an ordinary wire loop, and may be used to define its inductance. The strict identification of V depends on the detailed specification of the system, but is not a practical difficulty. Methods for the experimental measurement of inductance are mostly based on (4·70).

The *mutual* inductance of wire circuits can evidently be discussed without any need to take account of the thickness of the wire. The mutual inductance of two circuits C and C' is the flux of B through one due solely to unit current flowing round the other. A convenient formula is obtained in the following way.

The vector potential due to unit current in C is

$$\mathbf{A} = \frac{\mu_0}{4\pi}\oint \frac{\mathrm{d}\mathbf{s}}{R}, \tag{4·71}$$

where R is distance from the element $\mathrm{d}\mathbf{s}$ to the field point. The associated flux through C' can therefore be written down from (4·57), to give the mutual inductance in the form

$$M = \frac{\mu_0}{4\pi}\oint_C \oint_{C'} \frac{\mathrm{d}\mathbf{s}\,.\,\mathrm{d}\mathbf{s}'}{R}, \tag{4·72}$$

where R is now distance between the element $\mathrm{d}\mathbf{s}$ of C and the element $\mathrm{d}\mathbf{s}'$ of C'. This formula exhibits explicitly the reciprocity established in §4·3·4.

As an example suppose that C and C' are coplanar concentric circles of radii a and a' ($< a$) respectively. Then, as indicated by figure 4.11,

$$M = \frac{\mu_0 a a'}{4\pi}\iint \frac{\cos(\theta-\theta')}{[a^2+a'^2-2aa'\cos(\theta-\theta')]^{\frac{1}{2}}}\,\mathrm{d}\theta\,\mathrm{d}\theta', \tag{4·73}$$

where the integrals are each over an arbitrary 2π range. If, for each value of θ', the integration variable θ is changed to $\psi = \theta - \theta'$, and then the θ' integration performed, the result is

$$M = \mu_0 aa' \int_0^\pi \frac{\cos\psi}{(a^2 + a'^2 - 2aa' \cos\psi)^{\frac{1}{2}}}\, d\psi. \tag{4.74}$$

This can be evaluated exactly in closed form only in terms of the *complete elliptic integrals*, for which the standard representation is

$$E(k) = \int_0^{\frac{1}{2}\pi} f(\phi, k)\, d\phi, \quad K(k) = \int_0^{\frac{1}{2}\pi} d\phi / f(\phi, k), \tag{4.75}$$

where
$$f(\phi, k) = (1 - k^2 \sin^2\phi)^{\frac{1}{2}}. \tag{4.76}$$

In fact, by putting $\psi = 2\phi$ in (4.74), it is easy to check that

$$M = \mu_0 (aa')^{\frac{1}{2}} \left[\left(\frac{2}{k} - k\right) K(k) - \frac{2}{k} E(k)\right], \tag{4.77}$$

with
$$k = \frac{2(aa')^{\frac{1}{2}}}{a + a'}. \tag{4.78}$$

Simple approximations are available for $a'/a \ll 1$ and for $a'/a \simeq 1$.

When $a'/a \ll 1$ it is appropriate to express the integrand of (4.74) as a power series in a'/a and integrate term by term. The early stages of this procedure give

$$M = \tfrac{1}{2}\mu_0 \pi \frac{a'^2}{a}\left\{1 + \frac{3}{8}\left(\frac{a'}{a}\right)^2 + O\left[\left(\frac{a'}{a}\right)^4\right]\right\}, \tag{4.79}$$

which can be checked by integrating (4.27) over the portion of the surface $z = 0$ bounded by $\rho = a'$; and further terms can be readily obtained. The expansion converges for $a'/a < 1$, but not sufficiently rapidly to be useful when $a'/a \simeq 1$.

The derivation of (4.79) is equivalent to approximating $E(k)$ and $K(k)$ in (4.77) for small values of k, which is easily done by using the early terms of the expansion of $f(\phi, k)$ as a power series in k. When $a'/a \simeq 1$ an approximation to the elliptic integrals is required for $k \simeq 1$. As $k \to 1$, evidently $E(k) \to 1$, but $K(k)$ presents some difficulty since it diverges logarithmically. In fact

$$K(k) = \log\left(\frac{4}{\epsilon}\right) + O(\epsilon^2 \log\epsilon), \tag{4.80}$$

where
$$\epsilon = (1 - k^2)^{\frac{1}{2}}. \tag{4.81}$$

Since $k = 1 + O(\epsilon^2)$ an approximation to (4.77) for $\epsilon \ll 1$ is seen to be

$$M = \mu_0 (aa')^{\frac{1}{2}}\left[\log\left(\frac{4}{\epsilon}\right) - 2\right]; \tag{4.82}$$

and in terms of the parameter

$$b = a - a'$$

this approximation gives

$$M = \mu_0 a \left[\log\left(\frac{8a}{b}\right) - 2\right] \tag{4.83}$$

for $b/a \ll 1$.

The expression (4.83) is the same as the approximation previously quoted, (4.45), for the self-inductance of a circular loop of thin superconducting wire. The two problems involve identical fluxes.

4.4.2 *Solenoids*

A circuit of comparatively large self-inductance can be obtained by taking insulated wire round a number of turns. With wire carrying current I closely wound on the surface of a closed tube it is possible to confine the associated magnetic field to the interior of the tube, by adjusting the local density of the winding to give the desired surface current (cf. §4.2.4). Such a device, an ideal solenoid, generates the same field as an ideal inductor with surface current density nI, where n is the local number of turns per unit length. If the total number of turns is N the self-inductance of the solenoid is N^2 times that of the corresponding inductor; for N is the factor by which the total current in the inductor exceeds I, and is also the factor by which its flux must be multiplied to get the total flux through the wire circuit. For example, a narrow ideal solenoid, of uniform cross-sectional area A and length l, has inductance

$$L = \mu_0 N^2 A/l. \tag{4.84}$$

If $N = 200$, $A = 2 \times 10^{-4}\,\mathrm{m}^2$, $l = 0{\cdot}1\,\mathrm{m}$, then $L \simeq 10^{-4}$ henry.

An ideal, uniform, straight solenoid of infinite length contains uniform field of magnitude

$$B = \mu_0 nI. \tag{4.85}$$

In practice solenoids are commonly in the form of a circular cylinder, with the wire wound uniformly and lying perpendicular to the generators, and their finite length gives rise to field leakage, as indicated at the end of §4.2.5. If the length l is much greater than the radius a the leakage is small, and the inductance is approximately (4.84), where $N = nl$ is the total number of turns. But this formula is inaccurate when, as often happens, l/a is not large.

The general nature of the field of a circular cylindrical solenoid is depicted in figure 4.10. It is, in fact, easy to evaluate the field exactly at any point on the axis from a knowledge of the field on the axis of a circular current loop. For the magnitude of the field at a point P on the axis at distance ζ from one end is evidently

$$B = n\int_{-\zeta}^{l-\zeta} B'\,\mathrm{d}z, \tag{4.86}$$

where B' is the field of a circular loop, carrying current I, at a point on the

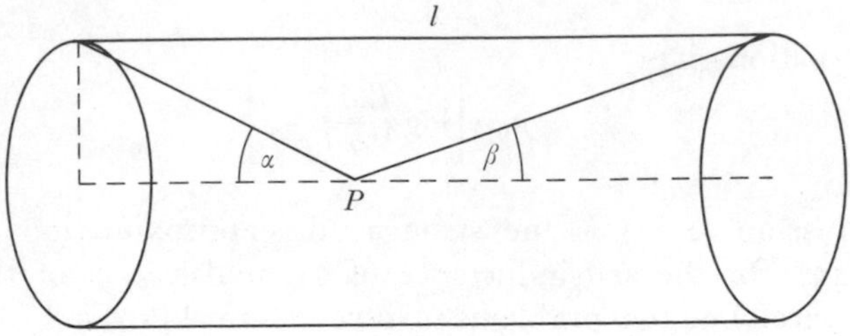

Figure 4.12

axis distance z from the centre of the loop; from which the formula (4.23) for B' gives

$$B = \tfrac{1}{2}\mu_0 nI(\cos\alpha + \cos\beta), \tag{4.87}$$

where α and β are the angles shown in figure 4.12.

Consider the case $l/a \gg 1$. At points well away from either end of the solenoid α and β are small and (4.85) is approximately achieved. On the other hand if β, but not necessarily α, is small the corresponding approximate result is

$$B = \tfrac{1}{2}\mu_0 nI(1 + \cos\alpha). \tag{4.88}$$

Thus the field falls off towards the mouth of the solenoid, dropping to

$$B = \tfrac{1}{2}\mu_0 nI \tag{4.89}$$

at the end $\zeta = 0$ ($\alpha = \frac{1}{2}\pi$). For points outside the solenoid α increases beyond $\frac{1}{2}\pi$ and the field continues to decrease monotonically.

4.4.3 *Elements of circuit theory*

In the analysis of circuits it is the quantity of current flowing between terminals, and its relation to the potential across them, that is of interest; details of the spatial flow pattern and potential distribution are of no concern. It is therefore legitimate to treat the contributions to resistance, capacitance and inductance in a circuit as though they were isolated in structureless elements; and the physical arrangement of actual elements and wires between terminals is only significant in so far as it affects the values of these quantities.

Indeed, in practice, resistance, capacitance and inductance in a circuit can often be taken to reside solely in 'lumped' elements specifically introduced; the contributions from other parts of the circuit, such as connecting wires, being negligible. Consider, for example, the resistance of a length l of uniform wire of cross-section A. The current is $I = AJ$, where $J = \sigma E$ is the current density, and the potential drop is $V = lE$. The resistance (4.7) is therefore

$$R = V/I = l/(A\sigma). \tag{4.90}$$

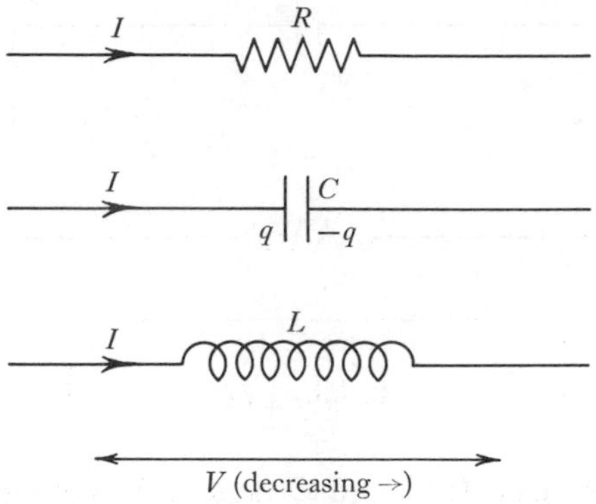

Figure 4.13

For copper $\sigma = 5.8 \times 10^7$, so that the resistance of one metre of copper wire of radius one millimetre is only about 0.005 ohm. The resistance of connecting wires is therefore unlikely to be significant except when it happens that no lumped resistors are present. Likewise 'stray' capacitance and inductance can often be ignored. When necessary, however, allowance can be made for extraneous effects; for instance, the resistance of the wire of a multi-turn inductor can be taken into account by including, in the representation of the actual inductor, resistance in series with inductance.

The essence of circuit theory is therefore the analysis of the flow of current in networks whose elements are pure resistors, capacitors and inductors.

The theory is based on the relations, for the respective elements, between the current I flowing through the element and the potential difference V between its terminals. For a resistor the relation is

$$V = IR. \tag{4.91}$$

For a capacitor it is

$$I = C\frac{\mathrm{d}V}{\mathrm{d}t}, \tag{4.92}$$

since $I = \mathrm{d}q/\mathrm{d}t$, where q is the charge on the plate at higher potential, and $q = CV$. For an inductor it is

$$V = L\frac{\mathrm{d}I}{\mathrm{d}t}, \tag{4.93}$$

this being the potential necessary to overcome the opposition of the voltage induced by changing current.

In each of these relations the positive sense of current flow is the same as that in which the potential decreases. This is indicated in figure 4.13, which also shows the conventional symbols for the circuit elements.

A simple illustration is afforded by the discharge of a capacitor through

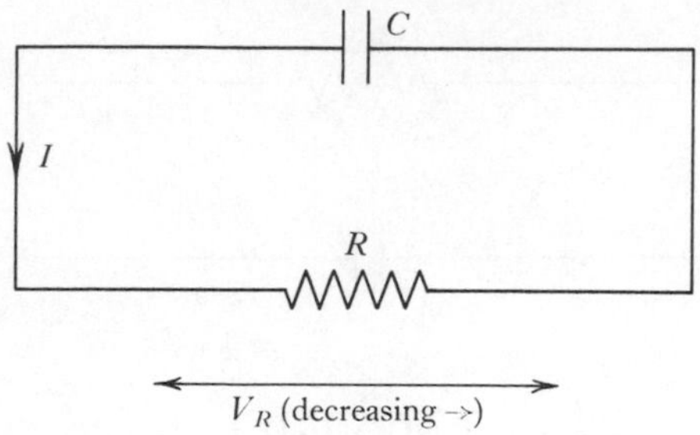

Figure 4.14

a resistor, as in figure 4.14. If V_R and V_C denote the potentials across resistor and capacitor respectively, then

$$V_R + V_C = 0, \tag{4.94}$$

the time derivative of which gives

$$R\frac{dI}{dt} + \frac{I}{C} = 0. \tag{4.95}$$

Hence $$I = I_0 e^{-t/(RC)}, \tag{4.96}$$

and correspondingly $$V_R = -V_C = V_0 e^{-t/(RC)}, \tag{4.97}$$

where $V_0 = I_0 R$ is the initial potential.

The rate of loss of energy, due to dissipation in the resistor, is IV_R (see §4.1.3). The total energy lost in this way during the discharge is therefore

$$I_0 V_0 \int_0^\infty e^{-2t/(RC)}\, dt, \tag{4.98}$$

which, as expected, has the same value, $\frac{1}{2}CV_0^2$, as the energy initially stored in the capacitor. In fact, the product of (4.94) with I yields the equation of energy balance

$$IV_R + \frac{d}{dt}(\tfrac{1}{2}CV_C^2) = 0. \tag{4.99}$$

It follows similarly that pure capacity and inductance together as in figure 4.15 give

$$L\frac{d^2I}{dt^2} + \frac{I}{C} = 0, \tag{4.100}$$

which implies that the circuit has a natural angular frequency of oscillation $\omega = (LC)^{-\frac{1}{2}}$. If, for example, $C = 10^{-10}$ farad, $L = 10^{-4}$ henry, then $\omega = 10^{-7}\,\text{sec}^{-1}$.

Consider now the circuit shown in figure 4.16, where the generator maintains a time varying voltage $V(t)$ between its terminals. Then

$$V_L + V_R + V_C = V, \tag{4.101}$$

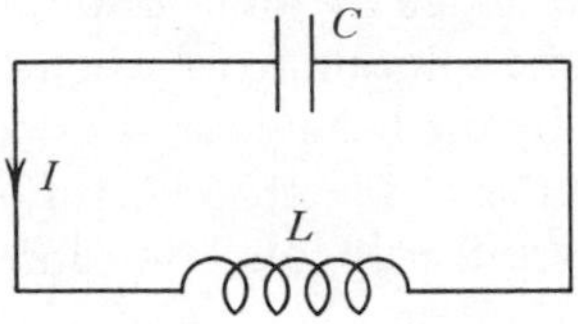

Figure 4.15

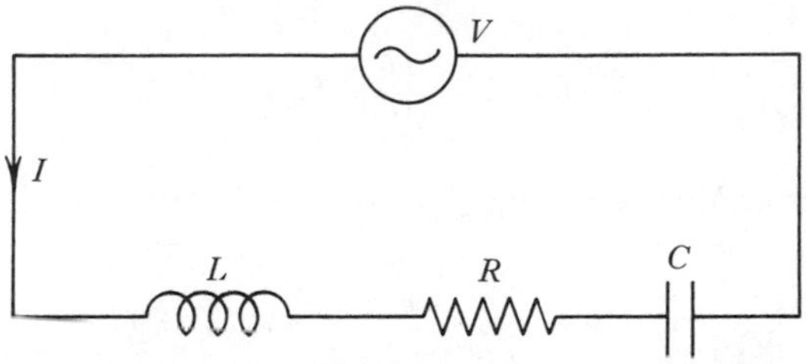

Figure 4.16

the time derivative of which gives

$$L\frac{\mathrm{d}^2 I}{\mathrm{d}t^2}+R\frac{\mathrm{d}I}{\mathrm{d}t}+\frac{I}{C}=\frac{\mathrm{d}V}{\mathrm{d}t}. \tag{4.102}$$

The general solution of (4.102) is the sum of any particular solution, called the particular integral, and the complementary function, the latter being the general solution of the corresponding homogeneous equation obtained by equating the right hand side to zero. Evidently exp (ipt) is a solution of the homogeneous equation if

$$p^2-\mathrm{i}\frac{R}{L}p-\frac{1}{LC}=0, \tag{4.103}$$

which gives

$$p=\frac{1}{2}\left[\mathrm{i}\frac{R}{L}\pm\left(\frac{4}{LC}-\frac{R^2}{L^2}\right)^{\frac{1}{2}}\right]. \tag{4.104}$$

The complementary function is therefore

$$\mathrm{e}^{-Rt/2L}[A\cos(\gamma t)+B\sin(\gamma t)], \quad \text{for} \quad R<2(L/C)^{\frac{1}{2}}, \tag{4.105}$$

and

$$\mathrm{e}^{-Rt/2L}(A\,\mathrm{e}^{\gamma t}+B\,\mathrm{e}^{-\gamma t}), \quad \text{for} \quad R>2(L/C)^{\frac{1}{2}}, \tag{4.106}$$

where

$$\gamma=\left|\frac{1}{LC}-\frac{R^2}{4L^2}\right|^{\frac{1}{2}}. \tag{4.107}$$

A particular solution can be obtained in the form of an integral for any generator voltage $V(t)$, but in specific cases it is often quicker to derive the result directly. Suppose, for example, that the circuit elements are connected across the terminals of a battery by throwing a switch at $t=0$. Then

$$V=\begin{cases}0, & \text{for} \quad t<0,\\ V_0, & \text{for} \quad t>0,\end{cases} \tag{4.108}$$

where V_0 is a constant, and hence $\mathrm{d}V/\mathrm{d}t = 0$ for all $t \neq 0$. Zero is therefore a particular integral, and the solution for I when $t > 0$ is (4.105) or (4.106) with A and B determined by the behaviour at $t = 0$. In fact I, which is zero for $t < 0$, must be continuous at $t = 0$, to keep V_L finite; and consequently (4.102) integrated over an arbitrarily small time interval spanning $t = 0$ gives

$$L\left[\frac{\mathrm{d}I}{\mathrm{d}t}\right]_{0-}^{0+} = [V]_{0-}^{0+}.$$

Thus, at $t = 0+$,

$$I = 0, \quad \frac{\mathrm{d}I}{\mathrm{d}t} = V_0/L; \tag{4.109}$$

from which it is at once apparent that, for $t > 0$,

$$I = \frac{V_0}{\gamma L}\,\mathrm{e}^{-Rt/2L}\sin(\gamma t) \tag{4.110}$$

in the case $R < 2(L/C)^{\frac{1}{2}}$. The current therefore performs a damped oscillation at angular frequency γ. That it is ultimately zero is, of course, due to the fact that the capacitor presents an open circuit to steady current.

4.4.4 *Time harmonic case: complex impedance*

Another example, most important because of its common occurrence in practice, is that in which the generator voltage is simple harmonic. A particular integral is then also simple harmonic at the same frequency, and this part of the complete solution soon predominates over the complementary function, since the latter contains the damping factor $\exp(-Rt/2L)$. Rarely, in such a context, is the transient behaviour of the current subsequent to switching on the generator of interest; rather it is the phase and amplitude of the current relative to that of the generator voltage in the steady state after the transients have died away that are required.

If the generator voltage is $V_0\cos(\omega t)$ the particular integral is most easily found by writing

$$V = \mathrm{Re}\,V_0\mathrm{e}^{\mathrm{i}\omega t}; \tag{4.111}$$

for it then follows at once that equation (4.102) is satisfied by

$$I = \mathrm{Re}\,I_0\mathrm{e}^{\mathrm{i}\omega t}, \tag{4.112}$$

where

$$ZI_0 = V_0, \tag{4.113}$$

with

$$Z = \mathrm{i}\omega L + R + \frac{1}{\mathrm{i}\omega C}. \tag{4.114}$$

Thus

$$I = \frac{V_0}{|Z|}\cos(\omega t - \phi), \tag{4.115}$$

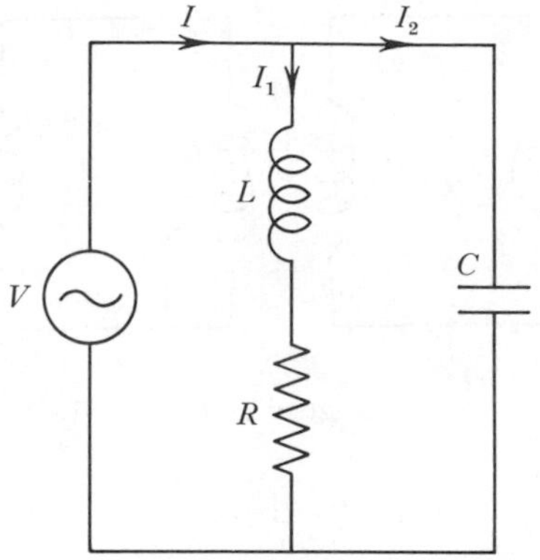

Figure 4.17

where $$\phi = \arg Z = \tan^{-1}\left[\left(\omega L - \frac{1}{\omega C}\right)\Big/ R\right], \tag{4.116}$$

and $$|Z| = \left[R^2 + \left(\omega L - \frac{1}{\omega C}\right)^2\right]^{\frac{1}{2}}. \tag{4.117}$$

It should be noted that the complex number Z contains all the information required. Its modulus gives the amplitude, and its argument the phase, of the applied voltage relative to the current. In particular, the current amplitude, regarded as a function of ω, has a single maximum at $\omega = (LC)^{-\frac{1}{2}}$; for which value of ω (4.116) shows that $\phi = 0$, and so $I = V_0/R$.

Equation (4.113) suggests that Z be regarded as a generalized resistance. It is called an *impedance*. Here, specifically, it is the impedance of a resistor, a capacitor and an inductor in series; and the expression (4.114) shows the explicit contribution of each element.

It is evident that, in the time harmonic case, the impedance concept can be readily applied to any circuit. For suppose now, with reference to equations (4.91), (4.92) and (4.93), that V is understood to be the *complex representation* of the actual voltage, in the sense that it has time dependence $\exp(\mathrm{i}\omega t)$ and its real part gives the actual voltage. Then, if I is likewise the complex representation of the current, the equations appear in the form

$$V = RI \tag{4.118}$$

from the resistor, $$I = \mathrm{i}\omega C V \tag{4.119}$$

for the capacitor, and $$V = \mathrm{i}\omega L I \tag{4.120}$$

for the inductor. The impedances of these elements are therefore R, $1/(\mathrm{i}\omega C)$ and $\mathrm{i}\omega L$, respectively.

Consider, for example, the circuit shown in figure 4.17. Evidently

$$V = I_1(R + \mathrm{i}\omega L) = I_2/(\mathrm{i}\omega C), \tag{4.121}$$

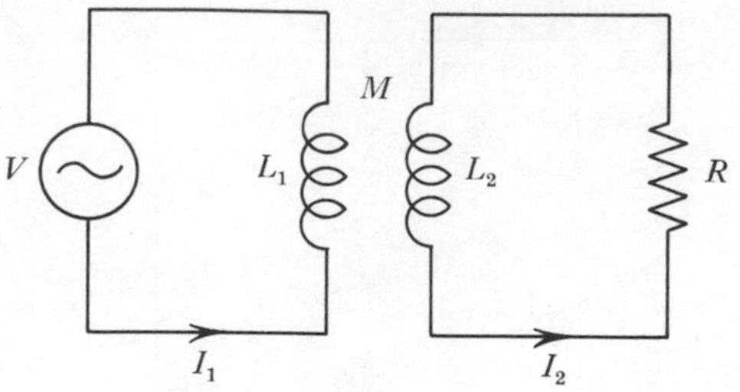

Figure 4.18

so that $$I = V/Z, \tag{4.122}$$

where $$\frac{1}{Z} = \frac{1}{R+i\omega L} + i\omega C. \tag{4.123}$$

The ratio of the amplitude of the current oscillation to that of the generator's voltage oscillation is therefore

$$|I/V| = |1/Z| = \left[\frac{(1-\omega^2 LC)^2 + \omega^2 C^2 R^2}{R^2+\omega^2 L^2}\right]^{\frac{1}{2}}, \tag{4.124}$$

and the phase of the current is in advance of that of the voltage by

$$\tan^{-1} \omega[CR + (\omega^2 LC - 1) L/R]. \tag{4.125}$$

Equation (4.123) illustrates the fact that impedances in series or in parallel have the same rules of combination as resistors.

As another example, consider circuits that are inductively coupled. For the arrangement of figure 4.18,

$$V = i\omega(L_1 I_1 + M I_2) \tag{4.126}$$

$$0 = i\omega M I_1 + (i\omega L_2 + R) I_2, \tag{4.127}$$

where M is the mutual inductance of the coils. Thus, in particular, the voltage developed across R through the coupling is

$$I_2 R = -\frac{MRV}{L_1(R+i\omega L_2) - i\omega M^2}. \tag{4.128}$$

Finally, a remark about energy in the time harmonic case is worth making. Reverting to the circuit of figure 4.16, the general statement of energy balance follows on multiplying (4.101) by I; for this gives

$$\frac{d}{dt}(\tfrac{1}{2}LI^2) + V_R I + \frac{d}{dt}(\tfrac{1}{2}CV_C{}^2) = VI, \tag{4.129}$$

in which the right hand side is the rate at which the generator works, and successive terms on the left hand side are the rate of change of energy stored in the inductor, the rate of dissipation of energy in the resistor, and the rate of change of energy stored in the capacitor.

In the time harmonic case it is chiefly the *time averaged* energy that is of interest, meaning the average over a period $2\pi/\omega$ or equivalently over an interval very much greater than the period. In this case, by integrating (4.129) over a period, the result is simply

$$\overline{V_R I} = \overline{VI}, \tag{4.130}$$

where the overline denotes the time average.

Suppose now that complex representations I, V are used, so that their time dependence is $\exp(\mathrm{i}\omega t)$ and their real parts give the actual current and voltage. Then the quadratic expression IV is of no significance, for in general, of course, $\mathrm{Re}\, IV \neq (\mathrm{Re}\, I)(\mathrm{Re}\, V)$. However, it is easy to confirm that

$$\mathrm{Re}\,\tfrac{1}{2}VI^* = \overline{(\mathrm{Re}\, V)(\mathrm{Re}\, I)}, \tag{4.131}$$

where the star denotes the complex conjugate; for if

$$V = V_0 \mathrm{e}^{\mathrm{i}(\omega t+\theta)}, \quad I = I_0 \mathrm{e}^{\mathrm{i}(\omega t+\phi)}, \tag{4.132}$$

where I_0, V_0, θ and ϕ are real, the left hand side of (4.131) is

$$\tfrac{1}{2}V_0 I_0 \cos(\theta - \phi), \tag{4.133}$$

whereas the right hand side is the average of

$$V_0 I_0 \cos(\omega t + \theta) \cos(\omega t + \phi), \tag{4.134}$$

which plainly agrees with (4.133) since the averages of $\cos^2(\omega t)$, $\sin^2(\omega t)$ and $\cos(\omega t)\sin(\omega t)$ are $\frac{1}{2}$, $\frac{1}{2}$ and 0, respectively.

The result (4.130) can therefore be deduced directly from the relation $V = IZ$, with Z given in (4.114), by noting that

$$\mathrm{Re}\,\tfrac{1}{2}VI^* = \mathrm{Re}\,\tfrac{1}{2}II^*Z = \tfrac{1}{2}II^*R = \mathrm{Re}\,\tfrac{1}{2}V_R I^*. \tag{4.135}$$

In general the time average of quadratic quantities such as energy is most quickly evaluated by using complex representations in the relation typified by (4.131).

4.4.5 *Forces*

Consider an infinitesimal, plane loop C of vector area $\mathbf{S}$ carrying steady current I. As explained in § 2.2.2, the loop is appropriately described as a magnetic dipole of moment

$$\mathbf{m} = I\mathbf{S}, \tag{4.136}$$

and when placed in a magnetic field $\mathbf{B}$ experiences a couple

$$\mathbf{m} \wedge \mathbf{B}. \tag{4.137}$$

By analogy with an electric dipole in an electric field (see (3.42)) it is to be expected that the loop also experiences a force $\mathrm{grad}(\mathbf{m}.\mathbf{B})$. This is now confirmed.

The force is
$$\mathbf{F} = I\oint d\mathbf{s}\wedge\mathbf{B}, \tag{4.138}$$
where the line integral is round the loop. The component in the direction of any unit vector $\mathbf{t}$ is
$$\mathbf{F}.\mathbf{t} = I\oint(\mathbf{B}\wedge\mathbf{t}).d\mathbf{s} = I\int[\operatorname{curl}(\mathbf{B}\wedge\mathbf{t})].d\mathbf{S}. \tag{4.139}$$
The second expression results from the application of Stokes' theorem, and reduces to
$$\mathbf{m}.\operatorname{curl}(\mathbf{B}\wedge\mathbf{t}), \tag{4.140}$$
since the area of the loop is infinitesimal. The expansion (A. 17) then gives
$$\mathbf{F}.\mathbf{t} = \mathbf{m}.[(\mathbf{t}.\operatorname{grad})\mathbf{B}] = \mathbf{t}.\operatorname{grad}(\mathbf{m}.\mathbf{B}), \tag{4.141}$$
because $\mathbf{t}$ and $\mathbf{m}$ are independent of position, and $\operatorname{div}\mathbf{B} = 0$. Hence
$$\mathbf{F} = \operatorname{grad}(\mathbf{m}.\mathbf{B}). \tag{4.142}$$

The formula (4.142) is evidently valid for any loop small enough for the variations of the gradients of the components of $\mathbf{B}$ over the loop to be negligible. It can, of course, be applied to an arbitrary current loop by treating the loop as a net of infinitesimal loops, the force on the equivalent magnetic shell then being given as an integral over a surface spanning the loop; but this approach is unlikely to be more expeditious than the direct application of (4.138). Incidentally, it is obvious from either formulation that the force is zero if $\mathbf{B}$ is uniform.

Suppose the field $\mathbf{B}$ were due to current I' in loop C'. Then
$$\mathbf{m}.\mathbf{B} = II'M, \tag{4.143}$$
where M is the mutual inductance of C and C'. But $II'M$ is the mutual energy, W_m say, of the two loop currents. Hence
$$\mathbf{F} = \operatorname{grad} W_m, \tag{4.144}$$
in which, of course, W_m is treated as a function of the position coordinates of C, all other parameters being kept constant.

Too hasty a comparison with (3.42) might suggest that a minus sign has gone astray. But this is not so. In giving C a small translation $\delta\mathbf{x}$ mechanical work $-\mathbf{F}.\delta\mathbf{x}$ is done against the force $\mathbf{F}$, and also, if I and I' are maintained constant, electrical work
$$I\delta\Phi + I'\delta\Phi' \tag{4.145}$$
is done, where $\delta\Phi$ and $\delta\Phi'$ are the respective changes in the magnetic fluxes through C and C' consequent on the translation of C. Evidently (4.145) is
$$2II'\delta M = 2\delta W_m,$$

and the equation of energy balance therefore reads

$$-\mathbf{F}.\delta\mathbf{x}+2\delta W_{\mathrm{m}} = \delta W_{\mathrm{m}}, \tag{4.146}$$

which is equivalent to (4.144).

This result is easily generalized. Suppose there are n current loops. Then the magnetic energy is

$$W = \frac{1}{2}\sum_{i=1}^{n} I_i\Phi_i, \tag{4.147}$$

and the increment associated with any small displacement is

$$\delta W = \frac{1}{2}\sum_{i=1}^{n} I_i\delta\Phi_i \tag{4.148}$$

if the current in each loop is kept constant. On the other hand the electrical work required to maintain the currents in the face of flux changes $\delta\Phi_i$ is

$$\sum_{i=1}^{n} I_i\delta\Phi_i = 2(\delta W)_{\text{constant currents}}. \tag{4.149}$$

The mechanical work required to effect the displacement against the magnetic force is therefore

$$-(\delta W)_{\text{constant currents}}. \tag{4.150}$$

If the displacement is made *without* the application of electrical work to maintain the currents, then the currents change in such a way that the magnetic flux through each loop is kept constant. The mechanical work effecting the displacement is therefore also

$$(\delta W)_{\text{constant fluxes}}. \tag{4.151}$$

Thus, for the same displacements, (4.150) and (4.151) have the same value. The situation is analogous to that in electrostatics when displacements of conductors are considered in which either the charges or the potentials of the conductors are constant (cf. § 3.2.5).

In terms of the inductances L_{ij} the expression (4.55) for W evidently gives

$$-(\delta W)_{\text{constant currents}} = -\tfrac{1}{2}\sum_{i,j} I_i I_j \delta L_{ij}. \tag{4.152}$$

If the loops are not distorted in the displacements there is a change only in the 'mutual' energy, defined by (4.147) with each Φ_i representing the flux through the ith loop due to the currents in all the other loops: the self-inductances L_{ii} are unaffected, and the terms with $i = j$ in (4.152) are zero. But for the general case in which the shape of the loops is altered the contributions $\frac{1}{2}I_i^2\delta L_{ii}$ must be included in (4.152).

These results are well illustrated by finding the force $\mathbf{F}$ and couple $\mathbf{G}$ on current I in a loop C due to current I' in a loop C'. Suppose C is given a small

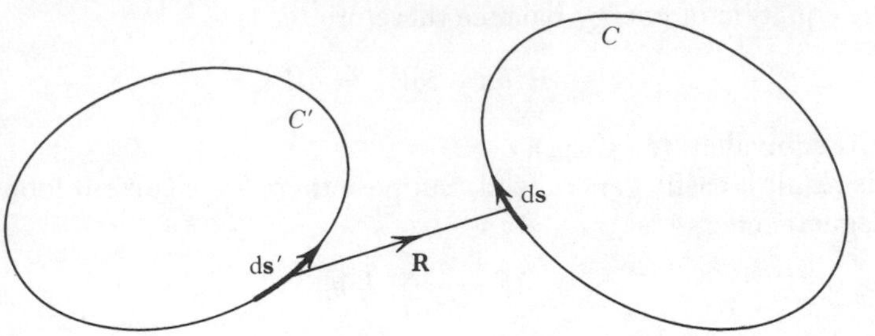

Figure 4.19

translation $\delta\mathbf{x}$. Then the statement that the work done against $\mathbf{F}$ is equal to (4.152) reads

$$-\mathbf{F}.\delta\mathbf{x} = -II'\delta M, \qquad (4.153)$$

where M is the mutual inductance of C and C', and is given by (4.72). Now

$$R\delta R = \mathbf{R}.\delta\mathbf{R} = \mathbf{R}.\delta\mathbf{x} \qquad (4.154)$$

if R is the vector from $\mathbf{ds}'$ to $\mathbf{ds}$ (see figure 4.19). Hence

$$\delta(1/R) = -\mathbf{R}.\delta\mathbf{x}/R^3, \qquad (4.155)$$

and the expression thereby obtained from (4.72) for δM, when substituted into (4.153), yields immediately

$$\mathbf{F} = -\frac{\mu_0 II'}{4\pi}\oint_C \oint_{C'} \frac{\mathbf{R}}{R^3}\,\mathbf{ds}.\mathbf{ds}'. \qquad (4.156)$$

This is evidently equivalent to the formula (2.60) which was obtained by an alternative method.

The couple $\mathbf{G}$ on C about some origin O can likewise be found, by supposing that C be given a small rotation $\delta\mathbf{\Omega}$ about O, and using

$$-\mathbf{G}.\delta\mathbf{\Omega} = -II'\delta M. \qquad (4.157)$$

To evaluate δM it is first noted that, if $\mathbf{r}$ and $\mathbf{r}'$ are the position vectors of $\mathbf{ds}$ and $\mathbf{ds}'$, respectively, then, as in (4.155),

$$\delta(1/R) = -\mathbf{R}.(\delta\mathbf{\Omega}\wedge\mathbf{r})/R^3 = \delta\mathbf{\Omega}.(\mathbf{r}\wedge\mathbf{r}')/R^3. \qquad (4.158)$$

Furthermore, in this case $\mathbf{ds}$ also has an increment, $\delta\mathbf{\Omega}\wedge\mathbf{ds}$, so that

$$\delta(\mathbf{ds}.\mathbf{ds}') = \delta\mathbf{\Omega}.(\mathbf{ds}\wedge\mathbf{ds}'). \qquad (4.159)$$

The expression for δM obtained from these results and substituted into (4.157) gives

$$\mathbf{G} = \frac{\mu_0 II'}{4\pi}\oint_C \oint_{C'} \left(\frac{\mathbf{r}\wedge\mathbf{r}'}{R^3}\,\mathbf{ds}.\mathbf{ds}' + \frac{\mathbf{ds}\wedge\mathbf{ds}'}{R}\right), \qquad (4.160)$$

which is equivalent to the result of problem 2.18.

4.5 Magnetic media

4.5.1 *Phenomenological theory*

It was seen in Chapter 3 that a material body placed in an electrostatic field may give rise to an additional field. If the body is a conductor the additional field is due to an induced surface charge density created by free electrons; if an insulator, it is due to an induced dipole moment density created by bound electrons.

Likewise a material body placed in a magnetostatic field may give rise to an additional field. So far in this connection only a superconductor has been considered, for which the additional field is due to a surface current density of a special kind associated with extremely low temperatures. But more commonly effects arise under ordinary conditions from an induced current density created by the 'circulation' of electrons in atoms, or by the intrinsic 'spin' of electrons. A simple, purely phenomenological, description of these latter effects is now given, based on experimental findings that show the inductance of a solenoid to depend on the substance with which it is filled.

Suppose that known current I is passed through a long cylindrical solenoid, and that the sensibly uniform magnetic field $\mathbf{B}$ thereby generated within the solenoid is measured, say by making I vary slowly and noting the voltage induced in a wire loop encircling the solenoid. Now imagine the experiment repeated, with the sole difference that the interior of the solenoid is filled uniformly with the substance under investigation. Then the factor μ/μ_0 by which the ratio B/I is changed by the presence of the medium is dependent on the medium, which is said to have *permeability* μ. The measurement of the magnetic field in the manner indicated presumes that it remains uniform within the medium, and refers, of course, to the macroscopic (or average) field there.

If μ differs from μ_0 the effect is naturally explained by the existence of currents additional to I, which must reside in the medium. Thus

$$\operatorname{div}\mathbf{B} = 0 \tag{4.161}$$

remains valid everywhere, but $\operatorname{curl}\mathbf{B} = \mu_0\mathbf{J}$ fails if the evaluation of $\mathbf{J}$ does not take account of the contribution from the (average) currents in the medium.

The phenomenological approach is to avoid explicit mention of the medium currents by following the reasoning of § 2.4.2 in introducing the vector $\mathbf{H}$, which satisfies

$$\operatorname{curl}\mathbf{H} = \mathbf{J} \tag{4.162}$$

and is related to $\mathbf{B}$ in a way that achieves agreement with experiment.

The relation

$$\mathbf{H} = \mathbf{B}/\mu \tag{4.163}$$

is clearly adequate to account for the experiment just described: for a given current I, the value of **H** in the region interior to the solenoid is independent of what homogeneous medium fills this region, and (4.163) implies that the value of **B** is proportional to μ.

Equations (4.161), (4.162) and (4.163) contain no explicit reference to any particular geometry, and it is natural to assume that they are the governing equations for any disposition of steady currents and magnetic media. If the permeability has the same value at all points then the problem differs from the corresponding vacuum problem only in the replacement of μ_0 by μ: but in general μ is recognized to be dependent on position, and then (4.163) is to be treated as a point relation between the vectors **B** and **H**. Note that the vector potential representation remains $\mathbf{B} = \text{curl}\,\mathbf{A}$, by virtue of (4.161), but that the scalar potential representation in regions where **J** is zero, is now $\mathbf{H} = -\text{grad}\,\phi$, from (4.162).

Much of what has been said so far runs parallel to the discussion of dielectrics in §3.4.1, and the constitutive relation (4.163) may be compared with (3.112). The characteristics of actual substances, however, throw the two relations into sharp contrast; whereas the latter is comparatively straightforward, the former is beset by various complications, some general appreciation of which is necessary before it can be properly understood. The main features of the dependence of **B** on **H** for magnetic media are now outlined in the broadest terms.

4.5.2 *The dependence of* **B** *on* **H**

There is a broad classification of magnetic substances into three categories; *diamagnetic*, *paramagnetic* and *ferromagnetic*. Specimens in the first two categories have values of μ/μ_0 which differ so little from unity that it would be out of place to elaborate on them here. Suffice it to say that for diamagnetic substances μ/μ_0 is *less than* unity, though only by a quantity of the order of 10^{-5}; whereas for paramagnetic substances μ/μ_0 exceeds unity by a quantity of the order of 10^{-3}, and is inversely proportional to the absolute temperature. That the induced currents in a diamagnetic medium are not significantly temperature dependent, and produce a magnetic field **B** that opposes the applied field, indicates their origin in a physical mechanism different from that of paramagnetism.

Ferromagnetic specimens are characterized by very large values of μ/μ_0, of the order of 1000, and are consequently of great technological value. At ordinary temperatures iron, nickel and cobalt are the only ferromagnetic elements; but numerous alloys, some of which do not even contain any of these metals, have been found to exhibit ferromagnetic behaviour, as have also some rare-earth elements at low temperatures.

A striking feature of ferromagnetic substances is that the ratio of B to H depends on the magnetic history of the specimen: the value of B associated with a given value of H is therefore not unique, and the relation (4.163) must be treated with due caution.

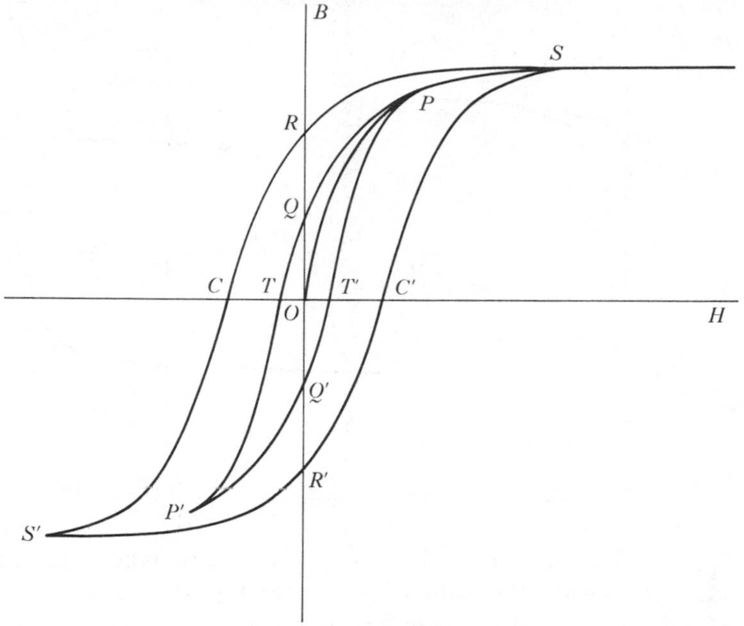

Figure 4.20

The situation is best described by considering again the ideal experiment referred to in §4.5.1. Suppose initially that the solenoid current I is zero, and that the ferromagnetic filling manifests no magnetic effect, so that $B = H = 0$. Now imagine that I, and therefore H, is made to increase slowly; then B also increases, and in this way a portion of the $B(H)$ curve is obtained, represented by OP in figure 4.20, say. If at this stage I, and with it H, is decreased, it is found that the curve is not retraced, but proceeds along PQ; B is still positive when H reaches zero at Q, and is not itself brought to zero until a *negative* value of H (associated with a reversal of I) is reached at T. Proceeding further, P' represents the point at which the value of H is the negative of its value at P, after which a continual increase of H takes the curve via Q' and T' back to P.

The cycle $PQTP'Q'T'P$, symmetric about O, is called a *hysteresis loop*. By stopping the so-called initial magnetization curve OP at different points P, different loops are obtained, except when saturation is reached. Saturation occurs when H is so large that the slope of the initial curve is μ_0, and remains so when H is further increased, indicating that the currents induced in the medium have reached their limit. All points P on this part of the curve have a common hysteresis loop, $SRCS'R'C'S$ in figure 4.20, whose tips S and S' represent the onset of saturation; this loop is referred to without qualification as *the* hysteresis loop of the specimen.

Several specific items in this general picture are now briefly mentioned.

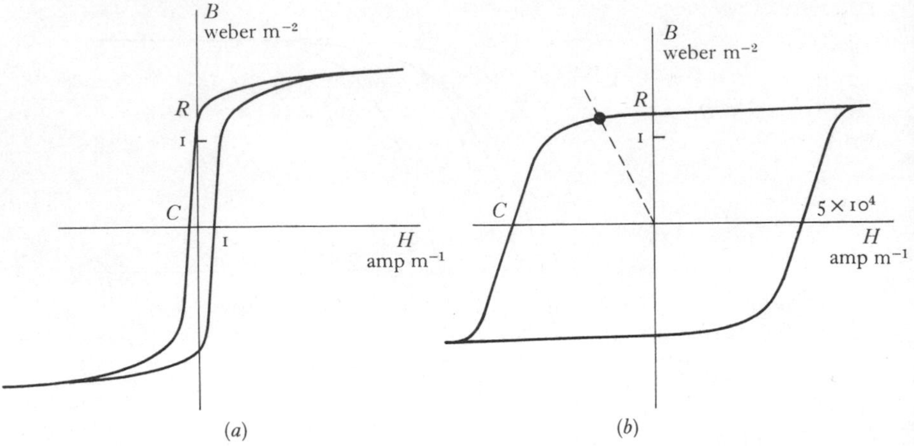

Figure 4.21

First, the scale. There are naturally considerable variations between different materials, but typically the value of H at saturation, H_S, is of the order of 10^4 amp m^{-1}, and B_S is of the order of 1 weber m^{-2}.

Secondly, the permeability. Defined as the ratio B/H, this can have any value, depending on how the fields are attained; and in such generality is not a particularly useful concept. However, it is important to recognize that if the specimen is operated sufficiently near to O then the magnetization curve remains close to the approximately linear section of the initial curve centred on O; in which case the permeability can be properly treated as a constant, characteristic of the substance, with value given by the slope of the initial curve at O. This situation is frequently met in practice, since H can be several amp m^{-1} without violating the linear relation. The value of μ/μ_0 so given is typically of the order of 1000.

Thirdly, the shape of the hysteresis loop. This can vary, depending on the substance, from very narrow to very broad. The *remanence*, that is, the value B_R of B when H is zero, is generally of the order of 1 weber m^{-2}; but the *coercivity*, that is, the value H_C of H required to reduce B to zero, may range from less than 1 to more than 10^5 amp m^{-1}. A substance is said to be magnetically soft or hard according as its coercivity is relatively low or high. Figure 4.21 shows (purely illustratively) hysteresis loops (*a*) for a soft and (*b*) for a hard material. Soft materials are required when hysteresis is an undesirable effect, as in the cores of circuit inductances. But the hardest materials are sought for making efficient *permanent* magnets.

The existence of permanent magnets is signalled by the presence of remanence: the material stays magnetized, giving rise to a magnetic field of its own accord, after the external current has been reduced to zero. However, only when the field is confined entirely within the substance does zero external current correspond to the physical state represented by the point R in figure 4.20. In practical circumstances in which the field is made available in a region outside the substance the operative conditions are represented by some point between R and C. As explained in §4.5.3, this point may be determined as the intersection of the hysteresis loop by a line through O of specified slope, like that indicated in figure 4.21 (*b*). The virtues of a hard substance, not easily demagnetized, are then apparent.

4.5.3 *Boundary conditions and simple problems*

The behaviour of **B** and **H** at an interface between two different magnetic media follows at once from (4.161) and (4.162). The normal component of **B** is continuous,

$$B_{n2} = B_{n1}; \tag{4.164}$$

and, assuming the interface carries no (external) surface current, the tangential components of **H** are continuous,

$$\mathbf{H}_{t1} = \mathbf{H}_{t2}; \tag{4.165}$$

these results being precisely analogous to (3.113) and (3.114) for a dielectric surface in electrostatics.

If the magnetic media can be treated as ideally soft, in the sense that they are characterized by a *linear* relation

$$\mathbf{B} = \mu\mathbf{H}, \tag{4.166}$$

they give rise to boundary value problems of the same nature as electrostatic problems involving dielectrics.

Suppose, for example, that current I flows in vacuum along an infinite straight line l parallel to the plane face of a half-space occupied by a homogeneous permeable medium. Then the solution can be obtained by images (cf. §3.4.5). For suppose that in the vacuum **H** has the value associated with the actual current I together with current I' flowing along the image of l in the interface; and in the medium **H** has the value associated with current I'' flowing along l; where in each case the currents are taken to exist in an unbounded vacuum. Then it is easy to see that the normal component of **B** and the tangential component of **H** are continuous at the interface if

$$I' = \frac{\mu-\mu_0}{\mu+\mu_0}I, \quad I'' = \frac{2\mu_0}{\mu+\mu_0}I. \tag{4.167}$$

Another simple example is that of a homogeneous permeable sphere placed in a previously uniform magnetic field $\mathbf{B}_0$. The solution can be written down at once in the form

$$\mathbf{H} = -\operatorname{grad}\phi, \tag{4.168}$$

where ϕ is given by the expression (3.136) after replacing E_0 by B_0/μ_0 and κ by μ/μ_0.

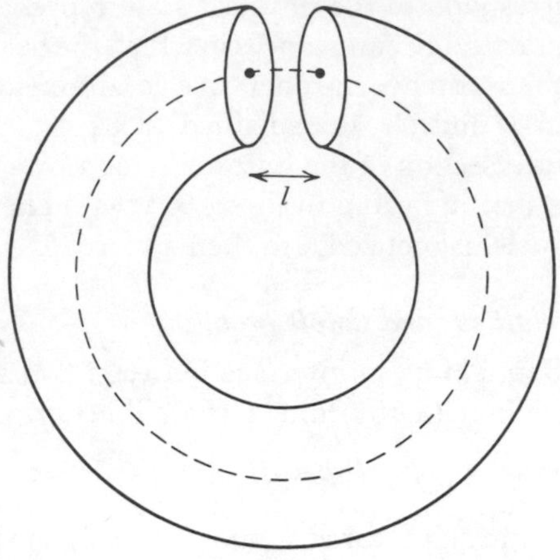

Figure 4.22

The boundary conditions are, of course, equally relevant in determining the fields of permanent magnets.

Consider a problem illustrative of the remarks in the final paragraph of §4.5.2. Imagine first a permanent magnet in the form of a thin torus. Then **B** exists only inside the magnet, its field lines being circles with centres on the axis of the torus, and its magnitude being sensibly uniform over the cross-section; and **H** is zero everywhere. How suppose the field is made accessible by creating a gap in the torus, as indicated in figure 4.22. This gives rise to a field **H** in the magnet, which can be evaluated approximately by the following argument.

If the width l of the gap is sufficiently small **B** is substantially uniform within it and must, by virtue of (4.164) have the same magnitude B_0 as it has in the magnet. Thus in the gap **H** has magnitude B_0/μ_0. But since there is no driving current ($\mathbf{J} = 0$ in (4.162)), the line integral of **H** round any closed path is zero. For the path shown as a dashed line in figure 4.22 this implies

$$B_0 l/\mu_0 + H_0 L = 0,$$

where H_0 is the negative of the magnitude of **H** in the magnet, and L is the length of the magnet. Thus the operative state of the magnet is represented by the point on the hysteresis loop where it is intersected by the line

$$H = -\frac{l}{L}\frac{B}{\mu_0}, \tag{4.169}$$

which is shown dashed in figure 4.21 (b).

In calculating the field due to a permanent magnet it may well be an adequate approximation to proceed to the theoretical limit of hardness and suppose that the currents in the medium remain saturated under the operative conditions. Then in the medium

$$\mathbf{B} = \mathbf{B}_\mathrm{R} + \mu_0 \mathbf{H}, \tag{4.170}$$

where $\mathbf{B}_\mathrm{R}$, the local remanence, is predetermined, and is usually, though not necessarily, independent of position.

A simple example is a uniformly magnetized sphere. With $\mathbf{B}_\mathrm{R}$ constant (4.170) gives

$$\operatorname{div} \mathbf{H} = \frac{1}{\mu_0} \operatorname{div} \mathbf{B} = 0, \tag{4.171}$$

so again $\mathbf{H}$ can be represented in terms of a potential ϕ, as in (4.168), which satisfies $\nabla^2 \phi = 0$. Now take spherical polar coordinates r, θ referred to the centre of the sphere O as origin, and the line through O parallel to $\mathbf{B}_\mathrm{R}$ as the axis from which θ is measured. Then it is clear that the solution is given by

$$\phi = \begin{cases} \dfrac{B_\mathrm{R}\, a^3}{3\mu_0} \dfrac{\cos\theta}{r^2}, & r > a, \\[2ex] \dfrac{B_\mathrm{R}}{3\mu_0} r \cos\theta, & r < a; \end{cases} \tag{4.172}$$

for the continuity of H_θ at $r = a$ is guaranteed by the continuity of ϕ, and the continuity of B_r is at once apparent from the expressions

$$B_r = \begin{cases} -\mu_0 \dfrac{\partial \phi}{\partial r} = \tfrac{2}{3} B_\mathrm{R} \left(\dfrac{a}{r}\right)^3 \cos\theta, & r > a, \\[2ex] B_\mathrm{R} \cos\theta - \mu_0 \dfrac{\partial \phi}{\partial r} = \tfrac{2}{3} B_\mathrm{R} \cos\theta, & r < a. \end{cases} \tag{4.173}$$

Two points are noteworthy. First, that inside the sphere $\mathbf{B}$, of magnitude $\frac{2}{3}B_\mathrm{R}$, and $\mu_0\mathbf{H}$, of magnitude $\frac{1}{3}B_\mathrm{R}$, are uniform and oppositely directed. Secondly, that outside the sphere the field is identical with that of a magnetic dipole of moment

$$\mathbf{m} = \tfrac{4}{3}\pi a^3 \mathbf{B}_\mathrm{R}/\mu_0. \tag{4.174}$$

In the absence of external currents the constitutive relation (4.170) for an ideally hard permanent magnet gives

$$\operatorname{curl} \mathbf{B} = \operatorname{curl} \mathbf{B}_\mathrm{R}. \tag{4.175}$$

The field to which the magnet gives rise is therefore identical with the vacuum field of a current density

$$(1/\mu_0) \operatorname{curl} \mathbf{B}_\mathrm{R}. \tag{4.176}$$

Likewise, since time independent magnetic forces satisfy Newton's third law (see § 2.2.4), a permanent magnet placed in an externally maintained magnetic field experiences the forces that act on the current density (4.176).

In particular, an infinitesimal permanent magnet itself produces a dipole field, and reacts as a dipole when placed in an external field.

If the magnetization of any permanent magnet is uniform, in the sense that $\mathbf{B}_R$ does not depend on position, (4.176) is only non-zero at the surface of the magnet. In fact, as Stokes' theorem shows, it consists solely of the surface current density

$$-\frac{1}{\mu_0}\mathbf{n}\wedge\mathbf{B}_R, \tag{4.177}$$

where $\mathbf{n}$ is the unit vector in the direction of the outward normal to the surface at each point.

Consider, for example, a permanent magnet in the shape of a cylinder, with plane ends, uniformly magnetized parallel to its generators. It is equivalent to the solenoid of the same shape carrying surface current density B_R/μ_0. Placed in a uniform external magnetic field $\mathbf{B}_{ext}$ it experiences a couple

$$\frac{1}{\mu_0}B_R V\mathbf{l}\wedge\mathbf{B}_{ext}, \tag{4.178}$$

where V is the volume of the cylinder and $\mathbf{l}$ is the unit vector parallel to its generators. A familiar illustration of the result is the couple exerted on a compass needle by the earth's magnetic field.

4.5.4 *Energy*

An adequate discussion of magnetostatic energy when magnetic media are present must take account of the fact that the relation between $\mathbf{B}$ and $\mathbf{H}$ is commonly non-linear.

Returning to the ideas developed in §4.3, consider first an ideal inductor carrying current I threaded by flux Φ. The work required to achieve an infinitesimal current increment δI in the face of the corresponding flux increment $\delta\Phi$ is

$$\delta W = I\delta\Phi. \tag{4.179}$$

Next, envisage a general steady current density $\mathbf{J}$, and apply the analogous result to each tube of flow of infinitesimal cross-section $\mathrm{d}S$, to get

$$\delta W = \Sigma J\,\mathrm{d}S\,\delta\Phi, \tag{4.180}$$

where the summation is over all tubes of flow. Now, as in (4.57) and the ensuing argument,

$$\delta\Phi = \oint \delta\mathbf{A}\,.\,\mathrm{d}\mathbf{s}, \tag{4.181}$$

which gives

$$\delta W = \int \mathbf{J}\,.\,\delta\mathbf{A}\,\mathrm{d}\tau, \tag{4.182}$$

where the integration is over all space. Since $\mathbf{J} = \operatorname{curl}\mathbf{H}$ the integrand is

$$\operatorname{div}(\mathbf{H}\wedge\delta\mathbf{A})+\mathbf{H}\,.\operatorname{curl}\delta\mathbf{A}. \tag{4.183}$$

The integral of the first term vanishes in the usual way, and so, from curl $\mathbf{A} = \mathbf{B}$,

$$\delta W = \int \mathbf{H} . \delta\mathbf{B} \, d\tau. \tag{4.184}$$

It may therefore be said that the local energy *density* of the magnetostatic field increases by

$$\mathbf{H} . \delta\mathbf{B}$$

when there is an increment $\delta\mathbf{B}$ in $\mathbf{B}$. However, it must be recognized that it is not possible to define magnetic energy density uniquely from this, because

$$\int_{\mathbf{B}_1}^{\mathbf{B}_2} \mathbf{H} . d\mathbf{B} \tag{4.185}$$

depends not only on the limits $\mathbf{B}_1$ and $\mathbf{B}_2$ but also, in general, in the way in which the transition from $\mathbf{B}_1$ to $\mathbf{B}_2$ was achieved. With reference, for example, to the magnetization curve shown in figure 4.20, if the specimen is taken along the initial curve from O to P the work done is given by the area between this curve and the B axis; but a traverse of the hysteresis loop $PQTP'Q'T'P$, which returns the specimen to the magnetic state represented by the point P, requires the expenditure of further work given by the area of the loop.

Only in circumstances in which a *linear* relation $\mathbf{B} = \mu\mathbf{H}$ holds can it be said that the magnetic field has energy density

$$\tfrac{1}{2}\mu\mathbf{H}^2. \tag{4.186}$$

Problems 4

4.1 Find the resistance between perfectly conducting spherical electrodes $r = a$ and $r = c\ (> a)$ when the medium between them has conductivity σ_1 for $a < r < b$ and σ_2 for $b < r < c$.

If the permittivity is ϵ_1 for $a < r < b$ and ϵ_2 for $b < r < c$, what charge density is associated with current flow I between the electrodes?

4.2 An interface separates homogeneous non-dielectric media of conductivities σ_1 and σ_2. If at a point on the interface the current density in the first medium is J making an angle θ with the normal to the interface, find the direction of the current in the second medium at that point, and also the surface charge density there.

4.3 A semi-infinite medium of uniform conductivity σ is bounded by a perfectly conducting plane, and contains a perfectly conducting sphere of radius a whose centre is distance d from the plane. Find an approximation to the resistance between the sphere and the plane when $a \ll d$.

4.4 A pair of perfectly conducting spherical electrodes, each of radius a, are at depth b in a sea of uniform conductivity σ, with their centres a distance $2b$ apart. Find an approximation to the resistance between them when $a \ll b$.

Evaluate the resistance when $a = 10^{-1}$ m, $b = 10$ m, $\sigma = 5$ mho m^{-1}.

4.5 Explain why a current flow problem in which perfectly conducting disc electrodes are embedded in a plane sheet of finite conductivity can be treated as a two-dimensional problem provided the sheet and discs are sufficiently thin.

An infinite plane sheet of uniform conductivity σ contains a pair of perfectly conducting electrodes in the form of circular discs of radius a, whose centres are distance $2b$ apart. Find the resistance between the electrodes on the assumption that the thickness h of the conducting sheet is small.

4.6 A thin circular sheet of radius d, thickness h, and uniform conductivity σ contains a pair of perfectly conducting electrodes in the form of small circular discs of radius a, whose centres are collinear with and distance b from the centre of the conducting sheet. Find an approximation to the resistance between the electrodes when a is much less than d and b.

4.7 Steady current I in the z direction flows uniformly in the region between the cylinders $x^2+y^2 = a^2$ and $(x+d)^2+y^2 = b^2$, where $a < d < b-a$. Show that the associated magnetic field $\mathbf{B}$ is uniform throughout the region $x^2+y^2 < a^2$, being in the y direction and of magnitude $\mu_0 Id/[2\pi(b^2-a^2)]$.

4.8 Show that the field of steady current that flows only in the z direction is independent of z and has the form $(B_x, B_y, 0)$. Show also that the field can be described, through $\mathbf{B} = \operatorname{curl} \mathbf{A}$, in terms of a vector potential $\mathbf{A} = [0, 0, A(x, y)]$; and that the field lines are given by $A =$ constant.

4.9 Current I flows in a wire loop that forms a rectangle whose corners are $(a, 0, -d)$, $(a, 0, d)$, $(-a, 0, d)$ and $(-a, 0, -d)$. Show that at $(x, y, 0)$ the vector potential is $(0, 0, A)$, where

$$A = \frac{\mu_0 I}{2\pi}\left\{\sinh^{-1}\frac{d}{[(x-a)^2+y^2]^{\frac{1}{2}}} - \sinh^{-1}\frac{d}{[(x+a)^2+y^2]^{\frac{1}{2}}}\right\}.$$

By letting $d \to \infty$ confirm that the magnitude of the vector potential of currents I flowing in opposite senses along a pair of parallel infinite straight lines is $(\mu_0 I/2\pi) \log (r_1/r_2)$, where r_1 and r_2 denote distances from the lines. Deduce that the field lines are circles.

4.10 Steady currents of the same magnitude I flow *towards* the origin along each half of the x axis, and *away from* the origin along each half of the y axis. Show that the x component of the vector potential at (x, y, z) may be written

$$A_x = -\frac{\mu_0 I}{4\pi}\int_{-x}^{x}\frac{d\xi}{(\xi^2+y^2+z^2)^{\frac{1}{2}}},$$

and derive similar integral expressions for A_y, A_z, B_x, B_y and B_z.
Show that $B = \mu_0 I/(2^{\frac{1}{2}}\pi a)$ at $(a, a, 0)$.

4.11 Show that if the centre and axis of a circular loop of radius a are taken as origin and axis of cylindrical polar coordinates r, θ, z, then the vector potential of current I flowing in the loop is given by $A_r = A_z = 0$,

$$A_\theta = \frac{\mu_0 Ia}{2\pi}\int_0^{\pi}\frac{\cos\phi\, d\phi}{(a^2+r^2+z^2-2ar\cos\phi)^{\frac{1}{2}}}.$$

Deduce that $B_\theta = 0$, and that for points close to the z axis,

$$B_r = \tfrac{3}{4}\mu_0 I\frac{a^2 rz}{(a^2+z^2)^{\frac{5}{2}}} + O(r^3/a^3), \quad B_z = \tfrac{1}{2}\mu_0 I\frac{a^2}{(a^2+z^2)^{\frac{3}{2}}} + O(r^2/a^2).$$

Compare with the results of §4.2.2.

Show that as the distance R of the field point from the centre of the loop tends to infinity, then $A_\theta \sim \frac{1}{4}\mu_0 Ia^2 r/R^3$. Confirm that this is the expected dipole potential.

4.12 A pair of coaxial circular loops, each of radius a, carry equal currents circulating in the same sense. Investigate the degree of uniformity of the field in the neighbourhood of the point midway between the centres of the loops, and show that the uniformity is enhanced when the distance between the centres is a.

4.13 Show that A_θ in problem 4.11 can be expressed exactly in terms of the complete elliptic integrals (4.75) as

$$A_\theta = \frac{\mu_0 I}{2\pi}\left(\frac{a}{r}\right)^{\frac{1}{2}}\left[\left(\frac{2}{k}-k\right)K(k)-\frac{2}{k}E(k)\right],$$

where $k^2 = 4ar/[(a+r)^2+z^2]$.

Use (4.80) to find the dominant term in A_θ when $z = 0$ and $r = a(1-2\epsilon)$ with $\epsilon \ll 1$. Show that the result agrees with that implicit in the second sentence of the penultimate paragraph of §4.2.2.

4.14 Generalize the formula (4.77) for the mutual inductance of a pair of coaxial circular loops to the case when the loops are not coplanar.

4.15 Write down, from (3.33) and (4.42), the capacitance C and inductance L per unit length of a pair of coaxial, circular cylindrical, superconducting surfaces, and verify that $LC = 1/c^2$. Show that this relation holds for any pair of parallel, cylindrical, superconducting surfaces, whose cross-sections do not intersect but are otherwise arbitrary.

4.16 Find the inductance of the superconducting surface formed by that part of the spherical surface $x^2+y^2+z^2 = a^2+b^2$ for which $x^2+y^2 > a^2$, together with that part of the cylindrical surface $x^2+y^2 = a^2$ for which $|z| < b$.

4.17 A toroidal tube is specified by the rotation of a circle of radius a about a fixed line in the plane of the circle and distance b ($> a$) from its centre. Show that the inductance of the solenoid formed by winding N turns of wire of total length $2\pi Na$ closely and uniformly on the tube is $\mu_0 N^2[b-(b^2-a^2)^{\frac{1}{2}}]$.

Consider the electric field that appears when the current in the solenoid is varied. Sketch the field lines both inside and outside the torus in the case when the variation is slow enough for the displacement current to be neglected, and $a \ll b$.

4.18 A homogeneous superconducting sphere of density ρ and radius a is held at its uppermost point, by a non-magnetic arm, in a uniform vertical magnetic field $\mathbf{B}$. If the sphere were bisected through the horizontal diametral plane find the least value of B that would maintain the lower hemisphere in position against gravity.

4.19 Calculate the total energy, per unit length in the z direction, of the magnetic field produced by steady current density $\mathbf{J} = (0, 0, J)$, with

$$J = J_0\frac{r_0}{r}\left(1-\frac{2r^2}{r_0{}^2}\right)e^{-r^2/r_0{}^2},$$

where r is distance from the z axis, and r_0, J_0 are constants.

4.20 (*a*) Uniform steady current in the surface of a long circular cylinder of radius r flows round the circumference, perpendicular to the axis of the cylinder. Show that there is an outward pressure $\frac{1}{2}\mu_0 j^2$ on the cylinder, where j is the surface current density.

Calculate the change in energy of the magnetic field (per unit length) when the radius of the cylinder is increased to R, the current being kept constant. Comment on the energy balance.

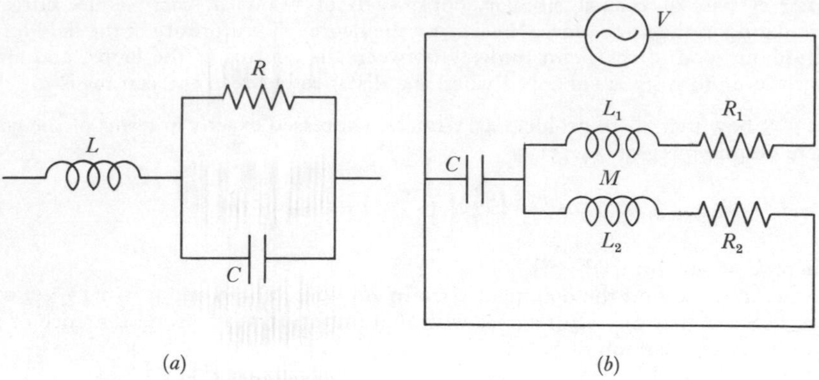

Figure 4.23

(*b*) A solenoid is constructed by winding a layer of copper strip of thickness 10^{-3} m round a plastic cylinder of diameter 10^{-1} m. If the tensile strength of the copper strip is 4×10^8 newton m^{-2}, what is the maximum magnetic field obtainable without damage to the solenoid by passing current pulses of short duration through the winding?

4.21 (*a*) Obtain the inductance per unit length of a pair of infinite, circular cylindrical, superconducting wires of radius a (carrying equal and opposite currents), whose axes are parallel and distance $2b$ ($> 2a$) apart (cf. problem 4.15).

(*b*) A pair of infinite, circular cylindrical wires of radius a, whose axes are parallel and distance $2b$ ($> 2a$) apart, carry equal and opposite currents uniformly distributed over their cross-sections. Show from energy considerations that the inductance per unit length is $(\mu_0/4\pi)\,[1+4\log(2b/a)]$.

4.22 (*a*) Find the complex impedance, at angular frequency ω, of the circuit elements shown in figure 4.23(*a*). Show that when $L = CR^2$ the impedance is approximately purely resistive for $\omega \ll 1/(LC)^{\frac{1}{2}}$.

(*b*) In the circuit shown in figure 4.23(*b*) the coils with self-inductance L_1 and L_2 have mutual inductance M, and the capacitance C is adjusted so that the application of the alternating voltage V at angular frequency ω produces no current through R_2. What, in this case, is the phase lag on V of the current through R_1?

4.23 (*a*) In the circuit shown in figure 4.24(*a*) the coils of self-inductance L have mutual inductance M. Show that the natural angular frequencies of the circuit are $[(L \pm M)\,C]^{-\frac{1}{2}}$.

(*b*) Alternating voltage V of angular frequency ω is applied to a primary circuit inductively coupled to a secondary circuit as shown in figure 4.24(*b*). Find the complex impedance $Z = V/I$, where I is the current in the primary circuit.

Show that if $\omega = (LC)^{-\frac{1}{2}}$, then $Z = M^2/(LCR)+R$. Show also that if

$$R^2 < 2[L-(L^2-M^2)]^{\frac{1}{2}}/C$$

there are two other values of the frequency for which Z is purely resistive; and that if R is sufficiently small these values are approximately $[(L \pm M)\,C]^{-\frac{1}{2}}$, and Z is then approximately $2R$.

4.24 In the bridge network shown in figure 4.25 there is a variable resistance R and a variable capacitance C. Show that on the application of the alternating voltage

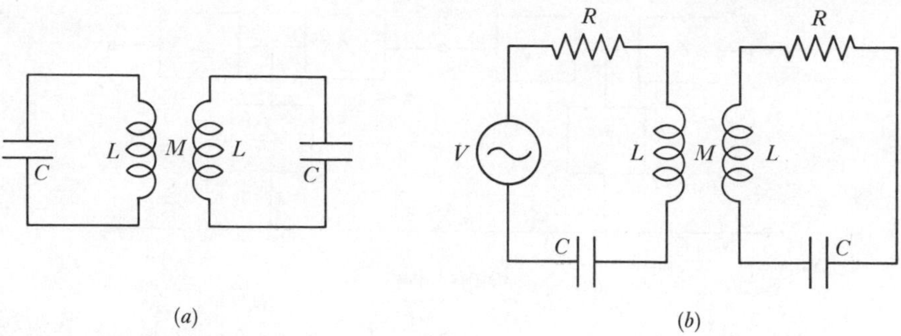

Figure 4.24

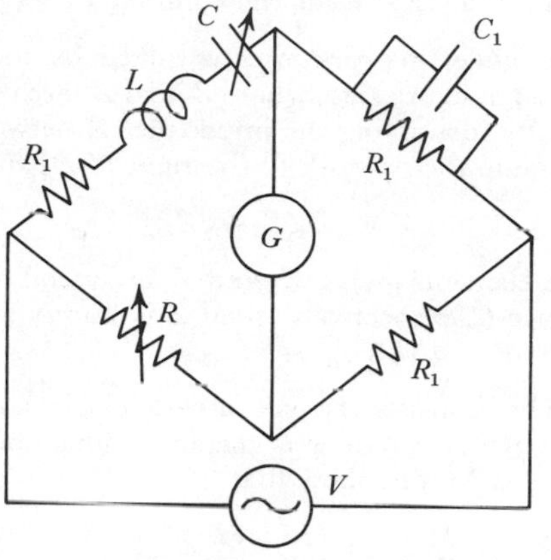

Figure 4.25

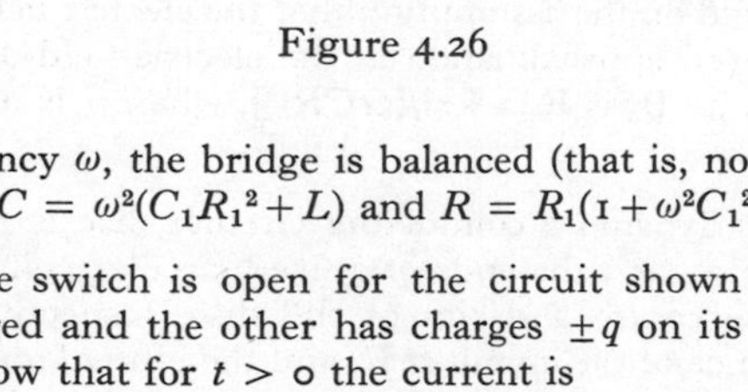

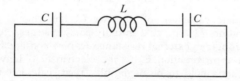

Figure 4.26

V of angular frequency ω, the bridge is balanced (that is, no current flows in the detector G) when $1/C = \omega^2(C_1R_1^2+L)$ and $R = R_1(1+\omega^2C_1^2R_1^2)$.

4.25 For $t < 0$ the switch is open for the circuit shown in figure 4.26, one capacitor is uncharged and the other has charges $\pm q$ on its plates. At $t = 0$ the switch is closed. Show that for $t > 0$ the current is

$$\frac{q}{(2LC)^{\frac{1}{2}}} \sin\left[\left(\frac{2}{LC}\right)^{\frac{1}{2}} t\right],$$

and obtain the charges on each capacitor.

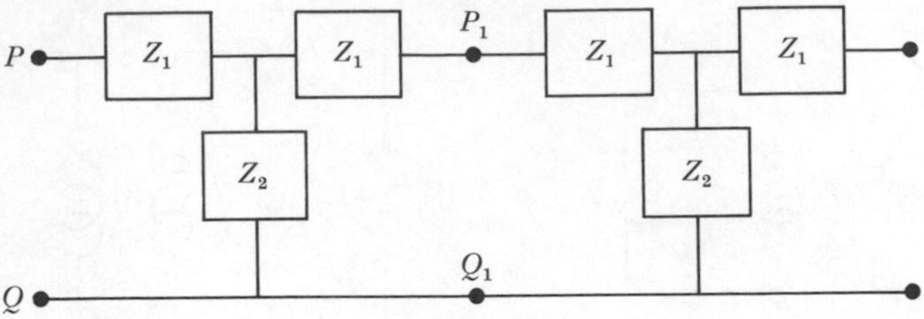

Figure 4.27

Find the energy in the inductance and in each capacitor. Check that the sum of the energies is that initially stored in the charged capacitor, and show that the energy in the inductance never exceeds the combined energy in the capacitors.

4.26 A semi-infinite ladder network is constructed by the repeated junction of 'T-sections' as indicated in figure 4.27, where Z_1 and Z_2 are complex impedances for angular frequency ω. By identifying the impedance Z between P and Q with that between P_1 and Q_1 due to the network to the right of P_1 and Q_1, show that

$$Z = [Z_1(Z_1+2Z_2)]^{\frac{1}{2}}.$$

Deduce that, if the elements giving Z_1 and Z_2 are a coil of inductance $\frac{1}{2}L$ and a capacitor of capacitance C, respectively, then Z is purely reactive if $\omega^2 > 4/(LC)$ and purely resistive if $\omega^2 < 4/(LC)$.

4.27 A transmission line consists of a pair of parallel wires, of negligible resistance, that carry currents $\pm I(x, t)$, where x is distance along the wires. If the voltage between the wires at x is $V(x, t)$, show that

$$\frac{\partial V}{\partial x} = -L\frac{\partial I}{\partial t}, \quad \frac{\partial I}{\partial x} = -C\frac{\partial V}{\partial t},$$

where L is the inductance and C the capacitance of the wires, per unit length.

Deduce that disturbances travel along the wires with speed $1/(LC)^{\frac{1}{2}}$.

4.28 A parallel plate capacitor, of capacity C, has coaxial circular plates, of radius a and separation d ($\ll a$), that initially carry charges $\pm q$. The capacitor is then discharged through an external resistance R. Neglecting edge effects throughout, find, first, an approximation $\mathbf{E}_0$ to the electric field between the plates by assuming the magnetic field to be negligible; next, calculate the approximation $\mathbf{B}_0$ to the magnetic field on the assumption that the electric field is $\mathbf{E}_0$; then, finally, show that the improved approximation to the electric field obtained by assuming the magnetic field to be $\mathbf{B}_0$ is $\mathbf{E}_0[1+r^2/(2cCR)^2]$, where r is distance from the axis of the plates.

4.29 In a primitive dynamo a conducting circular disc is rotated about its axis with angular velocity ω, and the ends of a fixed circular coil coaxial with the disc are connected to the centre and rim of the disc, respectively, through sliding contacts. The resistance of the circuit is R, and the mutual inductance between the coil and the perimeter of the disc is M. Show that any initial current in the circuit increases exponentially if ω exceeds a certain value, to be found.

4.30 A simple electric motor consists of a plane loop of wire, of area A, which is free to rotate about an axis in the plane of the loop, the axis being perpendicular to a constant, uniform magnetic field **B**. The angle between **n**, a unit vector normal to the plane of the loop, and **B** is θ; and the current $I(t)$ in the loop is measured in the sense of a right-handed screw about **n**. If a voltage $V(t)$ is applied across the ends of the loop, write down the equation relating $I(t)$ to θ, taking into account the resistance R of the loop, but neglecting its self-inductance.

Show that the equation of motion of the loop is

$$KR\ddot{\theta} + AB\sin\theta[AB\dot{\theta}\sin\theta + V(t)] = 0,$$

where K is the moment of inertia of the loop about its axis. Hence find the voltage $V(t)$ required to drive the motor with constant angular velocity ω. If, during such a motion, V is suddenly reduced to zero when $\theta = 0$ without breaking the circuit, show that, before stopping, the motor turns through a further angle θ_0 given by

$$4KR\omega/(AB)^2 = 2\theta_0 - \sin(2\theta_0).$$

4.31 A long circular cylindrical solenoid of radius a is held with its axis vertical. A small copper ring is supported with its plane horizontal and its centre on the axis and near the top of the solenoid. Show that when an alternating current is passed through the solenoid the lifting force on the ring is a maximum when the ring is distance $a/\sqrt{15}$ from the top of the solenoid.

4.32 An infinite, straight, thin wire, carrying a current I, lies parallel to and distance d from the axis of a superconducting cylinder of radius $a < d$. Find the force per unit length on the wire (*a*) when a current $-I$ flows along the cylinder, (*b*) when no current flows in the cylinder.

4.33 A toroidal tube is specified by the rotation of a square of side a about a fixed line l in the plane of the square, parallel to one of its sides and distance $b+\frac{1}{2}a$ $(b > 0)$ from its centre. The tube is filled with a medium of permeability $\mu(1+r/d)$, where μ and d are constants, and r is distance from the line l. Show that the inductance of the solenoid formed by winding N turns of wire of total length $4Na$ closely and uniformly on the tube is

$$\frac{1}{2\pi}\mu N^2 a[\log(1+a/b)+a/d].$$

4.34 Soft iron of uniform permeability μ occupying the region between concentric spherical surfaces of radii a and b $(> a)$ is placed in a uniform magnetic field $\mathbf{B}_0$. Find the resulting field within the inner spherical surface, showing that it is $\lambda\mathbf{B}_0$, where λ is a positive constant less than unity.

4.35 State and justify the image system that represents the effect due to a homogeneous permeable medium occupying the half-space $z < 0$ in the presence of an arbitrary steady current distribution confined to the half-space $z > 0$.

Deduce that the self-inductance of a plane wire loop is increased by the factor $2\mu/(\mu+\mu_0)$ if the loop is located in the plane face of a semi-infinite homogeneous medium of permeability μ.

4.36 Solve by images the problem of an infinite straight wire, carrying current I, lying parallel to an infinitely long circular cylinder of radius a and uniform permeability μ. Consider both the case $d > a$ and the case $d < a$, where d is the distance between the wire and the axis of the cylinder.

5

ELECTROMAGNETIC WAVES

5.1 Maxwell's equations

5.1.1 *The scope of the equations*

One of the objectives of § 2.5 was to give, at an early stage, some indication of the nature of solutions to the complete version of Maxwell's equations. The essential point is that the solutions have the character of electromagnetic waves, whose velocity in vacuum is c. This wave-like behaviour is suppressed under the special conditions considered in Chapters 2 and 3, even though some time varying situations were treated in the latter chapter. The possibility of time varying phenomena that are not wave-like was explained at the beginning of § 4.1.1. For example, the circuit theory of § 4.4.4 is valid provided only that the circuit dimensions are much less than the wavelength $2\pi c/\omega$ associated with angular frequency ω (cf. the inequality (4.2)). When $\omega = 2\pi \times 10^6\,\text{sec}^{-1}$ the wavelength is 300 m; it can therefore be appreciated that the restriction is not unduly severe in practice, but also that there are design problems for circuits operating at the very high frequencies now attainable. Such restricted time varying analyses, for instance, the treatment of inductance in circuit theory, involve the exact Maxwell equation (2.104), though only requiring the approximate version (4.1) of (2.106). It is the so-called displacement current term in the latter whose exclusion suppresses the realization of electromagnetic waves.

The purpose of this chapter is to discuss the exact equations, and to present simple solutions that typify some parts of the vast range of phenomena that they predict.

The equations set out in § 2.5.1 refer to the fields associated with charge and current in vacuum. There are many situations of practical importance in which it is valid to generalize these equations on the lines indicated in § 3.4 and § 4.5, so that with little outward change they apply to the average fields within a medium. Since the vacuum form of the equations is scarcely disturbed there is nothing to lose in admitting this generalization, provided its limitations are not forgotten.

Let ρ and $\mathbf{J}$ represent the charge and current densities associated with 'free' charged particles, such as conduction electrons, and introduce the vectors $\mathbf{D}$ and $\mathbf{H}$ in the manner of § 2.4.2, so that

$$\operatorname{curl} \mathbf{E} = -\dot{\mathbf{B}}, \tag{5.1}$$

$$\operatorname{curl} \mathbf{H} = \dot{\mathbf{D}} + \mathbf{J}, \tag{5.2}$$

$$\operatorname{div} \mathbf{B} = 0, \tag{5.3}$$

$$\operatorname{div} \mathbf{D} = \rho. \tag{5.4}$$

The effect of the average charge and current densities associated with the bound charged particles of the medium, which give rise to its electromagnetic properties, is to be allowed for by the formulation of appropriate *constitutive relations* between the field vectors.

For vacuum
$$\mathbf{D} = \epsilon_0 \mathbf{E}, \quad \mathbf{B} = \mu_0 \mathbf{H}. \tag{5.5}$$

For a medium the relations are of varying degrees of complexity depending on the molecular structure to be represented. Only the relations

$$\mathbf{D} = \epsilon \mathbf{E}, \quad \mathbf{B} = \mu \mathbf{H} \tag{5.6}$$

are considered in this book.

The simple phenomenological description embodied in (5.6) has its roots in the findings of § 3.4 and §4.5; but the justification for its adoption for time varying fields must be examined with care.

Only in specialist applications do ferromagnetic substances play a role in wave theory, and in the present context it is therefore supposed that μ is not significantly different from μ_0. Dielectrics, on the other hand, are ubiquitous, and important effects arise from the departure of ϵ/ϵ_0 from unity. For many media at ordinary fieldstrengths the relation between $\mathbf{D}$ and $\mathbf{E}$ is linear, and also isotropic in the sense that it is a single relation between corresponding components of the vectors. On the other hand (see § 6.3.5) $\mathbf{D}$ at time $t = t_0$ may well depend to some extent on $\mathbf{E}$ at prior times, $t \leqslant t_0$. Thus in using $\mathbf{D} = \epsilon \mathbf{E}$ it is essential to realize that the permittivity ϵ takes its static value (as introduced in § 3.4) only for fields that vary sufficiently slowly. Put more positively, though, it can be said that this constitutive relation has comparatively wide applicability for *monochromatic* fields, providing it is understood that the value of ϵ depends on the angular frequency ω. It may be noted in particular that, whatever the detailed mechanism, there is an upper limit to the frequency beyond which the bound charges in the medium are unable to respond, so that

$$\epsilon(\omega) \to \epsilon_0 \tag{5.7}$$

as $\omega \to \infty$.

A medium with dielectric properties may also be significantly conducting, and in the present context it is assumed that the relation between the corresponding conduction current and the field takes the simple form

$$\mathbf{J} = \sigma \mathbf{E}. \tag{5.8}$$

This form is the same as (4.3), previously introduced for steady currents, and again it is comparatively widely applicable on the understanding that the conductivity σ is frequency dependent.

Some substances are so highly conducting that they may often be treated as theoretically *perfect* conductors in the sense that σ is taken to be infinite.

There is thus no electric field inside a perfect conductor, and hence, from (5.1) and (5.2), no time varying magnetic field or current. For time varying fields the characterization of a perfect conductor is therefore analogous to that of a conductor in electrostatics and a superconductor in magnetostatics; its electromagnetic effect arises from charges and currents that exist solely on its surface.

5.1.2 *Boundary conditions*

The behaviour of the field vectors across surfaces of discontinuity has been obtained for static fields, and the same results continue to hold when the fields are time dependent. This is easily established on the reasonable assumption that the tangential components of **B** and **D** remain finite at such surfaces.

Suppose S is a surface of discontinuity, and apply the integral form of (5.1), namely (2.80), to an elementary rectangular circuit whose longer sides, of length ds, are on either side of S and effectively parallel to it (see figure 3.7). If the short sides of the rectangle are negligible compared with ds then the left hand side of (2.80) is in effect

$$(E_{t2} - E_{t1})\,\mathrm{d}s. \tag{5.9}$$

The right hand side of (2.80), however, being proportional to the area of the rectangle, is negligible compared with (5.9), provided only that the tangential components of **B** remain finite as assumed. The tangential components of **E** are therefore continuous across S; that is, as in (3.11),

$$[\mathbf{n} \wedge \mathbf{E}] = 0, \tag{5.10}$$

where **n** is the unit vector normal to S, and the square bracket signifies the jump in the quantity embraced on crossing S in the direction **n**.

A similar analysis applied to the integral form of (5.2) evidently gives (cf. (4.29))

$$[\mathbf{n} \wedge \mathbf{H}] = \mathbf{j}, \tag{5.11}$$

where the assumption that the tangential components of **D** remain finite is involved, and allowance is made for a surface current density **j** in S.

The equations (5.3) and (5.4) do not differ from their time independent forms, and the implication that

$$[\mathbf{n}.\mathbf{B}] = 0 \tag{5.12}$$

and

$$[\mathbf{n}.\mathbf{D}] = \sigma, \tag{5.13}$$

where σ is the surface charge density in S, therefore needs no further justification. However, it should be noted that, since the time derivatives of (5.3) and (5.4) are consequences of (5.1) and (5.2), the conditions (5.12) and (5.13) give information further to (5.10) and (5.11) only with respect to static fields. They are redundant, for example, when the field is monochromatic.

A particular case of great importance is that in which S is the surface of a perfect conductor. Since there is no electric field inside a perfect conductor it follows from (5.10) that *the tangential component of* **E** *is zero at the surface of a perfect conductor*. Moreover there is no time varying magnetic field inside a perfect conductor. Discounting magnetostatic fields, it follows from (5.11) that a perfect conductor carries a surface current density whose relation to the magnetic field at each point just outside its surface is expressed by

$$\mathbf{j} = \mathbf{n} \wedge \mathbf{H},$$

which has, of course, the same form as (4.34). The function of this surface current, with its associated charge, is analogous to that of surface charge on a conductor in electrostatics, in that the field it generates in the interior of the conductor exactly cancels the field there due to whatever external sources there may be.

As a simple illustration of (5.10) and (5.11), consider a uniform but variable surface current density

$$\mathbf{j} = (0, j(t), 0) \tag{5.14}$$

flowing in the plane $x = 0$. Since this source has no y or z dependence it must give rise to a field that depends only on x and t, and the field is therefore represented by disturbances of the type discussed in § 2.5.2. On physical grounds these disturbances are necessarily propagated away from the plane $x = 0$ into the respective half-spaces $x > 0$, $x < 0$; and in order to satisfy (5.10) and (5.11) the field is evidently

$$E_y = cB_z = -\tfrac{1}{2}(\mu_0/\epsilon_0)^{\frac{1}{2}} j(t - x/c), \quad \text{for} \quad x > 0, \tag{5.15}$$

$$E_y = -cB_z = -\tfrac{1}{2}(\mu_0/\epsilon_0)^{\frac{1}{2}} j(t + x/c), \quad \text{for} \quad x < 0, \tag{5.16}$$

all other components of **E** and **B** being identically zero. At each point, therefore, the variation of the field with time replicates exactly that of the current density, but with a time delay $|x|/c$.

Note that (5.12) and (5.13) are automatically satisfied, there being no variable charge density associated with a uniform current density. Note also that the boundary conditions at $x = 0$ are not alone sufficient to determine the solution; if incoming waves, proportional to $j(t + x/c)$ in $x > 0$ and to $j(t - x/c)$ in $x < 0$, are not rejected an infinity of 'solutions' is available.

The rejection on physical grounds of 'incoming' waves imposes on the mathematical solution a further condition, which in general can be regarded as a boundary condition at infinity. In wave boundary value problems a condition of this kind is necessary to ensure that the mathematical solution is unique. Uniqueness theorems, analogous to those of potential theory given in § 3.3.2, can be established, but are beyond the scope of this book.

5.1.3 *Poynting's theorem: field energy*

It is shown in § 3.2 that the work done in setting up, in vacuum, a static charge distribution from a state in which the charges are infinity dispersed can be expressed in the alternative forms

$$\frac{1}{2}\int \rho\phi\,\mathrm{d}\tau = \tfrac{1}{2}\epsilon_0\int \mathbf{E}^2\,\mathrm{d}\tau,$$

and that the right hand expression embodies the concept of field energy density

$$\tfrac{1}{2}\epsilon_0\mathbf{E}^2.$$

Likewise, in § 4.3 the same type of argument leads to the concept of energy density

$$\tfrac{1}{2}\mu_0\mathbf{H}^2$$

for the vacuum magnetic field of steady current. It is now to be established that the complete Maxwell equations yield the conclusion that the same expressions can legitimately be interpreted as energy densities in the general time varying case when both electric and magnetic fields are present.

Suppose that by 'mechanical' means a velocity $\mathbf{v}$, some function of position and time, is imposed on a charge density ρ in vacuum, giving rise to a current density $\rho\mathbf{v}$. The electromagnetic field $\mathbf{E}, \mathbf{B}\ (= \mu_0\mathbf{H})$ thereby produced reacts on the charge-current distribution through the Lorentz force density

$$\rho(\mathbf{E}+\mathbf{v}\wedge\mathbf{B}). \tag{5.17}$$

The resultant rate of change of kinetic energy of the charge distribution is the sum of the rates of working of the mechanical and the electromagnetic forces: in symbols,

$$\frac{\mathrm{d}\mathscr{E}}{\mathrm{d}t} = W_\mathrm{m}+W_\mathrm{e}, \tag{5.18}$$

where $\mathscr{E}$ is the kinetic energy of the charges in some volume V, W_m is the rate of working of the mechanical forces on these charges, and

$$W_\mathrm{e} = \int_V \rho(\mathbf{E}+\mathbf{v}\wedge\mathbf{B})\,.\,\mathbf{v}\,\mathrm{d}\tau = \int_V \rho\mathbf{E}\,.\,\mathbf{v}\,\mathrm{d}\tau \tag{5.19}$$

is the rate of working of the electromagnetic forces.

The idea now is to transform the integrand of (5.19) by discarding $\rho\mathbf{v}$ in favour of the field vectors $\mathbf{E}$ and $\mathbf{H}$, so that W_e in (5.18) can be interpreted in terms of energy stored in the field. Specifically, the Maxwell equation (5.2), which in the present context is

$$\operatorname{curl}\mathbf{H} = \epsilon_0\dot{\mathbf{E}}+\rho\mathbf{v}, \tag{5.20}$$

gives

$$W_\mathrm{e} = \int_V \mathbf{E}\,.\,(\operatorname{curl}\mathbf{H}-\epsilon_0\dot{\mathbf{E}})\,\mathrm{d}\tau.$$

Furthermore $\mathbf{E}.\operatorname{curl}\mathbf{H} = \mathbf{H}.\operatorname{curl}\mathbf{E} - \operatorname{div}(\mathbf{E}\wedge\mathbf{H})$

$$= -\mu_0\mathbf{H}.\dot{\mathbf{H}} - \operatorname{div}(\mathbf{E}\wedge\mathbf{H}), \tag{5.21}$$

using (5.1). Hence

$$W_e = -\int_V [\epsilon_0\mathbf{E}.\dot{\mathbf{E}} + \mu_0\mathbf{H}.\dot{\mathbf{H}} + \operatorname{div}(\mathbf{E}\wedge\mathbf{H})]\,d\tau. \tag{5.22}$$

The integral of the divergence term can be transformed to an integral over the surface S bounding the volume V, so that (5.18) can be written

$$W_m = \frac{d}{dt}\left[\mathscr{E} + \frac{1}{2}\int_V (\epsilon_0\mathbf{E}^2 + \mu_0\mathbf{H}^2)\,d\tau\right] + \int_S (\mathbf{E}\wedge\mathbf{H}).d\mathbf{S}. \tag{5.23}$$

The expression of W_m in the form (5.23) is the mathematical content of Poynting's theorem. To obtain the physical interpretation suppose, first, that prior to some past time all the charges were at rest in a bounded region. Then contemporaneously, because of the finite speed of propagation of electromagnetic effects, the electric field at sufficiently great distances r from some origin is $O(1/r^2)$ and the magnetic field is zero. Thus, for sufficiently remote surfaces S,

$$W_m = \frac{d}{dt}\left[\mathscr{E} + \frac{1}{2}\int_V (\epsilon_0\mathbf{E}^2 + \mu_0\mathbf{H}^2)\,d\tau\right]. \tag{5.24}$$

The natural interpretation of (5.25) is that

$$\frac{1}{2}\int_V (\epsilon_0\mathbf{E}^2 + \mu_0\mathbf{H}^2)\,d\tau \tag{5.25}$$

is the field energy in V, and this in turn suggests that

$$\tfrac{1}{2}\epsilon_0\mathbf{E}^2 + \tfrac{1}{2}\mu_0\mathbf{H}^2 \tag{5.26}$$

be regarded as field energy density. Thus the expressions originally derived for static fields do in fact apply to the general case.

Having identified field energy density in this way suppose next that W_m is zero and apply (5.23) to an arbitrary region. The requirement that energy be conserved implies that

$$\int_S (\mathbf{E}\wedge\mathbf{H}).d\mathbf{S}, \tag{5.27}$$

which is in general non-zero, must be interpreted as the rate of flow of field energy out of the arbitrary closed surface S, in order to balance the rate of decrease of energy within S. This in turn suggests that

$$(\mathbf{E}\wedge\mathbf{H}).d\mathbf{S} \tag{5.28}$$

be regarded as the rate of flow of field energy across each surface element $d\mathbf{S}$; in other words, the *Poynting vector*

$$\mathbf{E}\wedge\mathbf{H} \tag{5.29}$$

specifies the ***power flux density*** intrinsic to an electromagnetic field.

That there is a degree of arbitrariness in these interpretations need hardly be emphasized. Obviously the addition of any solenoidal vector to (5.29) cannot be rejected, since it would not affect (5.27). More generally, the addition of a scalar function f to (5.26) and a vector function $\mathbf{F}$ to (5.29) would not affect the argument provided only that

$$\operatorname{div}\mathbf{F}+\dot{f}=0.$$

Alternative expressions for the energy density and power flux density have in fact been proposed (see problem 5.5), but the reasons for advocating a departure from the conventional expressions (5.26) and (5.29) are too inconclusive to find a place here.

A simple example may serve immediately to illustrate the energy balance described by Poynting's theorem.

Let a parallel plate capacitor, with circular plates having separation d, be slowly charged. Then in the quasi-static approximation (cf. problem 4.28) the electric field between the plates is parallel to the axis l of the plates, with magnitude E say; and correspondingly the magnetic field is perpendicular to l, its lines being circles centred on l, and at distance r from l has magnitude $H=\frac{1}{2}\epsilon_0 r\dot{E}$. The Poynting vector is therefore $\frac{1}{2}\epsilon_0 rE\dot{E}$ towards l, giving the power flux across a cylindrical surface $r=a$ between the plates as $\pi\epsilon_0 a^2 dE\dot{E}$. This has the same value as the rate of change of the energy $\frac{1}{2}\epsilon_0 E^2\pi a^2 d$ stored by the capacitor in the region $r \leqslant a$ between the plates, the magnetic energy being negligible in the quasi-static approximation.

It should be noted that a more general form of Poynting's theorem is obtained if the derivation is carried through using equations (5.1) and (5.2) without specific reference to constitutive relations. The only change in the analysis is that $\dot{\mathbf{D}}$ replaces $\epsilon_0\dot{\mathbf{E}}$ in (5.20), and $\dot{\mathbf{B}}$ replaces $\mu_0\dot{\mathbf{H}}$ in (5.21). The result corresponding to (5.23) is therefore

$$W_{\mathrm{m}}=\frac{\mathrm{d}\mathscr{E}}{\mathrm{d}t}+\int_V(\mathbf{E}.\dot{\mathbf{D}}+\mathbf{H}.\dot{\mathbf{B}})\,\mathrm{d}\tau+\int_S(\mathbf{E}\wedge\mathbf{H}).\mathrm{d}\mathbf{S}. \qquad (5.30)$$

The interpretation now is that

$$\mathbf{E}.\dot{\mathbf{D}}+\mathbf{H}.\dot{\mathbf{B}} \qquad (5.31)$$

gives the rate of change of field energy density, so that the energy density itself is

$$\int\mathbf{E}.\mathrm{d}\mathbf{D}+\int\mathbf{H}.\mathrm{d}\mathbf{B}, \qquad (5.32)$$

where the integrations are taken from some zero energy state to the state under consideration. In making this assertion it must of course be remembered that, when media are present, the prescription yields a unique energy density associated with a given state of the field only if the media can be characterized by a rather restricted type of constitutive relation. The point has already been made, with respect to the second term in (5.32), in the discussion of magnetic hysteresis in §4.5.4.

If linear relations (5.6) hold to an adequate approximation for the field variations under consideration then the energy density is

$$\tfrac{1}{2}(\epsilon E^2 + \mu H^2) = \tfrac{1}{2}(\mathbf{E}.\mathbf{D} + \mathbf{H}.\mathbf{B}). \qquad (5.33)$$

This result is of very restricted validity in the time varying case, even when μ can in effect be equated to μ_0, because of the limitations on the relation $\mathbf{D} = \epsilon\mathbf{E}$ explained in §5.1.1. However, for monochromatic fields the relation with a frequency dependent ϵ has comparatively wide application, and there is then a result analogous to (5.33) which is derived in §6.2.4.

If conducting media are considered the further term

$$\int \mathbf{E}.\mathbf{J}\,\mathrm{d}\tau$$

appears in the right hand side of (5.30), where $\mathbf{J}$ is the conduction current density. This term gives the rate of dissipation of energy, as in (4.9) when it refers to steady currents. There is a simple illustration which exhibits the balance between the power flow into a region and the rate of dissipation of energy.

Let steady current I flow uniformly along an infinite cylindrical wire of radius a and conductivity σ. Then the electric field is parallel to the wire and its value just outside the wire is equal, by the boundary condition (5.10), to its value inside the wire, namely $E = I/(\pi a^2 \sigma)$. Also the magnetic field just outside the wire is perpendicular to the wire and tangent to its surface, with magnitude $H = I/(2\pi a)$. The Poynting vector (5.29) just outside the wire is therefore $I^2/(2\pi^2 a^3 \sigma)$ directed normally into the wire, so that the power flux into the wire per unit length is $I^2/(\pi a^2 \sigma)$; and this is indeed the rate of dissipation of energy $I^2 R$ per unit length of the wire, $R = 1/(\pi a^2 \sigma)$ being the resistance of unit length of the wire.

5.1.4 *Time harmonic fields*

If the field is monochromatic, with angular frequency ω, it is a great mathematical convenience to introduce the *complex representations* of the field vectors. These quantities have time dependence $\exp(\mathrm{i}\omega t)$, and their real parts give the values of the actual vectors. Thus, for example, the vector $\mathbf{E}'(\mathbf{r})$ defines the actual vector $\mathbf{E}(\mathbf{r}, t)$ through the relation

$$\mathbf{E} = \mathrm{Re}\,\mathbf{E}'\,\mathrm{e}^{\mathrm{i}\omega t}. \qquad (5.34)$$

In general $\mathbf{E}'$ is a *complex* vector function of position, its real and imaginary parts $\mathbf{E}_r'$ and $\mathbf{E}_i'$ being vectors that are not necessarily related in any particular way: the argument of each scalar component of $\mathbf{E}'$ gives the amount by

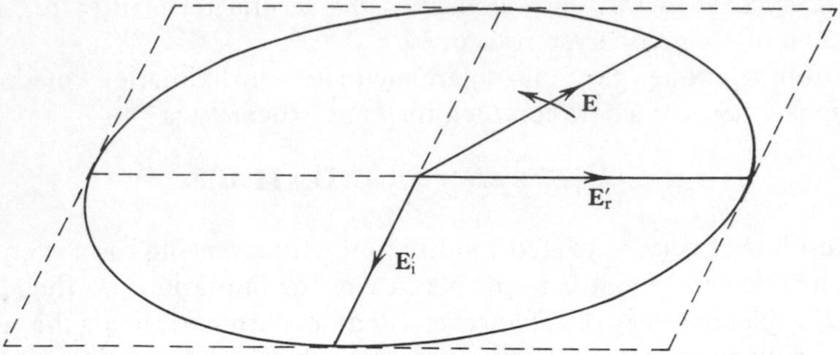

Figure 5.1

which the phase of the corresponding component of $\mathbf{E}$ is in advance of ωt. The relation (5.34) may be written

$$\mathbf{E} = \mathbf{E}_r' \cos(\omega t) - \mathbf{E}_i' \sin(\omega t), \tag{5.35}$$

from which it follows easily that, if the vector $\mathbf{E}$ at any specific position $\mathbf{r}$ is represented geometrically by an arrow drawn from a fixed point, then in general the tip of the arrow traces out an ellipse (figure 5.1). The ellipse is a circle only if $\mathbf{E}_r'$ and $\mathbf{E}_i'$ are mutually orthogonal and equal in magnitude. On the other hand it degenerates into a portion of a straight line only if $\mathbf{E}_r'$ and $\mathbf{E}_i'$ are parallel. The *polarization* of the field at $\mathbf{r}$ is said to be *elliptical circular* or *linear* in the respective cases.

Apart from the involvement of vectors the complex representation of the field is precisely the same as that of current and voltage introduced in § 4.4.4. In particular, the time averages of quadratic expressions may be calculated from the complex representations by applying the mathematical relation embodied in (4.131). Thus the time average of the field energy density in the special form (5.33) is

$$\mathrm{Re}\,\tfrac{1}{4}(\mathbf{E}'.\mathbf{D}'^* + \mathbf{H}'.\mathbf{B}'^*), \tag{5.36}$$

and the time averaged power flux density is

$$\mathrm{Re}\,\tfrac{1}{2}\mathbf{E}' \wedge \mathbf{H}'^*. \tag{5.37}$$

The satisfaction of the linear Maxwell equations by the actual vectors is synonymous with their satisfaction by the complex representations; and the latter quantities have the advantage that the action of the operator $\partial/\partial t$ merely multiplies them by $\mathrm{i}\omega$. Moreover, full information about the actual vectors can be read off from the complex representations themselves; nothing is gained by evaluating the real parts explicitly. For these reasons the discussion of monochromatic fields is almost always conducted solely in terms of the complex representations. This practice is so standard that a notational distinction (such as the dashes introduced above) is unnecessary: unless otherwise stated $\mathbf{E}, \mathbf{B}, \mathbf{D}, \mathbf{H}, \mathbf{J}, \rho$ are used to denote the complex representations, with the time factor exp (iωt) being understood but rarely included explicitly. This convention is adopted throughout the remainder of the present chapter.

In terms of complex representations Maxwell's equations (5.1) and (5.2) are

$$\operatorname{curl}\mathbf{E} = -\mathrm{i}\omega\mathbf{B}, \tag{5.38}$$

$$\operatorname{curl}\mathbf{H} = \mathrm{i}\omega\mathbf{D}+\mathbf{J}. \tag{5.39}$$

The charge conservation relation,

$$\operatorname{div}\mathbf{J}+\mathrm{i}\omega\rho = 0, \tag{5.40}$$

shows that there is no need to introduce ρ, since it is directly expressible in terms of **J**, and otherwise only appears in (5.4), which is a consequence of (5.39). Likewise, (5.3) follows from (5.38).

The key equations (5.38) and (5.39) must be supplemented by appropriate constitutive relations. In this book none more complicated than (5.6) and (5.8) are contemplated. With their introduction the right hand side of (5.39) becomes.

$$(\mathrm{i}\omega\epsilon_r+\sigma)\,\mathbf{E}, \tag{5.41}$$

where the permittivity has been written ϵ_r for a reason to be explained immediately. The fact that the parameters ϵ_r and σ enter in a specific combination suggests that they be replaced by a single, albeit complex, parameter. Since σ is zero for vacuum this parameter is written

$$\epsilon = \epsilon_r - \mathrm{i}\sigma/\omega \tag{5.42}$$

and called the *complex permittivity*. The parameter ϵ/ϵ_0 is called the *complex dielectric constant*.

The equations governing monochromatic fields in sufficiently simple media are therefore

$$\operatorname{curl}\mathbf{E} = -\mathrm{i}\omega\mu\mathbf{H}, \tag{5.43}$$

$$\operatorname{curl}\mathbf{H} = \mathrm{i}\omega\epsilon\mathbf{E}. \tag{5.44}$$

Before preceeding to discuss solutions of these equations it is useful to observe that they are invariant under the transformation

$$\mathbf{E}\rightarrow\mathbf{H},\quad \mathbf{H}\rightarrow-\mathbf{E},\quad \epsilon\leftrightarrow\mu. \tag{5.45}$$

Thus, for regions throughout which ϵ and μ are independent of position, from any known solution a second solution can be obtained merely by replacing **E** by **H** and **H** by $-$**E**, and interchanging ϵ and μ.

A further observation is that the equations as written are homogeneous, each term being linear in the dependent variables **E** and **H**. This is because all the currents are those associated with the media; they are accounted for in the complex permittitivity and in the permeability, the current densities being linear in **E** and **H**. A solution is sometimes termed 'self-consistent' because the current is driven by the field that it itself generates.

However, it is often the case that part of the current can be regarded as given, at least to an adequate approximation. Consider, for example, the vacuum field radiated by an isolated aerial. In a rigorous analysis the metal of the aerial is treated as a region of high conductivity, and a complicated boundary value problem is faced in which the aerial current, being the conduction current in the metal, appears as part of the solution. On the other hand it may well be practical to assume that the aerial current is known, perhaps from measurement, with the advantage that the calculation of the vacuum field of a given current distribution is a comparatively straightforward matter.

A given 'external' current density $\mathbf{J}_e$, if present, simply appears additionally on the right hand side of (5.44), which is then

$$\operatorname{curl}\mathbf{H} = \mathrm{i}\omega\epsilon\mathbf{E} + \mathbf{J}_e. \tag{5.46}$$

5.2 Monochromatic plane waves in vacuum

5.2.1 *Travelling waves*

Equations (5.43) and (5.44) for vacuum are

$$\operatorname{curl}\mathbf{E} = -\mathrm{i}\omega\mu_0\mathbf{H}, \quad \operatorname{curl}\mathbf{H} = \mathrm{i}\omega\epsilon_0\mathbf{E}. \tag{5.47}$$

The elimination of **H** and **E** in turn give

$$\nabla^2\mathbf{E} + k_0{}^2\mathbf{E} = 0, \quad \nabla^2\mathbf{H} + k_0{}^2\mathbf{H} = 0, \tag{5.48}$$

where

$$k_0 = \omega(\epsilon_0\mu_0)^{\frac{1}{2}} = \omega/c. \tag{5.49}$$

If a field is sought which depends only on the cartesian coordinate x each cartesian component of **E** and **H** satisfies

$$\frac{\mathrm{d}^2\chi}{\mathrm{d}x^2} + k_0{}^2\chi = 0. \tag{5.50}$$

A pair of independent solutions is

$$\mathrm{e}^{-\mathrm{i}k_0x}, \quad \mathrm{e}^{\mathrm{i}k_0x} \tag{5.51}$$

which represent plane waves travelling in the positive and negative x directions respectively. These travelling plane waves were discussed, without recourse to the complex representation, in §§ 2.5.2 and 2.5.3. It was shown that any such wave can be expressed as the superposition of two transverse, linearly polarized, travelling waves, one having non-zero field components E_y, H_z only, the other having non-zero components E_z, H_y only.

For the sake of definiteness consider the wave travelling in the positive x direction, polarized with **E** in the y direction. The non-zero components can be written

$$E_y = Z_0H_z = E_0\mathrm{e}^{-\mathrm{i}k_0x}. \tag{5.52}$$

Here E_0 is a complex constant whose modulus gives the amplitude of $\mathbf{E}$ and whose argument gives the phase of $\mathbf{E}$ at $x = t = 0$; and

$$Z_0 = (\mu_0/\epsilon_0)^{\frac{1}{2}} = 376.7\,\text{ohm}. \tag{5.53}$$

Z_0 is known as the *vacuum impedance*, and the fact that it is real indicates that at each point the electric and magnetic fields oscillate in phase; its reciprocal,

$$Y_0 = 1/Z_0, \tag{5.54}$$

is known as the *vacuum admittance*.

The significance of Z_0 may perhaps be brought out by imagining the plane wave field trapped between two perfectly conducting plates $y =$ constant, unit distance apart. This is permissible: just as in electrostatics equipotential surfaces can be treated as conducting surfaces, so in general any surface on which the tangential components of $\mathbf{E}$ vanish may be treated as a perfectly conducting surface, since no boundary condition is violated thereby. If the plane wave field is confined to the region $|y| < \frac{1}{2}$, and if the regions $|y| > \frac{1}{2}$ are field free, then, as explained in § 5.1.2, there are surface current densities

$$(\pm H_z, 0, 0) = (\pm Y_0 E_0 e^{-ik_0 x}, 0, 0)$$

in the respective planes $y = \pm\frac{1}{2}$. It is evident from symmetry that the individual current densities give rise to plane wave fields which are identical in $|y| < \frac{1}{2}$, but of opposite sign in $|y| > \frac{1}{2}$; when superposed these fields combine to give the field (5.52) between the plates, and annul one another elsewhere. Now although the electric field cannot, of course, be generally expressed as the gradient of a scalar function, it can be so expressed in this case if attention is confined to any plane $x =$ constant; and the potential difference, or voltage, between the plates is defined in this sense, being simply $E_0 \exp(-ik_0 x)$. Thus Z_0 appears as the ratio of the voltage between the plates to the current flowing in one of the plates per unit width in the z direction; the conductors are said to form a parallel plate transmission line, and Z_0 is its impedance.

Returning to the unbounded field (5.52), its time averaged energy density is

$$\text{Re}\,\tfrac{1}{4}(\epsilon_0 \mathbf{E}\,.\,\mathbf{E}^* + \mu_0 \mathbf{H}\,.\,\mathbf{H}^*) = \tfrac{1}{2}\epsilon_0\,|E_0|^2, \tag{5.55}$$

equal contributions coming from the electric and magnetic terms. (The actual energy density has a part which oscillates with angular frequency 2ω, being $\epsilon_0\,|E_0|^2 \cos^2(\omega t - k_0 x + \eta)$, where η is the argument of E_0.)

The time averaged power flux density is

$$\text{Re}\,\tfrac{1}{2}\mathbf{E} \wedge \mathbf{H}^* = \tfrac{1}{2} Y_0\,|E_0|^2 (1, 0, 0). \tag{5.56}$$

It is natural to define an *average velocity of energy flow* as the quotient of the time averaged power flux density and the time averaged energy density. For the plane wave considered this is

$$c(1, 0, 0), \tag{5.57}$$

which, in this case, happens also to be the quotient of the instantaneous quantities and moreover is identical with the *phase velocity* of the wave.

If a second wave, given by (5.52) but with complex amplitude E_0', were superposed on (5.52), the time averaged energy density and power flux density would be proportional to

$$|E_0|^2 + |E_0'|^2 + 2|E_0|\,|E_0'|\cos(\eta - \eta'), \tag{5.58}$$

where η and η' are the arguments of E_0 and E_0' respectively. Thus densities pertaining to the individual waves are additive only if the waves are in phase quadrature, that is, η and η' differ by an odd multiple of $\frac{1}{2}\pi$. For waves that are in phase, or out of phase, the densities of the combined field are the same as those for a single wave of amplitude $|E_0| + |E_0'|$, or $||E_0| - |E_0'||$, respectively.

A travelling plane wave with general *elliptic* polarization can be obtained by superposing on (5.52) the wave whose only non-zero field components are

$$E_z = -Z_0 H_y = \alpha E_0\,\mathrm{e}^{-\mathrm{i}k_0 x}, \tag{5.59}$$

where α is an arbitrary complex number. The combined field is

$$\mathbf{E} = E_0(0, 1, \alpha)\,\mathrm{e}^{-\mathrm{i}k_0 x}, \quad \mathbf{H} = Y_0 E_0(0, -\alpha, 1)\,\mathrm{e}^{-\mathrm{i}k_0 x}, \tag{5.60}$$

with time averaged energy and power flux densities differing from those of (5.52) by the factor

$$1 + \alpha\alpha^*, \tag{5.61}$$

so that the densities of the individual waves (5.52) and (5.59) are additive.

If $\alpha = \pm i$ the wave (5.60) is *circularly* polarized, with the actual electric vector rotating in one sense or the other according to the sign.

5.2.2 *Standing waves: reflection at a perfectly conducting plane*

The plane wave whose only non-zero field components are given by

$$-E_y = Z_0 H_z = E_0\,\mathrm{e}^{\mathrm{i}k_0 x} \tag{5.62}$$

has the same amplitude and polarization as (5.52), but travels in the opposite direction. The field resulting from the superposition of (5.52) and (5.62) is

$$E_y = -2\mathrm{i}E_0 \sin(k_0 x), \quad H_z = 2Y_0 E_0 \cos(k_0 x). \tag{5.63}$$

This is a *standing wave*. At each point the field vectors oscillate with a fixed amplitude whose value depends harmonically on the x coordinate of the point. Evidently $\mathbf{E}$ is zero in the planes

$$x = n\pi/k_0 = \tfrac{1}{2}n\lambda \quad (n = 0, \pm 1, \pm 2, \ldots), \tag{5.64}$$

where $\lambda = 2\pi c/\omega$ is the wavelength; and these *nodal planes* for the electric field are where the amplitude of the magnetic field attains its maxima.

Any of the planes (5.64) can be taken to be a perfectly conducting sheet. In particular, the field (5.63) can exist solely in $x > 0$, separated from the field free region $x < 0$ by the perfectly conducting plane $x = 0$. This evidently represents the solution to the problem of a plane wave, (5.62), normally incident, from $x > 0$, on a perfectly conducting plane $x = 0$. There is a *reflected* wave, (5.52), whose amplitude and phase are such that the *total* field, incident and reflected waves together, satisfies the boundary condition that the tangential component of **E** vanish on the surface of the conductor.

In the reflection problem the incident wave is supposed given, its source being irrelevant. The reflected wave, on the other hand, is generated by surface current in the conducting plane $x = 0$. The surface current density and the (total) magnetic field just outside the conductor satisfy the relation

$$\mathbf{j} = \mathbf{n} \wedge \mathbf{H}, \tag{5.65}$$

so that here

$$\mathbf{j} = -2Y_0 E_0(0, 1, 0). \tag{5.66}$$

Note that since **j** is uniform there is no associated surface charge density, in agreement with the fact that $\mathbf{n}.\mathbf{E} = 0$.

It should be emphasized that the total field is given *everywhere* by the superposition of the incident wave (5.62) and the field generated by the surface current density (5.66). In $x > 0$ the latter is the reflected wave (5.52), and in $x < 0$ it is the negative of (5.62) (cf. (5.15) and (5.16)): that the region $x < 0$ is field free is thus seen to be achieved by the exact cancellation of the incident wave.

It is apparent from (5.63) that $\mathbf{E} \wedge \mathbf{H}^*$ is purely imaginary, so that on a time average there is no power flux. The power fluxes in the two plane waves are algebraically additive if the waves travel in opposite directions.

The analysis can, of course, be applied to the case of a normally incident plane wave of arbitrary elliptical polarization, for example by regarding the plane wave as the superposition of two linearly polarized plane waves.

It is also easy to extend the analysis to the case of oblique incidence. Consider the plane wave

$$\left.\begin{aligned} \mathbf{E}^i &= (1, 0, 0)\, e^{ik_0(y \sin\alpha - z\cos\alpha)}, \\ \mathbf{H}^i &= Y_0(0, \cos\alpha, \sin\alpha)\, e^{ik_0(y\sin\alpha - z\cos\alpha)}, \end{aligned}\right\} \tag{5.67}$$

where, for convenience and without significant loss of generality, the amplitude of the electric field is taken to be unity. The wave travels in a direction making angle α with the positive z axis (see figure 5.2). If it is incident from $y > 0$ on a perfectly conducting plane in $y = 0$ there is a corresponding reflected wave

$$\left.\begin{aligned} \mathbf{E}^r &= (-1, 0, 0)\, e^{-ik_0(y\sin\alpha + z\cos\alpha)}, \\ \mathbf{H}^r &= Y_0(0, -\cos\alpha, \sin\alpha)\, e^{-ik_0(y\sin\alpha + z\cos\alpha)}. \end{aligned}\right\} \tag{5.68}$$

In $y > 0$ the total field $\mathbf{E} = \mathbf{E}^i + \mathbf{E}^r$, $\mathbf{H} = \mathbf{H}^i + \mathbf{H}^r$ is the sum of (5.67) and (5.68); in $y < 0$ it is zero. Thus, in $y > 0$,

$$\left.\begin{aligned} \mathbf{E} &= 2i(1, 0, 0) \sin(k_0 y \sin\alpha)\, e^{-ik_0 z\cos\alpha}, \\ \mathbf{H} &= 2Y_0[0, i\cos\alpha \sin(k_0 y\sin\alpha), \sin\alpha\cos(k_0 y\sin\alpha)]\, e^{-ik_0 z\cos\alpha}; \end{aligned}\right\} \tag{5.69}$$

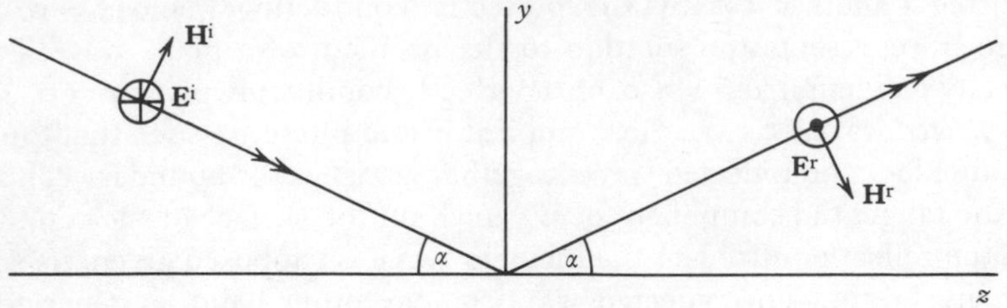

Figure 5.2

and the boundary condition, that the tangential component of **E** be zero on the surface $y = 0$, is palpably satisfied. The surface current density in $y = 0$ is

$$\mathbf{j} = -2Y_0 \sin\alpha(1, 0, 0)\, e^{-ik_0 z\cos\alpha}; \tag{5.70}$$

and since $\operatorname{div}\mathbf{j} = 0$ there is no charge density, in agreement with the fact that $E_y = 0$.

It should be noted that the angles of incidence and reflection are equal, and that the reflected wave has the same polarization (linear, with the electric vector parallel to the x axis) and amplitude as the incident wave. Only in this way can the boundary conditions be satisfied. In fact, the reflected wave is obtained from the negative of the incident wave on replacing α by $-\alpha$.

For an incident wave

$$\left.\begin{aligned}\mathbf{H}^{\mathrm{i}} &= (1, 0, 0)\, e^{ik_0(y\sin\alpha - z\cos\alpha)},\\ \mathbf{E}^{\mathrm{i}} &= -Z_0(0, \cos\alpha, \sin\alpha)\, e^{ik_0(y\sin\alpha - z\cos\alpha)},\end{aligned}\right\} \tag{5.71}$$

whose polarization is complementary to that of (5.67), the reflected wave is

$$\left.\begin{aligned}\mathbf{H}^{\mathrm{r}} &= (1, 0, 0)\, e^{-ik_0(y\sin\alpha + z\cos\alpha)},\\ \mathbf{E}^{\mathrm{r}} &= Z_0(0, -\cos\alpha, \sin\alpha)\, e^{-ik_0(y\sin\alpha + z\cos\alpha)}.\end{aligned}\right\} \tag{5.72}$$

The total field in $y > 0$ is

$$\left.\begin{aligned}\mathbf{H} &= 2(1, 0, 0)\cos(k_0 y\sin\alpha)\, e^{-ik_0 z\cos\alpha},\\ \mathbf{E} &= -2Z_0[0, \cos\alpha\cos(k_0 y\sin\alpha), i\sin\alpha\sin(k_0 y\sin\alpha)]\, e^{-ik_0 z\cos\alpha},\end{aligned}\right\} \tag{5.73}$$

showing that $E_x = E_z = 0$ on $y = 0$. The surface current density in $y = 0$ is

$$\mathbf{j} = -2(0, 0, 1)\, e^{-ik_0 z\cos\alpha}, \tag{5.74}$$

and in this case there is a surface charge density

$$\sigma = -\frac{2}{c}\cos\alpha\, e^{-ik_0 z\cos\alpha}, \tag{5.75}$$

which may be obtained by evaluating either $\epsilon_0 E_y$ at $y = 0$, or $(i/\omega)\operatorname{div}\mathbf{j}$.

The case of an obliquely incident plane wave with arbitrary polarization can be treated by expressing it as the appropriate combination of (5.67) and (5.71).

Finally, it should be noted that the reflection problems in this section have been discussed on the understanding that, in each case, there is, for a given incident wave, only one solution satisfying the boundary conditions at the conductor and the condition that the field additional to the incident wave be 'outgoing' in the sense that it travels away from the reflecting plane (cf. the remarks at the end of § 5.1.2).

5.2.3 *Evanescent waves*

Any vacuum plane wave of angular frequency ω can be written in vector form

$$\mathbf{E} = \mathbf{E}_0 \mathrm{e}^{-ik_0\mathbf{n}.\mathbf{r}}, \quad \mathbf{H} = Y_0 \mathbf{n} \wedge \mathbf{E}_0 \mathrm{e}^{-ik_0\mathbf{n}.\mathbf{r}}, \tag{5.76}$$

where $\mathbf{n}$ is the unit vector in the direction of propagation, and $\mathbf{E}_0$ satisfies

$$\mathbf{n}.\mathbf{E}_0 = 0. \tag{5.77}$$

For example, $\mathbf{n}$ is $(0, \sin\alpha, \cos\alpha)$ in both (5.68) and (5.72), whereas $\mathbf{E}_0$ is $(-1, 0, 0)$ in the former and $Z_0(0, -\cos\alpha, \sin\alpha)$ in the latter.

Since

$$\operatorname{curl}(\mathbf{A}\,\mathrm{e}^{-ik_0\mathbf{n}.\mathbf{r}}) = -ik_0\mathbf{n} \wedge \mathbf{A}\,\mathrm{e}^{-ik_0\mathbf{n}.\mathbf{r}} \tag{5.78}$$

for any vector $\mathbf{A}$ that is independent of position, inspection confirms that (5.76) satisfies the first Maxwell equation (5.47); and only the familiar formula for a triple vector product is needed to demonstrate that it also satisfies the second, provided that (5.77) holds and that

$$\mathbf{n}.\mathbf{n} = 1. \tag{5.79}$$

It is important to observe that, if relations (5.77) and (5.79) hold, then (5.76) is guaranteed to satisfy Maxwell's equations without any requirement that either $\mathbf{E}_0$ or $\mathbf{n}$ be real. A complex $\mathbf{E}_0$ is no novelty, having already been tacitly accepted; the arguments of its components determine the phases of the components of the electric field at $\mathbf{r} = 0$, and these are not even the same if the wave is elliptically polarized. So far, however, $\mathbf{n}$ has only appeared as a unit vector in a particular direction, and such a vector is necessarily real. The introduction of a complex $\mathbf{n}$ does yield a new type of wave.

Let

$$\mathbf{n} = \mathbf{n}_\mathrm{r} + i\mathbf{n}_\mathrm{i}, \tag{5.80}$$

where $\mathbf{n}_\mathrm{r}$ and $\mathbf{n}_\mathrm{i}$ are real vectors. Then, from the real and imaginary parts of (5.79),

$$n_\mathrm{r}^2 - n_\mathrm{i}^2 = 1, \tag{5.81}$$

$$\mathbf{n}_\mathrm{r}.\mathbf{n}_\mathrm{i} = 0. \tag{5.82}$$

In view of these relations, the latter of which states that $\mathbf{n}_\mathrm{r}$ and $\mathbf{n}_\mathrm{i}$ are orthogonal, the vectors can be represented as

$$\mathbf{n}_\mathrm{r} = (0, \cosh\beta, 0) \quad \mathbf{n}_\mathrm{i} = (0, 0, -\sinh\beta), \tag{5.83}$$

where β is a real parameter. Furthermore (5.77) now becomes

$$E_{0y}\cosh\beta = \mathrm{i}E_{0z}\sinh\beta, \tag{5.84}$$

so that

$$\mathbf{E}_0 = (E_1, \mathrm{i}E_2\sinh\beta, E_2\cosh\beta), \tag{5.85}$$

where E_1 and E_2 are complex parameters; and

$$\mathbf{n}\wedge\mathbf{E}_0 = (E_2, -\mathrm{i}E_1\sinh\beta, -E_1\cosh\beta). \tag{5.86}$$

Evidently, then, the general field of this type is an arbitrary linear combination of the wave (case $E_2 = 0$)

$$\left.\begin{aligned}\mathbf{E} &= (1,0,0)\,\mathrm{e}^{-\mathrm{i}k_0 y\cosh\beta}\,\mathrm{e}^{-k_0 z\sinh\beta},\\ \mathbf{H} &= -Y_0(0, i\sinh\beta, \cosh\beta)\,\mathrm{e}^{-\mathrm{i}k_0 y\cosh\beta}\,\mathrm{e}^{-k_0 z\sinh\beta},\end{aligned}\right\} \tag{5.87}$$

and the wave (case $E_1 = 0$) obtained from (5.87) by the transformation $\mathbf{E}\rightarrow\mathbf{H}$, $\mathbf{H}\rightarrow-\mathbf{E}$, $Y_0\leftrightarrow Z_0$. It is observed that (5.87) can be derived from (5.68) on replacing α by $\frac{1}{2}\pi+\mathrm{i}\beta$.

Mathematically all waves described by (5.76) may be classed together as plane waves. Physically, however, the distinction between waves typified by (5.68) and those typified by (5.87) is highly significant. The latter are sometimes called *inhomogeneous* plane waves; or, depending on the context in which they appear, *evanescent* waves or *surface* waves.

The prominent feature of an inhomogeneous plane wave in vacuum is that the amplitude decays exponentially in a direction at right angles to that of phase propagation. In (5.87), with $\beta > 0$, say, the decay is in the positive z direction, with decay coefficient

$$k_0\sinh\beta; \tag{5.88}$$

and the phase propagation is in the positive y direction, with phase velocity

$$c/\cosh\beta. \tag{5.89}$$

It may also be remarked that the wave is not purely transverse; there is a non-zero field component in the direction of propagation, H_y in (5.87). However, the longitudinal field component is in phase quadrature with the transverse components, so that the time averaged power flux density,

$$\mathrm{Re}\,\tfrac{1}{2}\mathbf{E}\wedge\mathbf{H}^* = \tfrac{1}{2}Y_0\cosh\beta(0,1,0)\,\mathrm{e}^{-2k_0 z\sinh\beta} \tag{5.90}$$

in (5.87), is strictly longitudinal. The time averaged energy density in (5.87) is

$$\tfrac{1}{4}(\epsilon_0\mathbf{E}.\mathbf{E}^*+\mu_0\mathbf{H}.\mathbf{H}^*) = \tfrac{1}{2}\epsilon_0\cosh^2\beta\,\mathrm{e}^{-2k_0 z\sinh\beta}, \tag{5.91}$$

the ratio of the magnetic to electric contribution being $\cosh^2\beta+\sinh^2\beta$. The average velocity of energy propagation, (5.90) divided by (5.91), is seen to equal the phase velocity.

In many cases, examples of which are met later on, inhomogeneous plane waves make an important contribution to the electromagnetic field. But the nature of their amplitude variation clearly restricts such waves to at most a half-space, even in an idealized model of a physical problem.

5.2.4 *Rectangular waveguides: TE_{01} modes*

In § 5.2.1 the idea that a travelling wave may be trapped between conducting plates was envisaged. This is indeed a most important way of transmitting electromagnetic energy, and the main features of the technique are readily analysed within the mathematical framework already introduced.

The case considered in § 5.2.1 was an idealized two-dimensional one in which a single plane wave travelling in the x direction is confined to the region between two perfectly conducting plates y = constant. This situation is theoretically possible if the wave is linearly polarized with $\mathbf{E}$ in the y direction. To be useful in practice, though, it is necessary to limit the field in the z direction as well. Clearly for the single plane wave the additional imposition of perfectly conducting planes z = constant is inadmissible, since E_y is not everywhere zero on any such plane.

On the other hand, if the two plane waves (5.67) and (5.68) are superposed, the resulting field (5.69) not only has E_y and E_z identically zero but also has a standing wave pattern in the y direction. Perfectly conducting sheets can therefore occupy the nodal planes y = constant of E_x, as well as the planes x = constant, without disturbing the field. The field can thus be confined to the interior of a hollow cylinder of rectangular cross-section with perfectly conducting walls.

The wave travels along the cylinder which acts as a ***rectangular waveguide***. In practice, of course, the walls are not perfectly conducting; but their conductivity is sufficiently high for a calculation based on perfect conductivity to be a good approximation. Broadly speaking, the losses associated with finite conductivity have negligible effect on the field pattern; they need only be taken into account in assessing the attenuation of the field along the guide, an effect which may be important if extensive lengths of guide are involved.

Circular waveguides are also used; their operation does not differ in principle from that of rectangular guides, but the theory involves Bessel functions and is therefore omitted here.

To analyse the action of the perfectly conducting rectangular guide in more detail suppose that the respective faces of the four walls are in the planes $x = 0$, $x = a$, $y = 0$, $y = b$. Then (5.69) is a possible field within the guide provided

$$k_0 b \sin \alpha = n\pi \quad (n = 1, 2, 3, \ldots). \tag{5.92}$$

If condition (5.92) holds, for one of the values of n, then the boundary conditions are satisfied; negative integral values of n are redundant, since they give fields differing only in sign from those characterized by the corresponding positive integral values. The field (5.69), with the factor 2i dropped, may now be written

$$\left.\begin{aligned} \mathbf{E} &= (1, 0, 0)\sin\left(\frac{n\pi y}{b}\right) e^{-ik_0\gamma z}, \\ \mathbf{H} &= Y_0\left[0, \gamma\sin\left(\frac{n\pi y}{b}\right), -i\frac{n\pi}{k_0 b}\cos\left(\frac{n\pi y}{b}\right)\right] e^{-ik_0\gamma z}, \end{aligned}\right\} \tag{5.93}$$

where
$$\gamma = \cos\alpha = \left(1 - \frac{n^2\pi^2}{k_0{}^2 b^2}\right)^{\frac{1}{2}} = \left(1 - \frac{n^2\lambda^2}{4b^2}\right)^{\frac{1}{2}}, \tag{5.94}$$

λ being the wavelength $2\pi c/\omega$.

The field (5.93) is called a *waveguide mode*. It is characterized by its particular polarization and by the integer n. Since $\mathbf{E}$ has no component along the guide the mode is described as *transverse electric*, abbreviated to TE, and is designated as a TE_{0n} mode. It is shown a little later that there is in fact a doubly infinite set of transverse electric modes, TE_{mn}, and an independent set of transverse magnetic modes TM_{mn}. Certain fundamental features common to all these modes are exemplified in the comparatively simple and practically important TE_{0n} mode, and are now discussed on the basis of (5.93).

The central observation is that γ, given by (5.94), is real if and only if

$$2b > n\lambda, \tag{5.95}$$

and is otherwise purely imaginary. If γ is real the mode travels along the guide (theoretically without attenuation) with phase speed

$$v = c/\gamma; \tag{5.96}$$

whereas if γ is purely imaginary the mode does not travel, but is simply an oscillatory disturbance whose amplitude decays exponentially with z. There is thus a clear cut distinction between so-called *propagated* modes and *evanescent* modes.

The condition (5.95) for the mode to be propagated may be regarded from several different viewpoints. As it stands it emphasizes that a TE_{0n} mode of given wavelength λ (or, equivalently, frequency ω) is propagated only if the guide dimension b exceeds $\frac{1}{2}n\lambda$. If on the other hand, λ and b are prescribed, the condition states that the propagated TE_{0n} modes are those (if any) for which $n < 2b/\lambda$. Yet again, it may be said that a TE_{0n} mode is propagated in a given guide only if

$$\lambda < 2b/n, \tag{5.97}$$

or, equivalently,
$$\omega > n\pi c/b. \tag{5.98}$$

It is demonstrated in (5.98) that there is a *cut-off* (or *critical*) angular frequency $n\pi c/b$ below which the TE_{0n} mode is not propagated. The corresponding cut-off wavelength is $2b/n$. Below the angular frequency

$$\omega_c = \pi c/b \tag{5.99}$$

there is no propagated mode at all. The corresponding wavelength is

$$\lambda_c = 2b. \tag{5.100}$$

Note also that, for a given wavelength λ, the waveguide can be constructed so that only the TE_{01} mode is propagated; this is achieved by making b less than λ but greater than $\frac{1}{2}\lambda$.

For TE_{0n} modes in an idealized waveguide with perfectly conducting walls the width a of the dimension parallel to **E** is irrelevant, since it has no effect on the field. However, in an actual case, in which the walls have finite conductivity, losses are increased by decreasing a. The obvious convenience of making a small is therefore limited by what losses are acceptable; in a practical waveguide, with $\frac{1}{2}\lambda < b < \lambda$, a is likely to be less than $\frac{1}{2}\lambda$, so that the orientation of the propagated wave is unambiguous, but not much less. Obviously a waveguide of convenient dimensions is operable only at wavelengths less than about 0·1 m, corresponding to frequencies greater than $3 \times 10^9 \text{ sec}^{-1}$.

The evanescent modes are those for which γ is purely imaginary. Suppose

$$\gamma = -\mathrm{i} \sinh \beta, \tag{5.101}$$

where β is real (and positive for decay in the positive z direction). Then from (5.94), $\alpha = \frac{1}{2}\pi + \mathrm{i}\beta$, and it is apparent that the individual waves (5.67) and (5.68) whose superposition constitutes the evanescent mode are precisely the evanescent plane waves discussed in § 5.2.3 (see the remark following (5.87)).

5.2.5 *TE_{mn} and TM_{mn} modes*

It has been shown that TE_{0n} modes can exist in a rectangular waveguide. What other modes are possible?

To examine this question it is helpful to revert to Maxwell's equations in the case when the field is time harmonic and also has z dependence specified by

$$\mathrm{e}^{-\mathrm{i}k_0\gamma z}, \tag{5.102}$$

where γ is an as yet undetermined constant.

The y component of the first equation in (5.47), and the x component of the second equation, are

$$-\mathrm{i}k_0\gamma E_x - \frac{\partial E_z}{\partial x} = -\mathrm{i}\omega\mu_0 H_y, \tag{5.103}$$

$$\frac{\partial H_z}{\partial y} + \mathrm{i}k_0\gamma H_y = \mathrm{i}\omega\epsilon_0 E_x, \tag{5.104}$$

respectively, since partial differentiation with respect to z is now synonymous with multiplication by $-\mathrm{i}k_0\gamma$. The elimination of H_y from (5.103) and (5.104) gives

$$\mathrm{i}k_0(1-\gamma^2)\,E_x = \gamma\frac{\partial E_z}{\partial x} + Z_0\frac{\partial H_z}{\partial y}, \tag{5.105}$$

and the elimination of E_x gives

$$ik_0(1-\gamma^2)H_y = Y_0\frac{\partial E_z}{\partial x}+\gamma\frac{\partial H_z}{\partial y}. \tag{5.106}$$

In a similar way the x component of the first equation in (5.47) and the y component of the second equation are equivalent to

$$ik_0(1-\gamma^2)H_x = -Y_0\frac{\partial E_z}{\partial y}+\gamma\frac{\partial H_z}{\partial x}, \tag{5.107}$$

$$ik_0(1-\gamma^2)E_y = \gamma\frac{\partial E_z}{\partial y}-Z_0\frac{\partial H_z}{\partial x}. \tag{5.108}$$

These latter four equations are prescriptions for the remaining field components if E_z and H_z are known. Moreover with their help the z components of (5.47) are seen to be the wave equation

$$\frac{\partial^2\psi}{\partial x^2}+\frac{\partial^2\psi}{\partial y^2}+k_0{}^2(1-\gamma^2)\psi = 0 \tag{5.109}$$

for H_z and E_z respectively. The problem therefore consists in choosing those solutions of (5.109) for E_z and H_z that yield fields satisfying the boundary conditions.

Now the boundary conditions are

$$E_y = E_z = 0 \quad \text{on} \quad x = 0, a, \tag{5.110}$$

$$E_x = E_z = 0 \quad \text{on} \quad y = 0, b. \tag{5.111}$$

Evidently (5.110) implies that $\partial E_z/\partial y = 0$ on $x = 0, a$; and (5.108) in turn implies that $\partial H_z/\partial x = 0$ on $x = 0, a$. Likewise (5.111) and (5.105) imply $\partial E_z/\partial x = \partial H_z/\partial y = 0$ on $y = 0, b$. Thus a field specified by $E_z = \psi_1$, $H_z = \psi_2$, where ψ_1 and ψ_2 are solutions of (5.109), can be regarded as the superposition of two fields, specified by

$$E_z = \psi_1, \quad H_z = 0, \tag{5.112}$$

and by

$$E_z = 0, \quad H_z = \psi_2, \tag{5.113}$$

respectively, each of which individually satisfies the boundary conditions.

In case (5.112), $\psi_1 = 0$ on all four walls, and appropriate solutions are

$$\psi_1 = \sin\left(\frac{m\pi x}{a}\right)\sin\left(\frac{n\pi y}{b}\right), \tag{5.114}$$

where

$$(m\pi/a)^2+(n\pi/b)^2-k_0{}^2(1-\gamma^2) = 0, \tag{5.115}$$

and m and n are integers.

In case (5.113), $\partial\psi_2/\partial x = 0$ on $x = 0, a$ and $\partial\psi_2/\partial y = 0$ on $y = 0, b$, so

$$\psi_2 = \cos\left(\frac{m\pi x}{a}\right)\cos\left(\frac{n\pi y}{b}\right), \tag{5.116}$$

where (5.115) holds, and m and n are integers.

Equations (5.112) and (5.114) specify the transverse magnetic modes, TM_{mn}. The field components, with convenient normalization, are

$$\left.\begin{aligned} E_x &= -\mathrm{i}\gamma\frac{m\pi}{a}\cos\left(\frac{m\pi x}{a}\right)\sin\left(\frac{n\pi y}{b}\right)\mathrm{e}^{-\mathrm{i}k_0\gamma z}, \quad H_x = -\frac{Y_0}{\gamma}E_y, \\ E_y &= -\mathrm{i}\gamma\frac{n\pi}{b}\sin\left(\frac{m\pi x}{a}\right)\cos\left(\frac{n\pi y}{b}\right)\mathrm{e}^{-\mathrm{i}k_0\gamma z}, \quad H_y = \frac{Y_0}{\gamma}E_x, \\ E_z &= k_0(1-\gamma^2)\sin\left(\frac{m\pi x}{a}\right)\sin\left(\frac{n\pi y}{b}\right)\mathrm{e}^{-\mathrm{i}k_0\gamma z}, \quad H_z = 0, \end{aligned}\right\} \quad (5.117)$$

with, from (5.115),

$$\gamma^2 = 1 - \tfrac{1}{4}\lambda^2\left(\frac{m^2}{a^2}+\frac{n^2}{b^2}\right), \quad (5.118)$$

and $m, n = 1, 2, 3, \ldots$. The field vanishes if either m or n is zero.

Equations (5.113) and (5.116) specify the transverse electric modes, TE_{mn}. The field components are

$$\left.\begin{aligned} H_x &= \mathrm{i}\gamma\frac{m\pi}{a}\sin\left(\frac{m\pi x}{a}\right)\cos\left(\frac{n\pi y}{b}\right)\mathrm{e}^{-\mathrm{i}k_0\gamma z}, \quad E_x = \frac{Z_0}{\gamma}H_y, \\ H_y &= \mathrm{i}\gamma\frac{n\pi}{b}\cos\left(\frac{m\pi x}{a}\right)\sin\left(\frac{n\pi y}{b}\right)\mathrm{e}^{-\mathrm{i}k_0\gamma z}, \quad E_y = -\frac{Z_0}{\gamma}H_x, \\ H_z &= k_0(1-\gamma^2)\cos\left(\frac{m\pi x}{a}\right)\cos\left(\frac{n\pi y}{b}\right)\mathrm{e}^{-\mathrm{i}k_0\gamma z}, \quad E_z = 0, \end{aligned}\right\} \quad (5.119)$$

where γ is given by (5.118), and $m, n = 0, 1, 2, \ldots$ (except that the field vanishes if m and n are both zero).

The observations made in § 5.2.4 about the distinction between propagated and evanescent modes clearly apply in general. For a given waveguide, with $b > a$, it is seen that of all the modes the TE_{01} has the lowest cut-off frequency, given by (5.99). Moreover this, and the physically equivalent TE_{10}, are the only modes that can be formed by the superposition of two crossing travelling waves. Clearly all other modes require four such waves for their construction.

Although a detailed investigation would be out of place here it is important to realize that the totality of modes (5.117) and (5.119) form a complete set in the sense that any field excited in the waveguide can be regarded as a superposition of such modes. If the excitation were a given surface current in the relevant region ($0 < x < a$, $0 < y < b$) of a plane $z =$ constant, a mode superposition could be matched uniquely to the source by using a double Fourier series representation of the dependence of the current on x and y, and including only modes that either travel or decay in the directions away from the source plane. The evanescent modes are essential to the matching process; their exclusion would mean the omission of terms in the Fourier series. On the other hand their contribution is, in general, insignificant at

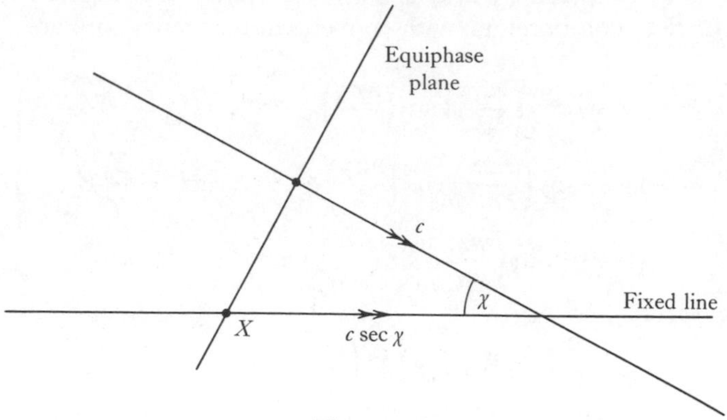

Figure 5.3

distances greater than a few wavelengths from the source plane; at such distances the field is in effect represented by a finite number of propagated modes.

5.2.6 *Energy and group velocity: dispersion*

A propagated waveguide mode has a standing wave pattern transverse to the direction of the guide and a travelling wave pattern along the guide. The travelling wave has phase speed

$$v = c/\gamma; \tag{5.120}$$

this was stated in (5.96) with reference to the TE_{0n} mode, but is of course true, with the appropriate value of γ, for any of the modes (5.117) or (5.119).

Now γ is less than unity, so v exceeds c. This is no cause of surprise, for the resolution of the mode field into purely travelling waves shows that the phase speed of the mode along the guide is derived from the speed c of a plane of constant phase of a pure travelling wave simply by following the point of intersection of the equiphase plane with a fixed line inclined to the normal to the plane at the angle $\chi = \cos^{-1}\gamma$. The point of intersection, indicated by X in figure 5.3, evidently has speed $c \sec \chi$.

There is, however, a physically more significant speed associated with each mode, namely

$$V = P/W, \tag{5.121}$$

where P is the time averaged power flux along the guide in a given mode, and W is the corresponding energy per unit length of the guide. V is called the speed of energy propagation, and in fact

$$V = \gamma c, \tag{5.122}$$

which is less than c (cf. the paragraph following (5.56)).

The result (5.122) is easily established by calculating P and W. If, for example, the TE_{0n} mode is considered, in the form (5.93), then, when γ is real

$$\text{Re}\,\tfrac{1}{2}\mathbf{E}\wedge\mathbf{H}^* = \tfrac{1}{2}Y_0\gamma(0, 0, 1)\sin^2\left(\frac{n\pi y}{b}\right), \tag{5.123}$$

the y component of $\mathbf{E}\wedge\mathbf{H}^*$ being purely imaginary; and also

$$\begin{aligned}\tfrac{1}{4}(\epsilon_0\mathbf{E}.\mathbf{E}^*+\mu_0\mathbf{H}.\mathbf{H}^*) &= \tfrac{1}{4}\epsilon_0\left[\sin^2\left(\frac{n\pi y}{b}\right)+\gamma^2\sin^2\left(\frac{n\pi y}{b}\right)+\frac{n^2\pi^2}{k_0{}^2b^2}\cos^2\left(\frac{n\pi y}{b}\right)\right]\\ &= \tfrac{1}{4}\epsilon_0\left[(1+\gamma^2)\sin^2\left(\frac{n\pi y}{b}\right)+(1-\gamma^2)\cos^2\left(\frac{n\pi y}{b}\right)\right],\end{aligned} \tag{5.124}$$

where (5.94) has been used. Now P is obtained by integrating the z component of (5.123) with respect to x and y over the range $0 < x < a, 0 < y < b$; and W is similarly obtained from (5.124). Since

$$\int_0^b \sin^2\left(\frac{n\pi y}{b}\right)dy = \int_0^b \cos^2\left(\frac{n\pi y}{b}\right)dy = \tfrac{1}{2}b,$$

it follows that

$$P = \tfrac{1}{4}Y_0\gamma ab, \tag{5.125}$$

$$W = \tfrac{1}{4}\epsilon_0 ab. \tag{5.126}$$

Hence $P/W = \gamma c$, as stated.

The same calculation for the TE_{mn} mode in the form (5.119) gives

$$P = \tfrac{1}{8}Z_0k_0{}^2\gamma(1-\gamma^2)\,ab; \tag{5.127}$$

again P/W has the value γc, as is also the case for each TM_{mn} mode. Incidentally, (5.127) is not applicable when m (or n) is zero; for if there is no x dependence the integration over x gives P a factor a instead of $\tfrac{1}{2}a$. This accounts for the factor $\tfrac{1}{4}$ in (5.125) as opposed to $\tfrac{1}{8}$ in (5.127), the remaining difference between the forms of the two expressions being solely due to the different way in which (5.119) with $m = 0$ and (5.93) are normalized.

Yet another velocity, intimately related to that of energy propagation, is the *group velocity*. This is also well illustrated by a waveguide mode.

The field of a single m, n mode is in no sense localized in any part of the waveguide. However, by superposing a number of modes of the same order m, n but having different frequencies ω, it is possible to form a more or less localized 'pulse' and to determine the speed with which this pulse travels along the guide.

To see how this can be achieved consider a group of modes (of specific order m, n) whose wave numbers

$$k = k_0\gamma \tag{5.128}$$

form a continuum between $K-\kappa$ and $K+\kappa$ $(0 < \kappa < K)$. From (5.118)

$$c^2k^2 = \omega^2-p^2, \tag{5.129}$$

where
$$p^2 = \pi^2 c^2 \left(\frac{m^2}{a^2} + \frac{n^2}{b^2}\right) \tag{5.130}$$
is fixed, so that the range of k is equivalent to the range of frequencies
$$[p^2 + c^2(K-\kappa)^2]^{\frac{1}{2}} < \omega < [p^2 + c^2(K+\kappa)^2]^{\frac{1}{2}}. \tag{5.131}$$
If the modes have electric fields of the same amplitude, and with the same phase at $z = t = 0$, the z dependence of the electric field formed by their superposition is described by
$$\frac{1}{2}\int_{K-\kappa}^{K+\kappa} e^{i(\omega t - kz)}\, dk, \tag{5.132}$$
where ω is the positive function of k implicit in (5.129), and the factor $\frac{1}{2}$ is merely included for future convenience.

Now suppose that κ is small enough to validate the linear approximation
$$\omega = \Omega + v_g(k-K), \tag{5.133}$$
where
$$\Omega = (c^2K^2 + p^2)^{\frac{1}{2}} \tag{5.134}$$
is the frequency corresponding to the central wave number K, and
$$v_g = d\omega/dk \tag{5.135}$$
evaluated at $k = K$. Then the corresponding approximation to (5.132) is
$$\tfrac{1}{2} e^{i(\Omega - v_g K)t} \int_{K-\kappa}^{K+\kappa} e^{i(v_g t - z)k}\, dk. \tag{5.136}$$
The integration is now trivial, and the expression can evidently be written
$$\frac{\sin[\kappa(z - v_g t)]}{z - v_g t} e^{i(\Omega t - Kz)}. \tag{5.137}$$

If $\kappa \ll K$ (5.137) is certainly a valid approximation to (5.132), and the broad features of the disturbance are readily discerned. The factor
$$e^{i(\Omega t - Kz)} \tag{5.138}$$
represents a sinusoidal travelling wave which is precisely the waveguide mode of frequency Ω, and this is *modulated* by the factor
$$\frac{\sin[\kappa(z - v_g t)]}{z - v_g t}. \tag{5.139}$$
Thus a snapshot picture of the z dependence at any particular time t shows a sinusoidal pattern, of wavelength $2\pi/K$, which has an amplitude (and sign) variation with z given by (5.139) and depicted in figure 5.4. This description is serviceable because the sinusoidal pattern varies much more rapidly than (5.139), having K/κ oscillations between the adjacent zeros of (5.139) on either side of $z = v_g t$.

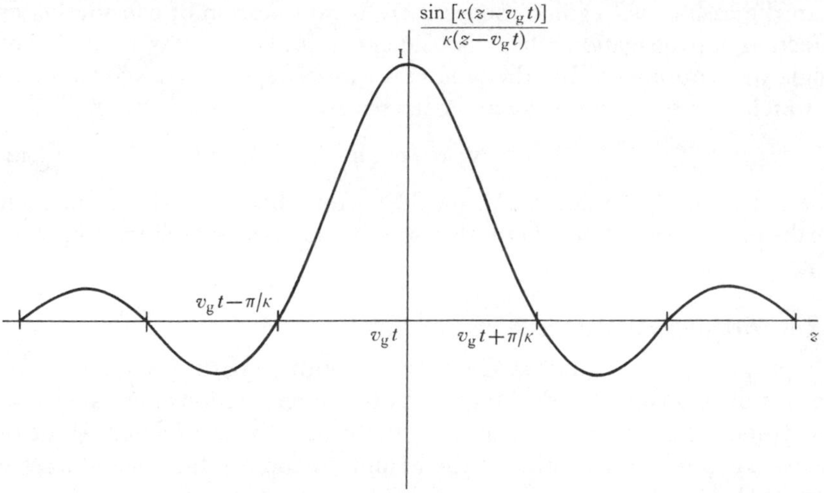

Figure 5.4

The sinusoidal pattern travels with phase speed Ω/K, but the 'pulse' created by its modulation (5.139) clearly has speed v_g. The quantity v_g defined by (5.135) is called the group speed. From (5.129)

$$c^2 k \, dk = \omega \, d\omega,$$

and so

$$v_g = c^2 K/\Omega. \tag{5.140}$$

The group speed associated with a given waveguide mode is therefore c^2 times the reciprocal of the phase speed of the mode, its value being identical with that of the speed of energy propagation defined by (5.121).

The analysis has indicated something of the physical significance of the group speed, but it should be appreciated that no statement to the effect that the group speed is the speed of travel of a pulse can be *precise*. That the pulse envelope was found to travel without change of form resulted from making the calculation on the basis of the approximation that ω depended linearly on k. It is clear that an exact calculation will yield a pulse envelope that changes shape as it travels, so that there is, at the least, some arbitrariness in the definition of its speed. The most that can be said is that under wide conditions the maximum of a pulse travels with approximately the speed v_g.

That a non-monochromatic disturbance in general fails to travel strictly without change of form is inevitable if the individual waves of different frequencies ω whose superposition constitutes the disturbance have different phase velocities ω/k. The effect is known as *dispersion*, and the relation between ω and k is called the *dispersion relation*. In the present discussion the concern is with dispersion in the slightly restricted sense that the pulse envelope does not travel without change of form; the dispersion relation is (5.129).

In the case of waveguide modes there is no question of considering any direction of propagation other than along the guide, and ω is a function of a single wave number k. In other contexts ω may depend on a wave vector $\mathbf{k}$, in which case the group velocity is defined by

$$\mathbf{v}_g = \partial\omega/\partial\mathbf{k}, \tag{5.141}$$

where the right hand side stands for the vector whose cartesian components are the partial derivatives of ω with respect to the corresponding components of $\mathbf{k}$.

5.2.7 *Rectangular cavities*

The propagated waveguide modes in (5.117) and (5.119) travel in the positive z direction. Consider the field formed by the superposition of one such mode and a mode of the same type travelling in the negative z direction. If the two modes have the same amplitude their combination produces standing wave patterns in the z direction for each field component. In particular E_x and E_y have common nodal planes z = constant on which they are zero, and any of these planes can therefore be regarded as perfectly conducting sheets. If $z = 0$ and $z = d$ are two such planes the implication is that the field can exist solely within the region $0 < x < a$, $0 < y < b$, $0 < z < d$, provided the boundaries are perfect conductors.

A necessary and sufficient condition for $z = 0$ and $z = d$ to be nodal planes of E_x and E_y is evidently

$$k_0\gamma d = p\pi, \tag{5.142}$$

where p is an integer. This can be written

$$\gamma = \tfrac{1}{2}\lambda p/d, \tag{5.143}$$

which in view of (5.118), and that fact that $\omega\lambda = 2\pi c$, requires that

$$\omega^2 = \pi^2 c^2\left(\frac{m^2}{a^2}+\frac{n^2}{b^2}+\frac{p^2}{d^2}\right). \tag{5.144}$$

The distinctive feature of the situation is embodied in the relation (5.144). With a given cavity in the shape of a rectangular box an electromagnetic field whose sole source is a surface current in the perfectly conducting walls of the box can exist, but only at the discrete frequencies specified by allocating integral values to m, n and p in (5.144). These are *resonant* frequencies of the cavity. In practice the walls of the cavity have finite conductivity, but if the conductivity is high the response of the cavity to excitation will be greatly enhanced when the frequency of the source of the excitation is close to one of the resonant frequencies. There is an obvious analogy with the natural frequencies of oscillation of a dynamical system.

Any enclosure with perfectly conducting walls has an infinite set of discrete resonant frequencies. These are determined by the shape of the enclosure, but their exact evaluation by analytic means is only possible for simple geometrical shapes, the rectangular box being much the easiest example. A comparison of measured and calculated resonant frequencies of a cavity affords an accurate method for the experimental determination of the speed of light c.

5·3 Monochromatic plane waves in media

5·3·1 *Travelling waves in dielectrics and conductors*

It is often necessary to take account of the electrical properties of the medium in which waves travel. For monochromatic waves the medium can in many cases be adequately characterized by a complex permittivity ϵ and a permeability μ, as explained in § 5.1.4. The governing equations are then (5·43) and (5·44).

If the medium is *homogeneous*, in the sense that ϵ and μ are independent of position, mathematical solutions of these equations are trivial generalizations of solutions of the corresponding vacuum equations (5·47), being obtained from them on replacing ϵ_0 and μ_0 by ϵ and μ respectively. The interpretation of travelling plane wave solutions exhibits the effect of the medium on wave propagation.

As in § 5.2.1, take a wave travelling in the positive x direction, polarized with $\mathbf{E}$ in the y direction. Corresponding to (5·52) the non-zero field components can be written

$$E_y = ZH_z = E_0 \mathrm{e}^{-\mathrm{i}kx}, \tag{5·145}$$

where

$$Z = (\mu/\epsilon)^{\frac{1}{2}}, \tag{5·146}$$

and

$$k = \omega(\epsilon\mu)^{\frac{1}{2}}. \tag{5·147}$$

Consider first the ideal case in which the medium is a lossless dielectric, so that ϵ and μ are real. Then the wave number k is real, the wavelength is

$$\lambda = \frac{2\pi}{k} = \frac{2\pi}{\omega(\epsilon\mu)^{\frac{1}{2}}}, \tag{5·148}$$

and the speed of phase propagation is

$$\omega/k = 1/(\epsilon\mu)^{\frac{1}{2}}. \tag{5·149}$$

Also the electric and magnetic fields are in phase, their ratio being the *intrinsic impedance* Z of the medium.

The relation (5·149) can be written

$$\omega/k = c/n, \tag{5·150}$$

where

$$n = \left(\frac{\epsilon\mu}{\epsilon_0\mu_0}\right)^{\frac{1}{2}}; \tag{5·151}$$

or, equivalently,

$$k = k_0 n. \tag{5·152}$$

It is a familiar result in elementary physics that the factor by which the phase speed differs from c is directly related to the degree of refraction experienced by a wave in passing between the medium and a vacuum (Snell's law; see § 5.3.2). Traditionally n is therefore known as the *refractive index* of the medium.

The dielectric constant $\kappa = \epsilon/\epsilon_0$ is commonly (though not necessarily) greater than unity, but not by an order of magnitude; for example, at the frequency $5 \times 10^{14}\,\text{sec}^{-1}$ (in the visible spectrum) κ is about 1.0006 for air and typically 2.5 for glass. On the other hand the permeability μ is not likely to be significantly different from μ_0, in which case

$$n = (\epsilon/\epsilon_0)^{\frac{1}{2}} = \kappa^{\frac{1}{2}}. \tag{5.153}$$

This expresses the famous prediction from Maxwell's theory, that the refractive index and the square root of the dielectric constant of a substance have the same value. It must be recognized that the comparison is not with the electrostatic value of the dielectric constant, as at first erroneously presumed, but with the value at the frequency at which the refractive index is measured.

Finally it may be observed that the time averaged power flux density, which is in the positive x direction, is

$$\tfrac{1}{2}Y\,|E_0|^2, \tag{5.154}$$

where

$$Y = 1/Z = (\epsilon/\mu)^{\frac{1}{2}} \tag{5.155}$$

is the *intrinsic admittance* of the medium.

Now consider the case in which the medium has conductivity σ. Then (5.145) still holds, but with ϵ given by (5.42), namely

$$\epsilon = \epsilon_r - i\sigma/\omega, \tag{5.156}$$

where ϵ_r is the (real) permittivity. In consequence k and Z are complex. Specifically

$$k = \omega\mu^{\frac{1}{2}}(\epsilon_r - i\sigma/\omega)^{\frac{1}{2}} = k_r - i\gamma, \tag{5.157}$$

say, where k_r and γ have the same sign, which without loss of generality can be taken positive, corresponding to propagation in the positive x direction when $\sigma = 0$.

The spatial dependence of the wave is therefore represented by

$$e^{-\gamma x}e^{-ik_r x}, \tag{5.158}$$

which exhibits the attenuation resulting from the dissipation associated with finite conductivity, γ being the *attenuation coefficient*. If $\gamma \ll k_r$ the wave can be pictured as that obtaining in the case $\sigma = 0$ modified to the extent that the amplitude decreases slowly in the direction of propagation; for the inequality is equivalent to

$$\sigma/\omega \ll \epsilon_r, \tag{5.159}$$

and this gives

$$k_r \simeq \omega(\epsilon_r \mu)^{\frac{1}{2}}, \quad \gamma \simeq \tfrac{1}{2}(\mu/\epsilon_r)^{\frac{1}{2}} \sigma. \tag{5.160}$$

At the other extreme is the perfect conductor for which σ and therefore γ is infinite, and no wave can exist.

The time averaged power flux density is

$$\tfrac{1}{2}Y_r |E_0|^2 e^{-2\gamma x}, \tag{5.161}$$

where Y_r is the real part of $Y = (\epsilon/\mu)^{\frac{1}{2}}$. Energy balance requires that the negative x derivative of this quantity equal the time averaged rate of dissipation of energy per unit volume

$$\text{Re}\,\tfrac{1}{2}\mathbf{J}.\mathbf{E}^* = \text{Re}\,\tfrac{1}{2}\sigma\mathbf{E}.\mathbf{E}^*. \tag{5.162}$$

This is easily confirmed, for the identification of the two forms of the imaginary part of k^2 implicit in (5.157) gives

$$\omega\mu\sigma = 2k_r\gamma;$$

that is, since $Y = k/(\omega\mu)$,

$$2Y_r\gamma = \sigma. \tag{5.163}$$

Hence the negative x derivative of (5.161) is

$$\tfrac{1}{2}\sigma |E_0|^2 e^{-2\gamma x}, \tag{5.164}$$

which is indeed the value of (5.162).

5.3.2 *Reflection and refraction at an interface between two media*

The phenomena of reflection and refraction, first studied in optics, are of importance throughout much of the electromagnetic spectrum. The fundamental calculation treats a model in which a plane wave is incident on an infinite plane boundary separating two isotropic homogeneous half-spaces. In practice one of the half-spaces may well be, in effect, a vacuum; but since the general case involves no loss of mathematical simplicity it is worth presenting the results for arbitrary media before discussing various physical phenomena which arise in particular circumstances.

Let the respective half-spaces be $y > 0$, with complex permittivity ϵ and permeability μ, and $y < 0$, with complex permeability ϵ' and permeability μ'. Let a monochromatic plane wave be incident from $y > 0$ on the interface $y = 0$, in the direction normal to the z axis that makes angle α with the positive x axis. Any such wave can be resolved into a pair of linearly polarized plane waves, one with the electric vector parallel to Oz, the other with the magnetic vector parallel to Oz. It is convenient to treat these cases separately, recognizing, of course, that any other case can be derived from them by superposition.

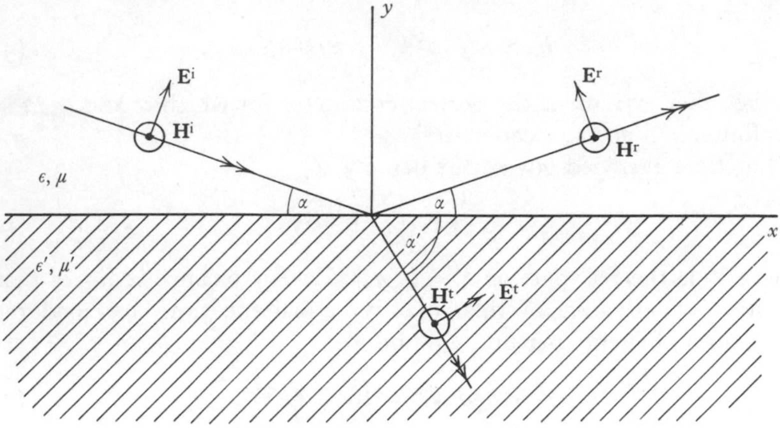

Figure 5.5

Suppose, then, that the field components of the incident wave in $y > 0$ are

$$\left.\begin{aligned} \mathbf{H}^{\mathrm{i}} &= (0, 0, 1)\, \mathrm{e}^{-\mathrm{i}k(x\cos\alpha - y\sin\alpha)}, \\ \mathbf{E}^{\mathrm{i}} &= Z(\sin\alpha, \cos\alpha, 0)\, \mathrm{e}^{-\mathrm{i}k(x\cos\alpha - y\sin\alpha)}, \end{aligned}\right\} \tag{5.165}$$

where $$k = \omega(\epsilon\mu)^{\frac{1}{2}}, \quad Z = (\mu/\epsilon)^{\frac{1}{2}}. \tag{5.166}$$

This gives rise to a reflected wave in $y > 0$ of the form

$$\left.\begin{aligned} \mathbf{H}^{\mathrm{r}} &= \rho(0, 0, 1)\, \mathrm{e}^{-\mathrm{i}k(x\cos\alpha + y\sin\alpha)}, \\ \mathbf{E}^{\mathrm{r}} &= Z\rho(-\sin\alpha, \cos\alpha, 0)\, \mathrm{e}^{-\mathrm{i}k(x\cos\alpha + y\sin\alpha)}; \end{aligned}\right\} \tag{5.167}$$

and a transmitted wave in $y < 0$ of the form

$$\left.\begin{aligned} \mathbf{H}^{\mathrm{t}} &= \tau(0, 0, 1)\, \mathrm{e}^{-\mathrm{i}k'(x\cos\alpha' - y\sin\alpha')}, \\ \mathbf{E}^{\mathrm{t}} &= Z'\tau(\sin\alpha', \cos\alpha', 0)\, \mathrm{e}^{-\mathrm{i}k'(x\cos\alpha' - y\sin\alpha')}, \end{aligned}\right\} \tag{5.168}$$

where $$k' = \omega(\epsilon'\mu')^{\frac{1}{2}}, \quad Z' = (\mu'/\epsilon')^{\frac{1}{2}}. \tag{5.169}$$

The *reflection coefficient* ρ, the *transmission coefficient* τ, and the angle of transmission α' are to be determined. The justification for taking the reflected and transmitted waves of the stated form, that is, linearly polarized in the same sense as the incident wave, with the angle of reflection being equal to the angle of incidence (see figure 5.5), is provided by the ensuing analysis, which shows that ρ and τ are in fact uniquely determined by the boundary conditions at the interface $y = 0$. The waves are, of course, physically acceptable in that they both travel away from the interface into the respective half-spaces: the solution would not be unique if incoming waves other than the incident wave were admitted (cf. the remarks at the end of § 5.2.2).

The total field $\mathbf{E}, \mathbf{H}$ is $\mathbf{E}^{\mathrm{i}} + \mathbf{E}^{\mathrm{r}}, \mathbf{H}^{\mathrm{i}} + \mathbf{H}^{\mathrm{r}}$ in $y > 0$, and is $\mathbf{E}^{\mathrm{t}}, \mathbf{H}^{\mathrm{t}}$ in $y < 0$. The boundary conditions are that the tangential components of **E** and **H**, namely E_x and H_z, be continuous across $y = 0$.

Since the boundary conditions hold for all x it is apparent that

$$k' \cos \alpha' = k \cos \alpha. \tag{5.170}$$

This is the general statement of Snell's law. It determines the angle of transmission α' in terms of the given parameters k, k' and α. The analogous result for the angle of reflection is simply its anticipated equality with the angle of incidence.

The reflection coefficient ρ can now be obtained most directly by considering E_x/H_z, since this *wave* (or *field*) impedance, being formed from the ratio of the tangential field components, is continuous across $y = 0$ and does not involve the transmission coefficient τ. The expression of the continuity of the impedance is evidently

$$Z\frac{1-\rho}{1+\rho}\sin\alpha = Z'\sin\alpha',$$

with solution

$$\rho_H = \frac{Z\sin\alpha - Z'\sin\alpha'}{Z\sin\alpha + Z'\sin\alpha'}, \tag{5.171}$$

where the suffix H is introduced to signify that this is the reflection coefficient for the polarization in which **H** is parallel to the interface.

Finally the continuity of H_z gives

$$1+\rho = \tau, \tag{5.172}$$

so that the transmission coefficient is

$$\tau_H = \frac{2Z\sin\alpha}{Z\sin\alpha + Z'\sin\alpha'}. \tag{5.173}$$

To complete the general solution the corresponding results for the case in which the field is polarized with **E** parallel to the interface must be given. There is no need to display the details, since they are obtained from those already set out simply by writing **E** and $-$**H** in place of **H** and **E**, respectively, and interchanging ϵ with μ and ϵ' with μ' (cf. the transformation (5.45)). Thus the impedances Z and Z' are replaced by the admittances Y and Y' (their reciprocals), the angle of transmission α' is again given by (5.170), and the reflection and transmission coefficients are

$$\rho_E = \frac{Y\sin\alpha - Y'\sin\alpha'}{Y\sin\alpha + Y'\sin\alpha'}, \quad \tau_E = \frac{2Y\sin\alpha}{Y\sin\alpha + Y'\sin\alpha'}. \tag{5.174}$$

It should be noted that ρ_E and τ_E are here defined as the respective ratios, at $y = 0$, of E_z^{r} and E_z^{t} to E_z^{i}, whereas ρ_H and τ_H are defined as the ratios of H_z^{r} and H_z^{t} to H_z^{i}. The precise definition of reflection and transmission coefficients is a matter of choice.

5·3·3 *Reflection by a good conductor: the skin effect*

Consider the idealized special case in which the half-space $y < 0$ is a perfect conductor. Then $|\epsilon'|$ is infinite, from which it follows that $|k'|$ is infinite, $Z' = 0$, $\alpha' = \frac{1}{2}\pi$, and

$$\rho_H = 1, \quad \rho_E = -1. \tag{5·175}$$

The reflection is therefore perfect, the amplitude of the reflected wave being equal to that of the incident wave. The modulus of each refraction coefficient is unity, and the difference in sign is associated with the fact that ρ_H refers to the tangential magnetic field and ρ_E to the tangential electric field. These results are equivalent to those already given in § 5.2.2.

There is, of course, no transmitted wave, for no time harmonic field can exist within a perfect conductor. This result is not dependent on the values of the transmission coefficients (in fact $\tau_E = 0$, $\tau_H = 2$) but simply arises through k' having a negative infinite imaginary part, so that $\exp(ik'y \sin\alpha')$ vanishes for any negative value of y (see the remark following (5.160)).

It is of interest to examine the less idealized case in which the half-space $y < 0$ is a good but not perfect conductor. If the conductivity σ satisfies the inequality

$$\sigma \gg \omega\epsilon_0 \tag{5·176}$$

then

$$\epsilon' \simeq -i\sigma/\omega. \tag{5·177}$$

If also $y > 0$ is vacuum, and $\mu' = \mu_0$, which would in effect commonly be the case, it follows that

$$k/k' = Z'/Z = \zeta, \tag{5·178}$$

where

$$\zeta = \left(\frac{\epsilon_0\,\omega}{2\sigma}\right)^{\frac{1}{2}}(1+i), \quad |\zeta| \ll 1; \tag{5·179}$$

and therefore, from (5·170), that $|\cos\alpha'| \ll 1$, or equivalently $\sin\alpha' \simeq 1$. Hence

$$\rho_H \simeq \frac{\sin\alpha - \zeta}{\sin\alpha + \zeta}, \quad \rho_E \simeq \frac{\zeta\sin\alpha - 1}{\zeta\sin\alpha + 1}. \tag{5·180}$$

The expressions (5.180) of course go over to (5.175) when $\zeta = 0$. Moreover ρ_E is close to -1 for any (real) value of the angle of incidence α. However ρ_H varies markedly with α as the angle approaches zero, swinging over from the approximate value $+1$ when α is appreciably greater than $|\zeta|$ to the value -1 when α is zero.

For all (real) angles α the transmitted wave travels in a direction close to the normal to the interface, since $\alpha' \simeq \frac{1}{2}\pi$. That α' is sensibly independent of α has the important implication that the field in $y > 0$ can in fact be calculated without reference to that in $y < 0$. For the boundary conditions that the tangential components of $\mathbf{E}$ and $\mathbf{H}$ are continuous across $y = 0$ imply the impedance conditions that

$$E_x = Z'H_z, \quad E_z = -Z'H_x \tag{5·181}$$

at $y = 0$. These, of course, yield (5.180), being equivalent, for example, to (5.171) with $\sin\alpha' = 1$. They can be regarded as generalizations of the conditions for the case of perfect conductivity ($Z' = 0$). When they hold, the plane $y = 0$ is said to act as a surface with *surface impedance* Z'.

The use of relations corresponding to (5.181) affords a comparatively simple method of estimating losses in waveguides and cavities due to the imperfect conductivity of the boundary walls. The expressions for the tangential components of **H** at the boundary are taken from the analysis using perfect conductivity (as in §§ 5.2.4–5.2.7), so that estimates of the (small) values of the tangential components of **E** are at once obtained. These in turn enable the Poynting flux normal to the boundary to be evaluated, and this gives the rate at which energy is dissipated in the walls.

Turning now to consider the interior of the conductor it has already been observed that, no matter what the direction or polarization of the incident wave, the field in $y < 0$ is approximately a plane wave travelling in the negative y direction. The attenuation, though, is severe. The y dependence is $\exp(ik'y)$, which from (5.178) and (5.179) is

$$\exp[(1+i)y/d], \tag{5.182}$$

where

$$d = 1/(\tfrac{1}{2}\mu_0\sigma\omega)^{\frac{1}{2}}. \tag{5.183}$$

The wave amplitude decreases by the factor e^{-1} in distance d, and this distance can be taken as a measure of the extent to which the field penetrates into the conductor.

Evidently d can be extremely small. The confinement of the field close to the surface of the conductor is known as the *skin effect*, and d is called the *skin depth*. As an example, for copper with $\sigma = 5.8\times10^7\,\text{mho}\,\text{m}^{-1}$ (5.183) gives roughly $d = 1/(6\omega^{\frac{1}{2}})$; so that at a microwave frequency

$$\omega \sim 3\times10^{10}\,\text{sec}^{-1}$$

d is about 10^{-6} m. It should not go unremarked that if the formula gives d a value comparable to or less than the mean free path of the conduction electrons, which can easily happen, then the simple macroscopic theory has no justification, and a vastly more sophisticated treatment is required to evaluate the field in the conductor.

5.3.4 *Total reflection and no reflection*

In the case in which the homogeneous half-spaces are both pure dielectrics, without conductivity, the dependence of the field on the angle of incidence exhibits some particularly striking features.

Consider, first, the implications of Snell's law. When k and k' are real it is convenient to introduce the refractive indices of the media, as in (5.152), and to write (5.170) in the form

$$\cos\alpha' = \frac{n}{n'}\cos\alpha. \tag{5.184}$$

Take α to be real and, without loss of generality, less than $\frac{1}{2}\pi$. Then, if $n' > n$, to each value of α there corresponds a real value of α'. More specifically, as the values of α go from $\frac{1}{2}\pi$ (normal incidence) to 0 (grazing incidence) the corresponding values of α' decrease monotonically from $\frac{1}{2}\pi$ to $\cos^{-1}(n/n')$; and $\alpha' > \alpha$, so that the refraction is towards the normal.

If, on the other hand, $n' < n$ the value of α' is real only when α exceeds the *critical angle* α_c, defined by

$$\cos\alpha_c = n'/n. \tag{5.185}$$

Refraction is away from the normal, and α' decreases from $\frac{1}{2}\pi$ more rapidly than α, reaching zero when $\alpha = \alpha_c$.

For $\alpha < \alpha_c$ (5.184) specifies a value of $\cos\alpha'$ greater than unity, showing that $\sin\alpha'$ is pure imaginary. The field in $y < 0$, given, for example, by (5.168), therefore has exponential dependence on y. Since it must decay (rather than increase) in the negative y direction, $\sin\alpha'$ is negative pure imaginary, and the field is evidently an evanescent surface wave precisely of the kind discussed (though specifically for a vacuum field) in § 5.2.3. On a time average this wave carries no energy away from the plane $y = 0$, which is consistent with the fact that when $\sin\alpha'$ is pure imaginary each reflection coefficient in (5.171) and (5.174) has modulus unity, the numerator being the complex conjugate of the denominator. The reflection is therefore *total*.

A common case is that in which $y < 0$ is in effect a vacuum and $y > 0$ a substance with a refractive index greater than unity. The phenomenon just described is then referred to as *total internal* reflection, since the wave incident on the interface and completely reflected originates in the medium. The effect is, of course, familiar in optics, where it is explained at an elementary level in terms of 'rays'. Simple ray theory includes Snell's law, in a restricted interpretation, but can say nothing about the field in the vacuum when it is evanescent.

Since in practice μ and μ' are not likely to differ significantly from μ_0 it is worth noting useful forms in which the reflection coefficients can be written when the permeabilities have the same value. Evidently $\mu = \mu'$ implies

$$\frac{Z'}{Z} = \left(\frac{\epsilon}{\epsilon'}\right)^{\frac{1}{2}} = \frac{n}{n'} = \frac{\cos\alpha'}{\cos\alpha}, \tag{5.186}$$

the last equality following from (5.184); and substituting for the impedance ratio in (5.171) and (5.174) gives

$$\rho_H = \frac{\sin(2\alpha) - \sin(2\alpha')}{\sin(2\alpha) + \sin(2\alpha')} = \cot(\alpha+\alpha')\tan(\alpha-\alpha'), \tag{5.187}$$

$$\rho_E = \frac{\sin(\alpha-\alpha')}{\sin(\alpha+\alpha')}. \tag{5.188}$$

From the remarks following (5.184) and (5.185) it is seen that, for $n' > n$, $\alpha+\alpha'$ decreases (monotonically) from π at normal incidence to $\cos^{-1}(n/n')$

at grazing incidence; and for $n' < n$, it decreases from π at normal incidence to α_c at critical incidence. Furthermore, α' never equals α, except for normal incidence. It is clear, therefore, that ρ_E cannot vanish, but that ρ_H is zero at one particular angle of incidence α, namely that for which

$$\alpha + \alpha' = \tfrac{1}{2}\pi.$$

This angle of incidence is called the *Brewster* angle. Its value, α_B say, can be obtained by setting $\alpha = \alpha_B$, $\alpha' = \frac{1}{2}\pi - \alpha_B$ in (5.184), which gives

$$\tan \alpha_B = n/n'. \tag{5.189}$$

If a plane wave falls on the interface at the Brewster angle the reflected wave (if any) is linearly polarized with electric field parallel to the interface, no matter what the polarization of the incident wave.

5.3.5 *Radiation pressure*

The problem of a plane wave incident from vacuum on the plane face of a homogeneous, conducting half-space can be used to exemplify the action of electro-magnetic forces in a time varying situation. Take the case of normal incidence, with the incident wave in $y > 0$

$$\mathbf{H}^i = (0, 0, 1)\, e^{ik_0 y}, \quad \mathbf{E}^i = Z_0(1, 0, 0)\, e^{ik_0 y} \tag{5.190}$$

giving rise to the reflected wave in $y > 0$

$$\mathbf{H}^r = \rho(0, 0, 1)\, e^{-ik_0 y}, \quad \mathbf{E}^r = -Z_0 \rho(1, 0, 0)\, e^{-ik_0 y} \tag{5.191}$$

and the transmitted wave in $y < 0$

$$\mathbf{H}^t = \tau(0, 0, 1)\, e^{iky}, \quad \mathbf{E}^t = Z\tau(1, 0, 0)\, e^{iky}, \tag{5.192}$$

where $$k = \omega(\epsilon\mu_0)^{\frac{1}{2}}, \quad Z = (\mu_0/\epsilon)^{\frac{1}{2}}, \quad \epsilon = \epsilon_0 - i\sigma/\omega,$$

$$\rho = \frac{Z_0 - Z}{Z_0 + Z}, \quad \tau = \frac{2Z_0}{Z_0 + Z}. \tag{5.193}$$

The current density in the conducting medium is $\mathbf{J} = \sigma\mathbf{E}^t$, and there is no charge density since $\operatorname{div}\mathbf{E} = 0$. The time averaged Lorentz force density is therefore

$$\mathbf{F} = \operatorname{Re} \tfrac{1}{2}\mathbf{J} \wedge \mathbf{B}^* = \operatorname{Re} \tfrac{1}{2}\mu_0 \sigma \mathbf{E}^t \wedge \mathbf{H}^{t*}, \tag{5.194}$$

giving $$\mathbf{F} = -\tfrac{1}{2}\mu_0 \sigma Z_r |\tau|^2 (0, 1, 0)\, e^{2\gamma y}, \tag{5.195}$$

where Z_r is the real part of Z, and γ the negative imaginary part of k (cf. (5.157)).

The integral of (5.195) over y from 0 to $-\infty$ gives the force exerted on the conductor per unit area of the xz plane. It is therefore in the negative y direction of amount

$$p = \tfrac{1}{4}\mu_0 \sigma Z_r |\tau|^2/\gamma. \tag{5.196}$$

It can be said that the irradiation of the conductor exerts on it a 'pressure' p, and the analogy with gas theory that this terminology suggests is of considerable significance. Some understanding of this interpretation is gained by casting (5.196) into an alternative form.

Note, first, that $Z_r = Y_r/|Y|^2$, and that

$$Y_r^2 - Y_i^2 = \operatorname{Re} Y^2 = Y_0^2, \tag{5.197}$$

where Y_r and Y_i are the real and imaginary parts of $Y = 1/Z = (\epsilon/\mu_0)^{\frac{1}{2}}$. Then, from (5.163),

$$\sigma Z_r/\gamma = 2Y_r^2/|Y|^2. \tag{5.198}$$

But $2Y_r^2 = Y_0^2 + |Y|^2$, from (5.197), so that (5.198) can be written

$$1 + |Z|^2/Z_0^2,$$

and

$$p = \tfrac{1}{4}\mu_0(1 + |Z|^2/Z_0^2)\,|\tau|^2. \tag{5.199}$$

Finally, if the expressions for ρ and τ in (5.193) are used, this can in turn be written

$$p = \tfrac{1}{2}\mu_0(1 + |\rho|^2). \tag{5.200}$$

The interest of the form (5.200) lies in the fact that it is $1/c$ times the sum of the magnitudes of the power flux densities of the incident and reflected waves, namely

$$\frac{1}{2c}(|\mathbf{E}^i \wedge \mathbf{H}^{i*}| + |\mathbf{E}^r \wedge \mathbf{H}^{r*}|). \tag{5.201}$$

The interpretation of this expression as a pressure is allied to the concept that an electromagnetic field has *momentum density*

$$\frac{1}{c^2}\mathbf{E} \wedge \mathbf{H}; \tag{5.202}$$

for then (5.201) represents, on a time average, the rate per unit area at which the momentum in the negative y direction of the incident wave is converted into the momentum in the positive y direction of the reflected wave.

A general analysis of field momentum, justifying the adoption of (5.202) as momentum density, can be based on Maxwell's equations in much the same way as the analysis of field energy through Poynting's theorem.

A special but important point arises incidentally out of this section. In the case of perfect conductivity the reflection problem is simple. The current in the conductor is then a surface current density $\mathbf{j} = (2, 0, 0)$ in $y = 0$, and the time averaged force density on it is

$$\operatorname{Re} \tfrac{1}{4}\mathbf{j} \wedge \mathbf{B}^* = (0, -\mu_0, 0), \tag{5.203}$$

$\mathbf{B} = \mu_0\mathbf{H}$ being evaluated just outside the surface of the conductor. A factor $\frac{1}{2}$ enters (5.203) because the effective field acting on $\mathbf{j}$ is the average of the fields on either side of $y = 0$, that in $y < 0$ being zero. The situation is much the same as in the time independent case (cf. (4.36)), since the field discontinuity is entirely a local effect; but it is interesting to confirm, as is easily done, that (5.203) is given by (5.196) in the limit $\sigma \to \infty$.

5.4 Radiation

5.4.1 *Potential representation: retarded potentials*

Many features of electromagnetic fields can be analysed without specific reference to the sources of the fields, with the advantage that the mathematics can be comparatively simple, as, for example, in the plane wave studies of §5.2 and §5.3. For the full development and application of the theory, however, it is essential to consider the field of localized sources. To this end the representation of the field in terms of scalar and vector potentials, which figures prominently in electrostatics and magnetostatics, can also be introduced into general vacuum fields governed by equations (2.103) to (2.107).

Since $\operatorname{div}\mathbf{B} = 0$ there is (§A. 2) a vector $\mathbf{A}$ such that

$$\mathbf{B} = \operatorname{curl}\mathbf{A}. \tag{5.204}$$

Substitution into (2.104) then gives

$$\operatorname{curl}\mathbf{E} = -\operatorname{curl}\dot{\mathbf{A}}, \tag{5.205}$$

which implies that there is a scalar ϕ such that

$$\mathbf{E} = -\dot{\mathbf{A}} - \operatorname{grad}\phi. \tag{5.206}$$

The representations (5.204) and (5.206) are generalizations of the familiar time independent representations. The next step is to find the equations satisfied by ϕ and $\mathbf{A}$. From (2.106), making use of (A. 5),

$$\nabla^2\mathbf{A} - \frac{1}{c^2}\ddot{\mathbf{A}} = \operatorname{grad}\left(\operatorname{div}\mathbf{A} + \frac{1}{c^2}\dot{\phi}\right) - \mu_0\mathbf{J}.$$

Now $\mathbf{A}$ is not as yet completely determined, only its curl being prescribed by (5.204), and it is convenient to choose it so that

$$\operatorname{div}\mathbf{A} + \frac{1}{c^2}\dot{\phi} = 0. \tag{5.207}$$

That this is possible is shown a little later. With this choice

$$\nabla^2\mathbf{A} - \frac{1}{c^2}\ddot{\mathbf{A}} = -\mu_0\mathbf{J}; \tag{5.208}$$

and furthermore, (2.107) gives

$$\nabla^2\phi - \frac{1}{c^2}\ddot{\phi} = -\rho/\epsilon_0. \tag{5.209}$$

Thus ϕ and the cartesian components of $\mathbf{A}$ satisfy inhomogeneous wave equations, as do the cartesian components of $\mathbf{E}$ and $\mathbf{B}$ (see §2.5.2), with the advantage that the source terms in the equations for ϕ and $\mathbf{A}$ are just proportional to the physical sources, charge and current density. Moreover (5.208) and (5.209) are formally simple generalizations of the corresponding time independent equations in (2.110) and (2.112).

The final step is to find the physical solutions of (5.208) and (5.209), namely the generalizations of (2.10) and (2.63).

To accomplish this consider first the time harmonic case, and work in terms of complex representations (time factor $\exp(i\omega t)$ understood). Then (5.209) is

$$\nabla^2\phi + k_0{}^2\phi = -\rho/\epsilon_0, \tag{5.210}$$

where, as heretofore, $k_0 = \omega/c$.

Were k_0 zero (5.210) would be Poisson's equation. The elementary solution of Poisson's equation corresponding to a unit point charge source is $1/(4\pi\epsilon_0 r)$, where r is distance from the point source; and the solution (2.10) for an arbitrary charge density ρ can be regarded as the superposition of such elementary solutions. It is possible to arrive at the solution of (5.210) in a similar manner, by finding first the solution for a unit point source. This solution is a function of r only, and satisfies the homogeneous equation, (5.210) with the right hand side zero, everywhere except at $r = 0$.

Now if ϕ depends on r only,

$$\frac{\partial\phi}{\partial x} = \frac{d\phi}{dr}\frac{\partial r}{\partial x} = \frac{x}{r}\frac{d\phi}{dr},$$

$$\frac{\partial^2\phi}{\partial x^2} = \frac{1}{r}\frac{d\phi}{dr} + \frac{x^2}{r}\frac{d}{dr}\left(\frac{1}{r}\frac{d\phi}{dr}\right),$$

so that

$$\nabla^2\phi = \frac{3}{r}\frac{d\phi}{dr} + r\frac{d}{dr}\left(\frac{1}{r}\frac{d\phi}{dr}\right) = \frac{1}{r}\frac{d^2}{dr^2}(r\phi). \tag{5.211}$$

The homogeneous equation can therefore be written

$$\frac{d^2}{dr^2}(r\phi) + k_0{}^2 r\phi = 0. \tag{5.212}$$

But (5.212) is a simple harmonic equation for $r\phi$, independent solutions of which are $\exp(\pm ik_0 r)$. Hence the solution sought for ϕ is some linear combination of

$$\frac{e^{-ik_0 r}}{r} \quad \text{and} \quad \frac{e^{ik_0 r}}{r}. \tag{5.213}$$

These spherically symmetric solutions of the wave equation may be compared with the plane wave solutions (5.51). That with the negative sign in the exponential represents a wave whose equiphase surfaces $r =$ constant travel outwards with speed c; that with the positive sign represents a similar wave, but travelling *inwards*, so that it is physically unacceptable as any part of the field of a source located at $r = 0$.

The required elementary solution of (5.210) is therefore proportional to the first expression in (5.213). Moreover, when $r \to 0$ the solution must agree with that for the case $k_0 = 0$, otherwise the differential operator $\nabla^2 + k_0{}^2$ acting on it will not match the source singularity. Hence the constant of proportionality is $1/(4\pi\epsilon_0)$.

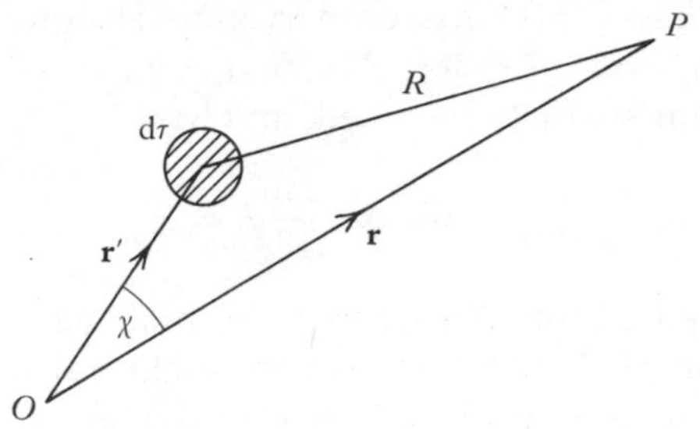

Figure 5.6

Since the solution for a unit point source at the origin is thus seen to be

$$\frac{1}{4\pi\epsilon_0}\frac{e^{-ik_0 r}}{r}, \tag{5.214}$$

that for any arbitrary charge density ρ is

$$\phi(\mathbf{r}) = \frac{1}{4\pi\epsilon_0}\int \rho(\mathbf{r}')\frac{e^{-ik_0 R}}{R}d\tau, \tag{5.215}$$

where the integration is over all space and R is the distance of the field point at $\mathbf{r}$ from the volume element $d\tau$ at $\mathbf{r}'$ (see figure 5.6).

Similarly the corresponding solution of (5.208) is

$$\mathbf{A}(\mathbf{r}) = \frac{\mu_0}{4\pi}\int \mathbf{J}(\mathbf{r}')\frac{e^{-ik_0 R}}{R}d\tau. \tag{5.216}$$

Evidently (5.215) and (5.216), in conjunction with the time harmonic forms of (5.204) and (5.206),

$$\mathbf{B} = \text{curl}\,\mathbf{A}, \quad \mathbf{E} = -i\omega\mathbf{A} - \text{grad}\,\phi, \tag{5.217}$$

give the field in terms of the charge and current densities. The formulae are generalizations of (2.10) and (2.63), respectively, to which they reduce when $k_0 = 0$.

It must not be forgotten, however, that the derivation of the formulae proceeded on the assumption that (5.207) was satisfied, and it remains to be confirmed that there is indeed this relation between (5.215) and (5.216). The relation is readily established in much the same way as it was shown in magnetostatics, by the argument preceding (2.66), that $\text{div}\,\mathbf{A} = 0$. Use a dash as before to distinguish differential operations with respect to the corrdinates (x', y', z') of the volume element $d\tau$, write

$$\psi = e^{-ik_0 R}/R, \tag{5.218}$$

and note that the divergence of the integrand of (5.216) can be expressed as

$$-\text{div}'(\mathbf{J}\psi) - i\omega\rho\psi, \tag{5.219}$$

because $\operatorname{div}' \mathbf{J} + i\omega\rho = 0$ (charge conservation) and $\operatorname{grad} \psi = -\operatorname{grad}' \psi$. Now integrate (5.219) over all space. The first term vanishes in the usual way, assuming that $\mathbf{J}$ is spatially bounded, and hence

$$\operatorname{div} \mathbf{A} + \frac{i\omega}{c^2} \phi = 0, \tag{5.220}$$

which is the required time harmonic version of (5.207).

The extension of these results for monochromatic fields to fields with arbitrary time variation can be achieved in the following manner. With the time factor $\exp(i\omega t)$ made explicit the integrand of (5.215), for example, is the product of $1/R$ with

$$\rho \, e^{i\omega(t-R/c)};$$

and since $\rho \exp(i\omega t)$ is the charge density evaluated at time t this is the charge density evaluated at time $t-R/c$. Thus (5.215) can be written

$$\phi(\mathbf{r}, t) = \frac{1}{4\pi\epsilon_0} \int \frac{\rho(\mathbf{r}', t-R/c)}{R} \, d\tau; \tag{5.221}$$

and likewise

$$\mathbf{A}(\mathbf{r}, t) = \frac{\mu_0}{4\pi} \int \frac{\mathbf{J}(\mathbf{r}', t-R/c)}{R} \, d\tau; \tag{5.222}$$

moreover the complex representation may here be dropped, and ϕ, $\mathbf{A}$, ρ and $\mathbf{J}$ be taken to stand for the actual quantities.

The significance of this way of expressing the formulae lies in the complete exclusion of the explicit appearance of ω (or, equivalently, of $k_0 = \omega/c$). The formal mathematical statements (5.221) and (5.222) are not one whit altered by altering ω. They are therefore valid no matter what the time variation, since any variation can be resolved by Fourier spectral analysis into its frequency components.

The physical interpretation of (5.221) and (5.222) is straightforward. They clearly reflect the fact that electromagnetic disturbances travel in vacuum with speed c, the effect of charge and current variations at each volume element taking time R/c to travel to a field point at distance R from $d\tau$. The instant $t-R/c$ is commonly called the *retarded time*, and the expressions for ϕ and $\mathbf{A}$ the *retarded potentials*.

It may be helpful to recall the chain of reasoning (see § 1.4.2) that has led to these formulae for the expression of the field in terms of charge and current in the general time varying case. The starting point was formulae for the field in terms of charge and current in the time independent case, taken as experimental laws. These results were then translated into differential equations, which, by adaption to the general case, yielded Maxwell's equations. Finally, the appropriate solution of Maxwell's equations was obtained in the form (5.204), (5.206), (5.221) and (5.222).

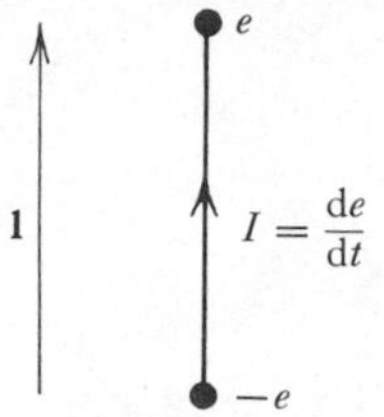

Figure 5.7

5·4.2 *Time dependent dipoles*

The simplest point source in electrostatics is a point charge, and in magnetostatics is an infinitesimal magnetic dipole. Have these sources counterparts in the time varying case?

A single point charge of changing magnitude is not, on its own, a realizable source, because it violates charge conservation. Consider, then, an electric dipole consisting of two variable point charges e and $-e$ at fixed locations (cf. § 2.1.5 for the static case). The total charge is always zero, and the charge must be conveyed from one station to the other by a current $\mathrm{d}e/\mathrm{d}t$. To get a point source (an infinitesimal dipole) it may be supposed, if $\mathbf{l}$ is the position vector of e relative to $-e$, that $l \to 0$ and $e \to \infty$ with the dipole moment

$$\mathbf{p} = e\mathbf{l} \tag{5.223}$$

remaining finite; and that a line current

$$I = \mathrm{d}e/\mathrm{d}t \tag{5.224}$$

runs directly from $-e$ to e (see figure 5.7).

The field of this point dipole is readily obtained from the formulae of § 5.4.1. It is convenient to treat first the time harmonic case, which is anyway of great practical importance; and the easiest procedure is to evaluate in turn $\mathbf{A}$, from (5.216), then

$$\mathbf{H} = \frac{1}{\mu_0} \operatorname{curl} \mathbf{A}, \tag{5.225}$$

and finally

$$\mathbf{E} = \frac{1}{\mathrm{i}\omega\epsilon_0} \operatorname{curl} \mathbf{H}. \tag{5.226}$$

Take the dipole at the origin, directed along the z axis. Then the volume integral of $\mathbf{J}$ has only a z component; and this is

$$Il = \mathrm{i}\omega p, \tag{5.227}$$

using (5.224) (here $I = \mathrm{i}\omega e$) and (5.223). From (5.216), therefore,

$$\mathbf{A} = A\hat{\mathbf{z}}, \tag{5.228}$$

where $\hat{\mathbf{z}}$ is the unit vector in the z direction, and

$$A = \frac{\mu_0}{4\pi} \mathrm{i}\omega p \frac{\mathrm{e}^{-\mathrm{i}k_0 r}}{r}, \tag{5.229}$$

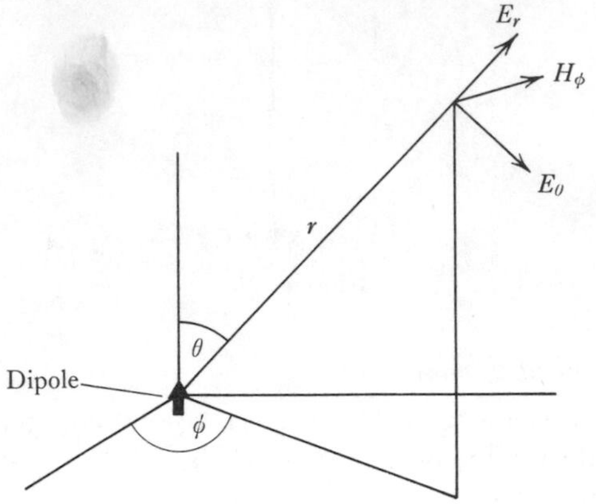

Figure 5.8

being simply the appropriate multiple of the elementary solution of the wave equation introduced in (5.213).

The application of (5.225) now gives, with the help of (A. 14),

$$\mathbf{H} = \frac{i\omega p}{4\pi}\frac{\mathrm{d}}{\mathrm{d}r}\left(\frac{\mathrm{e}^{-\mathrm{i}k_0 r}}{r}\right)\hat{\mathbf{r}}\wedge\hat{\mathbf{z}}, \tag{5.230}$$

where $\hat{\mathbf{r}}$ is the unit vector in the radial direction. The direction of the magnetic field at any point is therefore tangent to the circle, through that point, whose centre is on and whose plane is normal to the axis of the dipole.

To evaluate $\mathbf{E}$ introduce spherical polar coordinates r, θ, ϕ as in figure 5.8. Then $H_r = H_\theta = 0$, and

$$H_\phi = -\frac{pck_0^2}{4\pi}\sin\theta\left(1-\frac{\mathrm{i}}{k_0 r}\right)\frac{\mathrm{e}^{-\mathrm{i}k_0 r}}{r}. \tag{5.231}$$

Consequently (5.226) gives, from (A. 11),

$$E_r = \frac{1}{\mathrm{i}\omega\epsilon_0}\frac{1}{r\sin\theta}\frac{\partial}{\partial\theta}(\sin\theta H_\phi),\quad E_\theta = -\frac{1}{\mathrm{i}\omega\epsilon_0}\frac{1}{r}\frac{\partial}{\partial r}(rH_\phi),\quad E_\phi = 0; \tag{5.232}$$

from which, with (5.231),

$$E_r = \frac{pk_0^2}{4\pi\epsilon_0}2\cos\theta\frac{\mathrm{i}}{k_0 r}\left(1-\frac{\mathrm{i}}{k_0 r}\right)\frac{\mathrm{e}^{-\mathrm{i}k_0 r}}{r}, \tag{5.233}$$

$$E_\theta = -\frac{pk_0^2}{4\pi\epsilon_0}\sin\theta\left[1-\frac{\mathrm{i}}{k_0 r}\left(1-\frac{\mathrm{i}}{k_0 r}\right)\right]\frac{\mathrm{e}^{-\mathrm{i}k_0 r}}{r}. \tag{5.234}$$

These exact expressions for the field of an idealized dipole have comparatively simple approximations in each of the two cases $k_0 r \ll 1$ and $k_0 r \gg 1$.

In the former case the dominant terms are, in the electric field,

$$E_r \sim \frac{p}{4\pi\epsilon_0}\frac{2\cos\theta}{r^3}, \quad E_\theta \sim \frac{p}{4\pi\epsilon_0}\frac{\sin\theta}{r^3}; \tag{5.235}$$

and in the magnetic field

$$H_\phi \sim \frac{\mathrm{i}p\omega}{4\pi}\frac{\sin\theta}{r^2}. \tag{5.236}$$

Thus in the *near* field, namely, at distances from the dipole much less than $1/k_0$ (the wavelength divided by 2π) the electric field has the spatial dependence of a static dipole (see (2.29) and (2.30)), and the magnetic field follows the Biot–Savart law (see (2.56)).

When $k_0 r \gg 1$ the dominant terms are

$$E_\theta \sim Z_0 H_\phi, \quad H_\phi \sim -\frac{pck_0^{\,2}}{4\pi}\sin\theta\frac{\mathrm{e}^{-\mathrm{i}k_0 r}}{r}; \tag{5.237}$$

and E_r is of order $1/r^2$. The terms of order $1/r$ as $r \to \infty$ constitute what is called the *far field*, or *radiation field*. In the radiation field the polarization is linear, with **E** and **H** being mutually orthogonal, and orthogonal also to the radius vector from the dipole. The local field structure is that of a linearly polarized plane wave travelling radially outwards.

The electric dipole is approximately realized in practice by a linear aerial that is very much shorter than a wavelength, and its radiation characteristics are of considerable interest. A closer examination of this aspect is left to the next section, and meanwhile two results are interposed which follow directly from the present analysis.

First, the generalization of the electric dipole to the case of arbitrary time variation. This is immediately achieved by writing the field components (5.231), (5.233) and (5.234) in a form which removes the explicit appearance of ω, the argument being the same as that used to justify the general applicability of the retarded potentials (5.221) and (5.222). The result for the non-zero field components at (r, θ, ϕ) at time t is

$$H_\phi = \frac{1}{4\pi}\sin\theta\left(\frac{\ddot{p}}{cr}+\frac{\dot{p}}{r^2}\right), \tag{5.238}$$

$$E_\theta = \frac{Z_0}{4\pi}\sin\theta\left(\frac{\ddot{p}}{cr}+\frac{\dot{p}}{r^2}+\frac{cp}{r^3}\right), \tag{5.239}$$

$$E_r = \frac{Z_0}{4\pi}2\cos\theta\left(\frac{\dot{p}}{r^2}+\frac{cp}{r^3}\right), \tag{5.240}$$

where p and its time derivatives (signified by the dots) are evaluated at the retarded time $t-r/c$. The radiation field is therefore proportional to the second derivative of the (retarded) dipole moment.

The other result is the evaluation of the field of a time varying infinitesimal *magnetic* dipole. If the dipole has moment m its field is obtained from the electric dipole field by the transformation $\mathbf{E} \to \mathbf{H}$, $\mathbf{H} \to -\mathbf{E}$, $\epsilon_0 \leftrightarrow \mu_0$, and, ultimately, $p \to \mu_0 m$. The transformation for p, and indeed the identification of the source of the field, comes from comparison of the near field with the magnetostatic field (2.45) of a constant dipole.

5.4.3 *Dipole radiation: radiation resistance*

A spatially bounded time harmonic current distribution radiates, on a time average, a certain amount of power. This is the average rate at which the field energy crosses any closed surface that contains the entire current distribution, and is given by the flux of $\mathrm{Re}\,\frac{1}{2}\mathbf{E} \wedge \mathbf{H}^*$ out of the surface. That the flux is independent of the surface, so that the concept of radiated power is unambiguous, is essentially contained in the analysis of Poynting's theorem (§ 5.1.3), but is worth demonstrating explicitly in terms of the complex representation.

The appropriate form of Maxwell's equations is

$$\mathrm{curl}\,\mathbf{E} = -\mathrm{i}\omega\mu_0\mathbf{H}, \quad \mathrm{curl}\,\mathbf{H} = \mathrm{i}\omega\epsilon_0\mathbf{E} + \mathbf{J}, \tag{5.241}$$

where $\mathbf{J}$ is the given current density. From these equations

$$\begin{aligned} \mathrm{div}\,(\mathbf{E} \wedge \mathbf{H}^*) &= \mathbf{H}^*.\mathrm{curl}\,\mathbf{E} - \mathbf{E}.\mathrm{curl}\,\mathbf{H}^* \\ &= \mathrm{i}\omega(\epsilon_0\mathbf{E}.\mathbf{E}^* - \mu_0\mathbf{H}.\mathbf{H}^*) - \mathbf{E}.\mathbf{J}^*. \end{aligned} \tag{5.242}$$

Taking the real part of (5.242) and integrating over some volume V gives

$$\mathrm{Re}\,\frac{1}{2}\int_S (\mathbf{E} \wedge \mathbf{H}^*).\mathrm{d}\mathbf{S} = \mathrm{Re} - \frac{1}{2}\int_V \mathbf{E}.\mathbf{J}^*\,\mathrm{d}\tau, \tag{5.243}$$

where S is the surface bounding V.

The left hand side of (5.243) therefore has the same value for all surfaces S that completely enclose the current. This value is the average radiated power. It is seen also to be the average rate at which the field does work against the current, which confirms that the energy 'lost' in radiation is drawn from the work done by the external forces that maintain the current.

The application of this theory to a dipole source is straightforward. If S is taken to be a spherical surface (centred on the dipole) whose radius becomes indefinitely great, only the radiation field, for which the Poynting flux decreases as the inverse square of the distance, contributes to the radiated power. For the electric dipole with radiation field given by (5.237) the time averaged radiated power is evidently

$$\tfrac{1}{2}Z_0\left(\frac{ck_0^2}{4\pi}\right)^2 |p|^2 \int_{\phi=0}^{2\pi}\int_{\theta=0}^{\pi} \sin^3\theta\,\mathrm{d}\theta\,\mathrm{d}\phi; \tag{5.244}$$

and since the θ integral is $\frac{4}{3}$, this is

$$\frac{Z_0}{12\pi}c^2k_0^4|p|^2. \tag{5.245}$$

In experimental work the current flowing in a short aerial is a more practical parameter than the dipole moment. The two quantities are related by (5.227), so that (5.245) can be written

$$\frac{Z_0}{12\pi}(k_0 l)^2\,|I|^2. \tag{5.246}$$

This form is reminiscent of the time average power $\frac{1}{2}R\,|I|^2$ dissipated when current I flows through a resistance R, and the result is sometimes expressed by the statement that the dipole has *radiation resistance*

$$R = \frac{Z_0}{6\pi}(k_0 l)^2 = 20(k_0 l)^2\,\text{ohm}. \tag{5.247}$$

The statement simply means that in maintaining the current in the dipole, and hence the radiation, the generator works at the rate required to drive the same current through resistance R.

5.4.4 *The radiation field: gain*

The electrostatic field of a spatially bounded stationary charge distribution ultimately decreases at least as rapidly as the inverse square of the distance, and the magnetostatic field of a steady current distribution as the inverse cube. But for currents that vary with time the influence of the field is much more extensive: both the electric and the magnetic field fall off only as the inverse distance, a feature which is, of course, intimately linked with the radiation of energy. The truth of this statement has been established explicitly for the case of an infinitesimal dipole, and may therefore be expected to apply to any localized current density $\mathbf{J}$, since the current in each volume element $d\tau$ represents an electric dipole of moment $\mathbf{p}$, where $\dot{\mathbf{p}} = \mathbf{J}\,d\tau$.

It is simplest to give the formal mathematical analysis for the time harmonic case. The current density is supposed confined to a region within some sphere centred on the origin O, and the distance r of the field point from O is allowed to become infinitely great. Then, with the notation of figure 5.6, it follows from

$$R^2 = r^2 - 2rr'\cos\chi + r'^2$$

that

$$R = r\left[1 - \frac{r'}{r}\cos\chi + O\left(\frac{r'^2}{r^2}\right)\right], \tag{5.248}$$

and therefore that

$$k_0 R = k_0(r - r'\cos\chi) + O(k_0 r'^2/r) \tag{5.249}$$

and

$$1/R = 1/r + O(r'/r^2). \tag{5.250}$$

Moreover, the unit vector in the direction of $\mathbf{R}$ is

$$\hat{\mathbf{R}} = (\mathbf{r} - \mathbf{r}')/R = \hat{\mathbf{r}} + O(r'/r). \tag{5.251}$$

Now from (5.237) the far field of a dipole of moment p located at $\mathbf{r}'$ can be written

$$\mathbf{E} \sim Z_0 \mathbf{H} \wedge \hat{\mathbf{R}}, \quad \mathbf{H} \sim \frac{ik_0}{4\pi}(\mathrm{i}\omega\mathbf{p} \wedge \hat{\mathbf{R}})\frac{\mathrm{e}^{-\mathrm{i}k_0 R}}{R}; \tag{5.252}$$

and the application of (5.249), (5.250) and (5.251) gives

$$\mathbf{E} \sim Z_0 \mathbf{H} \wedge \hat{\mathbf{r}}, \quad \mathbf{H} \sim \frac{\mathrm{i}k_0}{4\pi}(\mathrm{i}\omega\mathbf{p} \wedge \hat{\mathbf{r}})\frac{\mathrm{e}^{-\mathrm{i}k_0(r-r'\cos\chi)}}{r}. \tag{5.253}$$

In (5.253) the radiation field is referred to the origin rather than to the point $\mathbf{r}'$ at which the dipole is located, and the result can be adapted to a general current distribution by replacing $\mathrm{i}\omega\mathbf{p}$ by $\mathbf{J}\,\mathrm{d}\tau$ and integrating over the distribution. In this way the radiation field of a current density $\mathbf{J}$ is seen to be

$$\mathbf{E} \sim Z_0 \mathbf{H} \wedge \hat{\mathbf{r}}, \quad \mathbf{H} \sim \mathbf{F} \wedge \hat{\mathbf{r}}\frac{\mathrm{e}^{-\mathrm{i}k_0 r}}{r}, \tag{5.254}$$

where

$$\mathbf{F} = \frac{\mathrm{i}k_0}{4\pi}\int \mathbf{J}\,\mathrm{e}^{\mathrm{i}k_0 r'\cos\chi}\,\mathrm{d}\tau \tag{5.255}$$

is independent of distance r but is a function of $\hat{\mathbf{r}}$ by virtue of the $\cos\chi$ dependence of the integrand.

In terms of spherical polar components (5.254) is

$$\mathbf{E} \sim -Z_0(0, F_\theta, F_\phi)\frac{\mathrm{e}^{-\mathrm{i}k_0 r}}{r}, \quad \mathbf{H} \sim (0, F_\phi, -F_\theta)\frac{\mathrm{e}^{-\mathrm{i}k_0 r}}{r}. \tag{5.256}$$

Evidently $\mathbf{E}$, $\mathbf{H}$ and $\hat{\mathbf{r}}$ are mutually orthogonal, and in general the polarization is elliptic. The time averaged power flux density is radial, with magnitude $P(\theta, \phi)/r^2$, where

$$P(\theta, \phi) = \tfrac{1}{2}Z_0(|F_\theta|^2 + |F_\phi|^2); \tag{5.257}$$

and the radiated power is

$$P_0 = \int_{\phi=0}^{2\pi}\int_{\theta=0}^{\pi} P(\theta, \phi)\sin\theta\,\mathrm{d}\theta\,\mathrm{d}\phi. \tag{5.258}$$

A highly important feature of the radiation, peculiar to each current source, is its directional dependence. Since the function $P(\theta, \phi)$ gives the time averaged power radiated per unit solid angle in the direction θ, ϕ, a normalized dimensionless measure of the directional dependence is

$$G(\theta, \phi) = 4\pi P(\theta, \phi)/P_0. \tag{5.259}$$

In aerial theory G is called the *gain* of the aerial; and the normalization is such that, were it possible (which it is not for electromagnetic waves) to have an isotropic radiator, meaning one that radiates equally in every direction, its gain would be unity.

For an infinitesimal electric dipole the function $\mathbf{F}$ is just a constant times the vector dipole moment. Hence $\mathbf{F}_\theta$ is proportional to $\sin\theta$, θ being measured

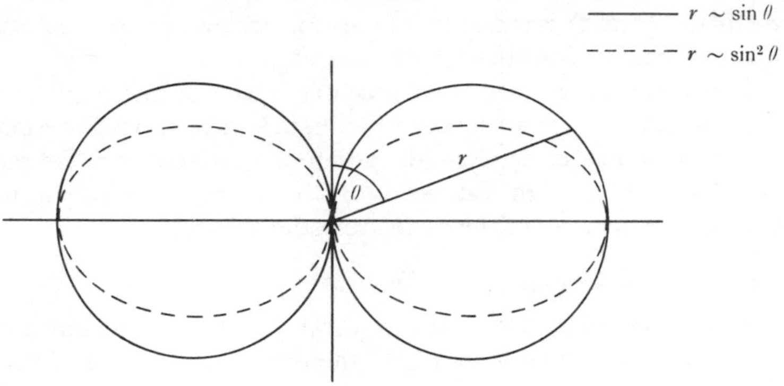

Figure 5.9

from the dipole axis, and F_ϕ is zero, so that $P(\theta, \phi)$ is proportional to $\sin^2\theta$, and

$$G = \tfrac{3}{2}\sin^2\theta. \tag{5.260}$$

A polar plot of the gain, that is, a surface the length of whose radius vector in each direction θ, ϕ is proportional to $G(\theta, \phi)$, is called the ***power polar diagram.*** A similar plot of the amplitude of the far field is called the amplitude polar diagram. The length of the radius vector of the former surface is, of course, proportional to the square of that of the latter. For a dipole the polar diagrams are symmetric about the dipole axis, and their cross-sections in a plane through the axis are shown in figure 5.9.

Power calculations can be made from a knowledge of the radiation field only, since the surface across which the power flux is evaluated can be taken arbitrarily far from the current distribution. However, in considering the field *per se* it is pertinent to ask under what conditions the radiation field represents a good approximation to the complete field. Note that the problem involves two length parameters with which the distance r of the field point from the origin may be compared, namely the wavelength $2\pi/k_0$, and the radius, a say, of the smallest spherical surface, with centre at the origin, that encloses all the current. Necessary conditions for the radiation field to be a good approximation are obviously, in general,

$$r \gg 1/k_0, \tag{5.261}$$

from the dipole analysis (see (5.237)), and

$$r \gg a, \tag{5.262}$$

from the expansion (5.248). Another necessary condition, which together with (5.261) and (5.262) makes a sufficient set, is seen from (5.249) to be

$$r \gg k_0 a^2. \tag{5.263}$$

The right hand side of (5.263) is called the *Rayleigh distance.* Since the Rayleigh distance is $(k_0 a)\,a$, or, equally, $(k_0 a)^2/k_0$, it is evident that the latter condition, (5.263), is redundant to the former two, (5.261) and (5.262), when $k_0 a < 1$, but contains them both when $k_0 a > 1$.

In practice the Rayleigh distance may give a surprisingly large value for the least distance at which the radiation field takes over. For example, in the reception of 10^{-1} m wavelength radiation scattered from the moon (radius $\sim 10^6$ m) the Rayleigh distance is of the order of $2\pi \times 10^{13}$ m, which vastly exceeds the distance of the earth from the moon.

5.4.5 *Interference: the linear antenna*

The exponential factor in the integrand of (5.255) merely expresses the phase difference corresponding to the different distances of the origin and the volume element $d\tau$ from a point in the far field. Stated physically, these phase differences give rise to *interference* between the waves generated by the current in the various volume elements, and it is this interference that largely determines the polar diagram of any extended source.

A simple example is afforded by two identical time harmonic dipoles with a common axis. If they are distance d apart, and oscillate in phase, the far field of the dipole pair is that of one of the dipoles multiplied by the factor

$$1 + e^{ik_0 d \cos\theta} = 2\cos(\tfrac{1}{2}k_0 d\cos\theta)\, e^{\frac{1}{2}ik_0 d\cos\theta}, \tag{5.264}$$

where θ is measured from the common axis. The modulus of this factor is

$$2\left|\cos\left(\frac{\pi d}{\lambda}\cos\theta\right)\right|, \tag{5.265}$$

the θ dependence of which is sharply contrasted in the two cases $d/\lambda \ll 1$ and $d/\lambda \gg 1$. In the former case the difference in the paths from the dipoles to any distant field point can only represent a small phase difference, and (5.265) is approximately 2: the interference is everywhere constructive. In the latter case the path difference decreases monotonically from much greater than the wavelength at $\theta = 0$ to zero at $\theta = \frac{1}{2}\pi$; the interference is constructive at those values of θ for which the path difference is an integral number of wavelengths, and destructive, giving zeros in the polar diagram, when the path difference is an odd number of half wavelengths.

Thus when the dipoles have a separation of many wavelengths the polar diagram, which is, of course, symmetric about their common axis, has a lobe structure modulating the pattern of a single dipole. An axial plane cross-section of the amplitude polar diagram for the case $\mathrm{d}/\lambda = 4$ is shown in figure 5.10.

As another illustration of the results of § 5.4.4, consider their application to the elementary theory of the linear antenna. It is supposed that a time harmonic current $I(\xi)$ flows in a straight wire of length l, ξ being the coordinate along the wire measured from its centre. If θ is measured from the line of the wire (see figure 5.11) (5.255) gives $F_\phi = 0$ and

$$F_\theta = -\frac{ik_0}{4\pi}\sin\theta \int_{-\frac{1}{2}l}^{\frac{1}{2}l} I(\xi)\, e^{ik_0\xi\cos\theta}\, d\xi. \tag{5.266}$$

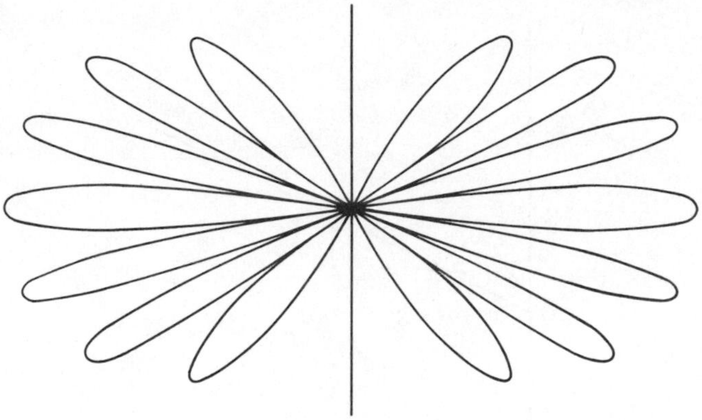

Figure 5.10

Strictly the dependence of the current on ξ is not known, and a full theory involves a sophisticated boundary value problem. However, in the case where the antenna is an integral number of half wavelengths, so that

$$k_0 = m\pi/l, \tag{5.267}$$

where m is an integer, a reasonable assumption is

$$I(\xi) = I_0 \sin[m\pi(\tfrac{1}{2}+\xi/l)]. \tag{5.268}$$

This vanishes at the ends $\xi = \pm\frac{1}{2}l$, as the current must if the wire is thin, and is in fact known to be a reasonably good approximation.

If (5.268) is adopted the integration in (5.266) is straightforward and a little tidying up gives

$$F_\theta = \frac{I_0}{4\pi\sin\theta}[(-)^{m+1}\,e^{\frac{1}{2}im\pi\cos\theta} + e^{-\frac{1}{2}im\pi\cos\theta}]$$

$$= \begin{cases} -\dfrac{iI_0}{2\pi}\sin(\tfrac{1}{2}m\pi\cos\theta)/\sin\theta, & m \text{ even,} \\[2ex] \dfrac{I_0}{2\pi}\cos(\tfrac{1}{2}m\pi\cos\theta)/\sin\theta, & m \text{ odd.} \end{cases} \tag{5.269}$$

Again it is evident that the polar diagram has a lobe structure.

5.4.6 *Reciprocity*

In electrostatics and magnetostatics there are reciprocal relations (3.103) and (4.65) between, respectively, scalar potentials due to each of two charge densities and vector potentials due to each of two current densities. It is now shown that there is an analogous relation between electric fields $\mathbf{E}$ and $\mathbf{E}'$ due to respective time harmonic current densities $\mathbf{J}$ and $\mathbf{J}'$, namely

$$\int \mathbf{J}.\mathbf{E}'\,d\tau = \int \mathbf{J}'.\mathbf{E}\,d\tau, \tag{5.270}$$

where the integrals are over all space.

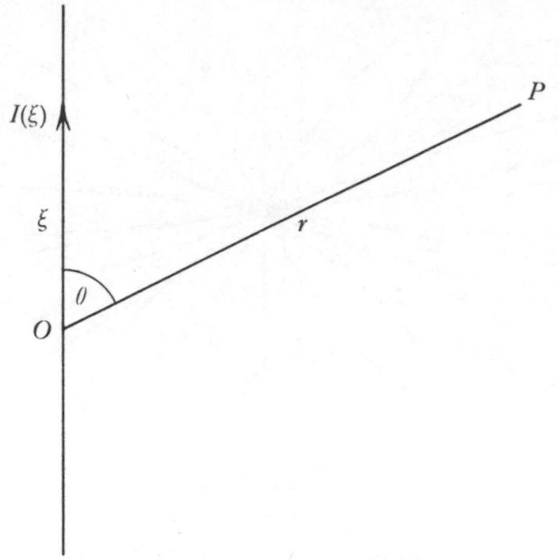

Figure 5.11

In establishing this result it is enough to suppose that the medium in which the fields are generated can be characterized, in the complex representation, by constitutive relations $\mathbf{D} = \epsilon\mathbf{E}$, $\mathbf{B} = \mu\mathbf{H}$. The medium may be inhomogeneous, with permittivity, permeability and conductivity depending arbitrarily on position provided that they behave suitably as the distance from the current distribution tends to infinity. In this latter respect the natural assumption that there is vacuum everywhere outside some sufficiently large sphere will serve.

The result is proved by expressing the current densities in terms of the field through Maxwell's equation

$$\mathbf{J} = \operatorname{curl}\mathbf{H} - \mathrm{i}\omega\epsilon\mathbf{E}. \tag{5.271}$$

For then

$$\begin{aligned}\mathbf{J}.\mathbf{E}' &= \mathbf{E}'.\operatorname{curl}\mathbf{H} - \mathrm{i}\omega\epsilon\mathbf{E}.\mathbf{E}' \\ &= -\operatorname{div}(\mathbf{E}'\wedge\mathbf{H}) + \mathbf{H}.\operatorname{curl}\mathbf{E}' - \mathrm{i}\omega\epsilon\mathbf{E}.\mathbf{E}';\end{aligned}$$

and with the help of Maxwell's other equation

$$\operatorname{curl}\mathbf{E} = -\mathrm{i}\omega\mu\mathbf{H}, \tag{5.272}$$

it follows that

$$\int(\mathbf{J}.\mathbf{E}' - \mathbf{J}'.\mathbf{E})\,\mathrm{d}\tau = \int(\mathbf{E}\wedge\mathbf{H}' - \mathbf{E}'\wedge\mathbf{H}).\mathrm{d}\mathbf{S}, \tag{5.273}$$

where the surface integral is over the sphere at infinity.

Now sufficiently distant points are in the vacuum radiation field of $\mathbf{J}$ and $\mathbf{J}'$, and of the current densities associated with the medium. Hence as $r \to \infty$ in the direction of the unit vector $\hat{\mathbf{r}}$, $\mathbf{E}$ and $\mathbf{H}$ (and likewise $\mathbf{E}'$ and $\mathbf{H}'$) are of order $1/r$, and to that order satisfy $\mathbf{E} = Z_0\mathbf{H}\wedge\hat{\mathbf{r}}$, with $\mathbf{H}$ being orthogonal to $\hat{\mathbf{r}}$ (see § 5.4.4). Consequently

$$\mathbf{E}\wedge\mathbf{H}' = Z_0(\mathbf{H}.\mathbf{H}')\hat{\mathbf{r}}+O(1/r^3) = \mathbf{E}'\wedge\mathbf{H}+O(1/r^3). \qquad (5.274)$$

It follows that the right hand side of (5.273) is zero, and (5.270) is established.

As an example, suppose the current systems are simply infinitesimal dipoles at A and B respectively, with moments $p\mathbf{l}$ and $p\mathbf{l}'$, where $\mathbf{l}$ and $\mathbf{l}'$ are real unit vectors. Then (5.270) states that

$$(\mathbf{l}.\mathbf{E}')_A = (\mathbf{l}'.\mathbf{E})_B. \qquad (5.275)$$

In particular, by taking $\mathbf{l} = \mathbf{l}'$, it is seen that, if $\mathbf{E}(\mathbf{r}_0, \mathbf{r})$ is the field at $\mathbf{r}$ due to a dipole along $\mathbf{l}$ located at $\mathbf{r}_0$, then

$$\mathbf{l}.\mathbf{E}(\mathbf{r}_0, \mathbf{r}) = \mathbf{l}.\mathbf{E}(\mathbf{r}, \mathbf{r}_0). \qquad (5.276)$$

Problems 5

[If the text does not imply otherwise the fields in these problems are monochromatic, at angular frequency ω, and exist in a vacuum region free of charge and current. Also $k_0 = \omega/c$, $Z_0 = 1/Y_0 = (\mu_0/\epsilon_0)^{\frac{1}{2}}$.]

5.1 Show that two-dimensional fields that are independent of z can be expressed as the superposition of the following fields:

(*a*) one in which $\mathbf{E} = (0, 0, E_z)$, $\mathbf{H} = (H_x, H_y, 0)$, where

$$H_x = \frac{\mathrm{i}Y_0}{k_0}\frac{\partial E_z}{\partial y}, \quad H_y = -\frac{\mathrm{i}Y_0}{k_0}\frac{\partial E_z}{\partial x},$$

$$\frac{\partial^2 E_z}{\partial x^2}+\frac{\partial^2 E_z}{\partial y^2}+k_0^2E_z = 0;$$

(*b*) one in which $\mathbf{H} = (0, 0, H_z)$, $\mathbf{E} = (E_x, E_y, 0)$, where

$$E_x = -\frac{\mathrm{i}Z_0}{k_0}\frac{\partial H_z}{\partial y}, \quad E_y = \frac{\mathrm{i}Z_0}{k_0}\frac{\partial H_z}{\partial x},$$

$$\frac{\partial^2 H_z}{\partial x^2}+\frac{\partial^2 H_z}{\partial y^2}+k_0^2H_z = 0.$$

5.2 A plane wave travelling in the x direction has $E_z = Z_0 \exp(-\mathrm{i}k_0x)$, $H_z = \lambda \exp(-\mathrm{i}k_0x)$. Describe, precisely, its polarization in each of the cases $\lambda = 0, 1, \mathrm{i}, -\mathrm{i}, 2, \exp(\frac{1}{4}\mathrm{i}\pi)$.

5.3 Surface current density $j_0\,(1, 0, \mathrm{i}\sin\alpha)\exp(-\mathrm{i}k_0x\cos\alpha)$, where j_0 and α are constants, flows in the plane $y = 0$. Show that it generates a pair of circularly polarized plane waves, and find the time average of the power radiated across unit area of any plane $y =$ constant.

5.4 A monochromatic field consists of two linearly polarized plane waves, of arbitrary amplitudes, travelling either in the same direction or in opposite directions. Establish the circumstances in which the time averaged power flux density is the sum or difference of the time averaged power flux densities of the individual plane waves. Investigate similarly the time averaged energy density.

5.5 Show that, for the vacuum field of charge and current with arbitrary time dependence,

$$\operatorname{div}\mathbf{S}+\frac{\partial\mathscr{E}}{\partial t}+\mathbf{E}.\mathbf{J}=0$$

in each of the cases

(*a*) $\mathscr{E}=\frac{1}{2}(\epsilon_0\mathbf{E}^2+\mu_0\mathbf{H}^2)$, $\mathbf{S}=\mathbf{E}\wedge\mathbf{H}$;

(*b*) $\mathscr{E}=\frac{1}{2}(\epsilon_0\mathbf{E}^2+\mathbf{A}.\operatorname{curl}\mathbf{H})$, $\mathbf{S}=\mathbf{E}\wedge\mathbf{H}+\frac{1}{2}\frac{\partial}{\partial t}(\mathbf{A}\wedge\mathbf{H})$;

(*c*) $\mathscr{E}$ as in (*b*), $\mathbf{S}=\frac{1}{2}(\mathbf{A}\wedge\dot{\mathbf{H}}-\dot{\mathbf{A}}\wedge\mathbf{H})+\phi\operatorname{curl}\mathbf{H}$;

where ϕ and $\mathbf{A}$ are scalar and vector potential.

Evaluate $\mathbf{S}$, in each case, for the plane wave given by $\phi=0$,

$$\mathbf{A}=(0,0,1)\cos(\omega t-k_0x).$$

5.6 A two-dimensional waveguide is formed by the pair of infinite, perfectly conducting planes $y=0$ and $y=a$. Show that a mode independent of z and with x dependence $\exp(-ik_0\gamma x)$ has $\gamma^2=1-(n\pi/k_0a)^2$, where n is a positive integer or zero.

Evaluate the power flux along the guide for a TE mode, and show that the speed of energy propagation is γc.

5.7 The waveguide of problem 5.6 is excited by the surface current density $\mathbf{j}=(0,j,0)$ flowing in the region $0<y<a$ of the plane $x=0$, where

$$j=j_0\sin^2(\pi y/a)$$

and j_0 is a constant. Obtain the field in the waveguide, and show that if $k_0a>2\pi$ the time averaged emitted power is

$$\tfrac{1}{16}j_0^2Z_0(1+\tfrac{1}{2}\gamma)\,a,$$

per unit length in the z direction, where $\gamma^2=1-(2\pi/k_0a)^2$.

5.8 Show that throughout a source free vacuum region in which $E_z=H_z=0$ a monochromatic field has z dependence $\exp(\pm ik_0z)$, and can be represented through

$$\mp Z_0H_y=E_x=-\frac{\partial\phi}{\partial x},\quad \pm Z_0H_x=E_y=-\frac{\partial\phi}{\partial y}$$

in terms of a function ϕ that satisfies

$$\frac{\partial^2\phi}{\partial x^2}+\frac{\partial^2\phi}{\partial y^2}=0.$$

Deduce that no such field can exist inside a waveguide formed by a perfectly conducting cylindrical surface parallel to the z axis whose cross-section is a single closed curve.

5.9 Two perfectly conducting, non-intersecting, infinitely long cylindrical surfaces have their generators parallel to the z axis. Show that they can support a TEM mode propagated in the z direction (one that has $E_z=H_z=0$); and that, for this mode, $\mathbf{E}$ in any plane $z=$ constant is a potential field for which the cross-section of each cylinder is an equipotential.

Show also that, in any plane $z=$ constant, the ratio of the voltage between the cylinders to the total current flowing along one of them is $(L/C)^{\frac{1}{2}}$, where L and C are, respectively, the inductance and capacitance per unit length of the pair of cylinders (cf. problems 4.15 and 4.27).

5.10 A monochromatic field is confined within an evacuated cavity enclosed by perfectly conducting walls. Show that the time averages of the electric and magnetic energies are equal.

5.11 A cavity with perfectly conducting walls is formed between coaxial cylinders $r = a$ and $r = b\,(> a)$ by inserting two plane cross-sections, normal to the axis and distance d apart. Show that the cavity can support modes that have

$$E_z = H_z = 0$$

everywhere; give the frequencies of these modes, and obtain expressions for the field components.

5.12 Show that

$$\nabla^2(\mathbf{r}U) = 2\,\mathrm{grad}\,U + \mathbf{r}\nabla^2 U.$$

Deduce that

$$\mathbf{E} = -ik_0 Z_0\,\mathrm{curl}\,(\mathbf{r}U), \quad \mathbf{H} = \mathrm{curl}\,\mathrm{curl}\,(\mathbf{r}U)$$

satisfies Maxwell's equations if $\nabla^2 U + k_0{}^2 U = 0$.

Show that a possible field within a spherical cavity of radius a with perfectly conducting walls is given by taking

$$U = \frac{\partial}{\partial z}\left[\frac{\sin(k_0 r)}{r}\right],$$

provided the frequency satisfies $\tan(k_0 a) = k_0 a$.

5.13 A two-dimensional waveguide, symmetrical about the plane $y = \frac{1}{2}a$, consists of a vacuum region $0 < y < a$ between semi-infinite media whose properties are independent of x and z. If the medium in $y > a$ were absent the plane wave with magnetic vector $\mathbf{H}_1 = (0, 0, 1)\exp[-ik_0(x\cos\alpha - y\sin\alpha)]$ incident from $y > 0$ on the medium in $y < 0$ would give rise to a reflected wave

$$\mathbf{H}_2 = (0, 0, \rho)\exp[-ik_0(x\cos\alpha + y\sin\alpha)],$$

where the reflection coefficient ρ is some function of the angle of incidence α. Show that the field specified by $\mathbf{H} = \mathbf{H}_1 + \mathbf{H}_2$ $(0 < y < a)$ is a possible mode in the waveguide provided α satisfies

$$\rho(\alpha) = \pm e^{-ik_0 a \sin\alpha}.$$

Deduce that in the case when the planes $y = 0$ and $y = a$ have surface impedance ζZ_0 the equation for $s = \sin\alpha$ is

$$\frac{1 \mp e^{ik_0 as}}{1 \pm e^{ik_0 as}}\,s = \zeta.$$

Confirm that the equation gives the correct values of s when the planes $y = 0$ and $y = a$ are perfect conductors.

5.14 Consider problem 5.13 in the case $|\zeta| \ll 1$. Show that the value of s that corresponds to the value zero for the case $\zeta = 0$ is given by

$$s^2 = 2i\zeta/(k_0 a),$$

approximately. Show further that this implies that the mode suffers attenuation specified by the factor $\exp(-\gamma x)$, where the attentuation coefficient γ is approximately the real part of ζ/a.

Deduce that if the regions $y < 0$ and $y > a$ are homogeneous conductors with high conductivity σ the attenuation coefficient is approximately $k_0 d/(2a)$, where d is the skin depth (5.183).

5.15 Homogeneous, lossless, non-magnetic dielectrics, with dielectric constants κ_1, κ_2, κ_3, occupy the regions $z < 0$, $0 < z < a$, $z > a$, respectively. A plane wave in $z < 0$ is normally incident on the interface $z = 0$. Show that if $a = \pi c/(2\omega\kappa_2^{\frac{1}{2}})$ the ratio of the amplitudes of the reflected and incident waves is $|\kappa-1|/(\kappa+1)$, where $\kappa^2 = \kappa_1\kappa_3/\kappa_2^2$.

5.16 The plane wave $\mathbf{E} = (0, 0, 1)\,\mathrm{e}^{-\mathrm{i}k_0 y}$ is incident from the vacuum half-space $y < 0$ on a lossless non-magnetic medium consisting of a succession of homogeneous slabs with interfaces $y = y_r$ ($r = 0, 1, 2, 3, \ldots; y_{r+1} > y_r, y_0 = 0$). The slab $y_r < y < y_{r+1}$ has refractive index n_{r+1}. By adopting the exact transmission coefficient at each interface and subsequently neglecting each reflected wave obtain the approximation

$$E_z = \exp\left\{-\sum_{r=0}^{m}\log\left(1+\frac{n_{r+1}-n_r}{2n_r}\right)-\mathrm{i}k_0\sum_{r=1}^{m}n_r(y_r-y_{r-1})-\mathrm{i}k_0 n_{m+1}(y-y_m)\right\} \quad (n_0 = 1),$$

for the z component of the electric field at y when $y_m < y < y_{m+1}$.

Show that in the limit as the thickness of each slab tends to zero, so that the refractive index n can be regarded as a continuous function of y, the approximation is

$$E_z = n^{-\frac{1}{2}}\exp\left[-\mathrm{i}k_0\int_0^y n(y')\,\mathrm{d}y'\right].$$

5.17 A homogeneous slab of lossless, non-magnetic dielectric of permittivity ϵ occupies the region $-a < y < 0$, the regions $y > 0$ and $y < -a$ being vacuum. A linearly polarized plane wave is incident on the slab from $y > 0$. Show that the wave transmitted into the region $y < -a$ has (with appropriate phase reference) transmission coefficient

$$\tau = \frac{4Y_0 Y\,\mathrm{e}^{\mathrm{i}k_0 a}}{(Y+Y_0)^2\,\mathrm{e}^{\mathrm{i}ka}-(Y-Y_0)^2\,\mathrm{e}^{-\mathrm{i}ka}},$$

where $Y = (\epsilon/\mu_0)^{\frac{1}{2}}$, $k = \omega(\epsilon\mu_0)^{\frac{1}{2}}$. What values of ω give $\tau = \exp(\mathrm{i}k_0 a)$?

Show that, if $\epsilon = \epsilon_0(1+2\delta)$, where $\delta \ll 1$, then, to the first order in δ,

$$\tau = 1-\mathrm{i}k_0 a\delta.$$

5.18 A homogeneous slab of lossless, non-magnetic dielectric of permittivity ϵ occupies the region $-a < y < 0$, and is backed by a perfectly conducting sheet $y = -a$. A plane wave, linearly polarized with $\mathbf{H}$ parallel to the z axis, is incident on the slab from the vacuum region $y > 0$, the direction of propagation making an angle α with the x axis. Show that the wave reflected into $y > 0$ has (with appropriate phase reference) reflection coefficient

$$\rho = \frac{Z_0\sin\alpha-\mathrm{i}Z\sin\beta\tan(ka\sin\beta)}{Z_0\sin\alpha+\mathrm{i}Z\sin\beta\tan(ka\sin\beta)},$$

where $k_0\cos\alpha = k\cos\beta$, $k = \omega(\epsilon\mu_0)^{\frac{1}{2}}$, $Z = (\mu_0/\epsilon)^{\frac{1}{2}}$.

If $ka \ll 1$ and $\epsilon/\epsilon_0 \gg 1$, obtain the approximation

$$\rho = (\sin\alpha-\mathrm{i}k_0 a)/(\sin\alpha+\mathrm{i}k_0 a),$$

and interpret this result in terms of surface impedance.

5.19 A two-dimensional waveguide has perfectly conducting plane walls $y = 0$ and $y = a$. Investigate the possible z independent modes if the medium inside the guide, $0 < y < a$, is a homogeneous lossy dielectric of permittivity ϵ and conductivity σ. Show that if $\sigma \ll \omega\epsilon$ the least attenuated mode has attenuation coefficient $\frac{1}{2}(\mu_0/\epsilon)^{\frac{1}{2}}\sigma$.

5.20 Show that in a homogeneous non-magnetic medium of conductivity σ, in which the displacement current can be neglected, the magnetic vector **B** satisfies the equation

$$\nabla^2\mathbf{B} = \mu_0\sigma\dot{\mathbf{B}}.$$

What corresponding equation is satisfied by the electric vector?

A slab of the medium occupies $-a < y < a$, and an electric field

$$\mathbf{E} = [0, 0, \cos(\omega t)]$$

is maintained both in $y = -a$ and in $y = a$. Show that the mean square current density at any point in the slab is

$$\tfrac{1}{2}\sigma^2 \frac{\cosh(2\kappa y) + \cos(2\kappa y)}{\cosh(2\kappa a) + \cos(2\kappa a)},$$

where $\kappa^2 = \frac{1}{2}\mu_0\sigma\omega$.

5.21 The region $0 < z < b$ between perfectly conducting planes $z = -a$ and $z = b$ is occupied by a uniform, lossless, non-magnetic dielectric of refractive index n, and the region $-a < z < 0$ is a vacuum. Show that a monochromatic field consisting of plane waves travelling parallel to the z axis can exist between the conductors if the wave number satisfies the equation

$$n \tan(k_0 a) + \tan(k_0 nb) = 0.$$

What are the possible frequencies (*a*) if $n = 1$, (*b*) if $nb = a$?

5.22 The plane wave $\mathbf{H}^{\mathrm{i}} = (0, 0, 1) \exp[-ik_0(x\cos\alpha - y\sin\alpha)]$ is incident from $y > 0$ on a homogeneous conductor, of permittivity ϵ_0, permeability μ_0 and conductivity σ, that occupies $y < 0$. Obtain the current and charge densities within the conductor, and also the surface charge density on it. Show that the results in the limit $\sigma \to \infty$ agree with those obtained by a direct calculation of reflection at a perfectly conducting plane.

5.23 Write down the field of a plane surface wave that is propagated in the x direction and attenuated in the y direction, and is linearly polarized with either the electric or the magnetic vector in the z direction. Show that such a wave in the half-space $y > 0$ can be supported by a plane $y = 0$ that has a purely reactive surface impedance $Z = \zeta Z_0$, where ζ is pure imaginary; and find the attenuation coefficient in terms of ζ.

Confirm that the analysis is equivalent to that for the reflection at $y = 0$ of a plane wave whose angle of incidence is so chosen that the reflection coefficient vanishes (cf. (5.180)).

5.24 A linearly polarized plane surface wave in the region $y > 0$ has $H_x = H_y = 0$, and is supported by the surface $y = 0$ which has impedance iZ_0/γ, where γ is real. Show that the time averaged power flux in the x direction, per unit length in the z direction, is

$$H_0^2(1+\gamma^2)^{\frac{1}{2}}/(4\omega\epsilon_0),$$

where H_0 is the amplitude of **H** at $y = 0$.

5.25 A corner reflector consists of two semi-infinite surfaces, $y = 0$, $x > 0$ and $x = 0$, $y > 0$. The surfaces have uniform impedances $\zeta_1 Z_0$ and $\zeta_2 Z_0$, respectively. Incident on the reflector from the quarter space $x > 0$, $y > 0$ is a linearly polarized plane wave, with electric vector parallel to the z axis, whose direction of propagation makes an angle α with the negative x axis. Explain how to find the complete field in $x > 0$, $y > 0$, and obtain the amplitude of the reflected wave that travels in the opposite direction to the incident wave.

5.26 A linearly polarized plane wave is incident obliquely on the infinite plane face of a perfect conductor. Calculate the radiation pressure from the Lorentz force in the way indicated in the final paragraph of §5.3.5, treating separately the cases (*a*) **E** parallel to the interface, (*b*) **H** parallel to the interface.

Show in each case that the result can be interpreted in terms of rate of change of momentum if (5.202) is adopted as field momentum density.

5.27 A spherical charge, of uniform volume charge density, is in translational motion with constant velocity **v**, where $v \ll c$. Obtain an approximation to the magnetic field by using the form of the vector potential applicable to steady current distributions (which is equivalent to (1.6)).

By adopting also the approximation that, at each instant, the electric field is the appropriate static field, calculate (from (5.202)) the total field momentum, and show that it is $4v/(3c^2)$ times the electrostatic energy of the spherical charge.

5.28 In a system of cylindrical polar coordinates r, θ, z a monochromatic line source lying along the entire z axis generates a field independent of z whose magnetic vector **H** has only a z component. For large values of r the dominant contribution to this component is

$$p \sin\theta \frac{e^{-ik_0 r}}{r^{\frac{1}{2}}},$$

where p is a complex constant. Find the time averaged power radiated per unit length of the source.

5.29 A uniform line current I, of angular frequency ω, flows along the z axis. Show that, in cylindrical polar coordinates r, θ, z, the scalar and vector potentials are $\phi = 0$, $\mathbf{A} = (0, 0, A)$, where

$$A = \frac{\mu_0 I}{2\pi}\int_0^\infty e^{-ik_0 r \cosh\psi}\, d\psi;$$

and that $\mathbf{E} = (0, 0, -i\omega A)$, $\mathbf{H} = \left(0, -\frac{1}{\mu_0}\frac{\partial A}{\partial r}, 0\right)$.

Note that, apart from the implicit time factor $\exp(i\omega t)$, A depends on ω and r only through $k_0 r$. It may therefore be expected that as $r \to 0$ A approaches $-(\mu_0 I/2\pi)\log r$, the value it would have were ω zero. On this basis deduce that, as $r \to 0$,

$$E_z \sim \frac{i\mu_0 I\omega}{2\pi}\log r, \quad H_\theta \sim \frac{I}{2\pi r}.$$

Write down, to within a constant factor, the dominant contribution to the field as $r \to \infty$.

5.30 Show that the retarded potentials of the field of a constant current I that starts to flow uniformly along the z axis at time $t = 0$ give $\phi = 0$, $\mathbf{A} = (0, 0, A)$, where

$$A = \begin{cases} \dfrac{\mu_0 I}{2\pi}\log\left[\dfrac{ct + (c^2t^2 - r^2)^{\frac{1}{2}}}{r}\right] & \text{for } r \leqslant ct, \\ 0 & \text{for } r \geqslant ct, \end{cases}$$

r being distance from the z axis.

Obtain the corresponding field components, and discuss the behaviour of the field as $t \to \infty$.

5·31 Describe the image system for time dependent charge and current densities ρ and $\mathbf{J}$ in $z > 0$ in the presence of the perfectly conducting plane $z = 0$, and write down the scalar and vector potentials of the total field.

Obtain the far field in the case when the source is a time harmonic dipole of moment $(p, 0, 0)$ located at $(0, 0, a)$.

5·32 Show that a monochromatic current I flowing in a small loop of vector area $\mathbf{A}$ represents a magnetic dipole of moment $\mu_0 I\mathbf{A}$. Write down the electric vector in the far field, and show that the radiation resistance is $Z_0(k_0{}^2A)^2/(6\pi)$.

5·33 Two identical electric dipoles have moments $p\mathbf{s}$, and the vector distance between them is $n\lambda\mathbf{s}$, where $\mathbf{s}$ is a unit vector, λ is the wavelength, and n an integer. Discuss the radiation pattern, and show that the ratio of the total power radiated to twice that radiated by a single dipole of moment p is $1-3/(4n^2\pi^2)$.

5·34 An aerial consists of n identical, collinear dipoles, with uniform spacing $2a$. The dipoles oscillate in phase and are directed along their common line. Show that the dependence of the far field amplitude on the angle θ from the line of dipoles is given by the modulus of

$$\frac{\sin\theta\,\sin\,(nk_0a\cos\theta)}{\sin\,(k_0a\cos\theta)}.$$

Describe the general features of the amplitude pattern in the case when $2a$ is much less than a wavelength but $2na$ is much greater than a wavelength.

5·35 A linear antenna lies along the portion $-\frac{1}{2}l < z < \frac{1}{2}l$ of the z axis; $k_0 l \ll 1$, and the current may be taken to be

$$I(z) = \begin{cases} I_0(1-2z/l) & \text{for} \quad 0 < z < \frac{1}{2}l, \\ I_0(1+2z/l) & \text{for} \quad -\frac{1}{2}l < z < 0, \end{cases}$$

where I_0 is a constant. Find the far field to the lowest order in $k_0 l$, and show that the corresponding expression for the radiation resistance is $Z_0(k_0 l)^2/(24\pi)$.

5·36 Let $\mathbf{F}(\mathbf{r}, \mathbf{l}, \mathbf{n})$ be the total resulting electric vector when the plane wave with electric vector $\mathbf{l}\exp(ik_0\mathbf{n}.\mathbf{r})$, where $\mathbf{l}$ and $\mathbf{n}$ are mutually orthogonal unit vectors, falls on an obstacle. Show from the reciprocity theorem that if, in place of the incident plane wave, a dipole of moment $\mathbf{p}$ is placed at $\mathbf{r}_0$ in the presence of the obstacle, the resulting electric vector $\mathbf{E}$ in the *far* field at $r\mathbf{n}$ is given by

$$\mathbf{l}.\mathbf{E} \sim \frac{k_0{}^2}{4\pi\epsilon_0}\,\mathbf{p}.\mathbf{F}(\mathbf{r}_0, \mathbf{l}, \mathbf{n})\,\frac{e^{-ik_0 r}}{r}.$$

Illustrate this result by using it to calculate the far field at $(0, 0, z)$ of a dipole of moment p placed at $(0, 0, a)$ in the presence of a perfectly conducting plane $z = 0$.

6

THE ELECTROMAGNETIC PROPERTIES OF MEDIA

6.1 Polarization and magnetization

6.1.1 *The vectors* **P** *and* **M**

The discussion so far has for the most part covered vacuum fields whose sources are either prescribed charges or currents, or else charges or currents on the surfaces of conductors. Where media have been more specifically introduced they have been treated on a simple phenomenological basis, using, in the appropriate context, the constitutive relations $\mathbf{J} = \sigma\mathbf{E}$, $\mathbf{D} = \epsilon\mathbf{E}$, $\mathbf{B} = \mu\mathbf{H}$. The main task of this chapter is to go a little way towards explaining the manner in which media give rise to electromagnetic effects.

The starting point is the recognition that a medium is an aggregate of elementary charged particles to which all electromagnetic effects are due. Most of these particles are likely to be attached to groups forming atoms or molecules; others may not be so attached, such as the free electrons in conductors and ionized gases. It is therefore commonly possible to distinguish between free charge and so-called bound charge; and the distinction is significant since it is essentially the existence of free charge which accounts for the passage of steady current. Included in free charge is the net charge, if any, of a molecule; thus a singly ionized molecule contributes the charge of one proton to the free charge.

Maxwell's equations (2.104)–(2.107) apply in the presence of media, provided ρ and $\mathbf{J}$ take account of all charge and current, including the charged particles constituting the media. But as they stand these are the (classical) *mi*croscopic equations. The actual ρ and $\mathbf{J}$ fluctuate wildly in distances of the order of the spacing between adjacent particles, and so correspondingly do $\mathbf{E}$ and $\mathbf{B}$. To go over to the *ma*croscopic equations the actual charge and current densities must be replaced by their space averages, as explained in §§ 1.2.1, 1.2.2. Let the symbols now be supposed to stand for the space averaged quantities, and furthermore use suffixes b and f to distinguish between bound and free charge. Then Maxwell's equations are

$$\operatorname{curl}\mathbf{E} = -\dot{\mathbf{B}}, \quad \operatorname{div}\mathbf{B} = 0, \tag{6.1}$$

$$\frac{1}{\mu_0}\operatorname{curl}\mathbf{B} = \epsilon_0\dot{\mathbf{E}} + \mathbf{J}_b + \mathbf{J}_f, \quad \epsilon_0 \operatorname{div}\mathbf{E} = \rho_b + \rho_f. \tag{6.2}$$

Consider an ordinary conductor. The conduction current is $\mathbf{J}_f$, and it is known from experiment that $\mathbf{J}_f = \sigma\mathbf{E}$. This can be explained by a simple theory which also provides an expression for σ in terms of the free electron model (see §6.2.2).

For dielectric and magnetic media, however, the connection between ρ_b, $\mathbf{J}_b$, and $\mathbf{E}$ and $\mathbf{B}$ is not quite so direct. The phenomenological discussion proceeded via the vectors $\mathbf{D}$ and $\mathbf{H}$, and to see the link it is helpful to compare (6.2) with the corresponding macroscopic equations previously obtained, which in effect are

$$\operatorname{curl}\mathbf{H} = \dot{\mathbf{D}} + \mathbf{J}_f, \quad \operatorname{div}\mathbf{D} = \rho_f. \tag{6.3}$$

The identification of (6.2) and (6.3) implies

$$\rho_b = -\operatorname{div}\mathbf{P}, \tag{6.4}$$

$$\mathbf{J}_b = \dot{\mathbf{P}} + \operatorname{curl}\mathbf{M}, \tag{6.5}$$

where

$$\mathbf{D} = \epsilon_0\mathbf{E} + \mathbf{P}, \tag{6.6}$$

$$\mathbf{B} = \mu_0(\mathbf{H} + \mathbf{M}). \tag{6.7}$$

In a strictly logical derivation of the macroscopic equations as a space average of the microscopic equations the relations (6.4)–(6.7) could be formally introduced in the following way. No matter what ρ_b may be there is some vector $\mathbf{P}$ that satisfies (6.4): then

$$\operatorname{div}(\mathbf{J}_b - \dot{\mathbf{P}}) = \operatorname{div}\mathbf{J}_b + \dot{\rho}_b = 0,$$

from charge conservation, so that there is also some vector $\mathbf{M}$ that satisfies (6.5). Thus (6.2) appears in the form

$$\frac{1}{\mu_0}\operatorname{curl}\mathbf{B} = \epsilon_0\dot{\mathbf{E}} + \dot{\mathbf{P}} + \operatorname{curl}\mathbf{M} + \mathbf{J}_f, \quad \epsilon_0 \operatorname{div}\mathbf{E} = -\operatorname{div}\mathbf{P} + \rho_f. \tag{6.8}$$

The vectors $\mathbf{D}$ and $\mathbf{H}$ are then *defined* by (6.6) and (6.7), and substitution into (6.8) gives (6.3).

Thus far the exercise is purely formal, since the constitutive relations have not been considered. Indeed, mathematically $\mathbf{P}$ and $\mathbf{M}$ are not completely specified, for given ρ_b, $\mathbf{J}_b$, by (6.4) and (6.5). However, consideration of their physical interpretation leads naturally to a precise specification of these vectors, and moreover to one which is directly related to those individual and collective properties of molecules that come under experimental and theoretical scrutiny. In this way macroscopic constitutive relations can be traced to their origins in the particular molecular structure of the substance concerned. A linear relation between $\mathbf{D}$ and $\mathbf{E}$ is synonymous with a linear relation between $\mathbf{P}$ and $\mathbf{E}$; and one between $\mathbf{H}$ and $\mathbf{B}$ with one between $\mathbf{M}$ and $\mathbf{B}$.

In brief, the interpretation of $\mathbf{P}$ and $\mathbf{M}$ is that they represent, respectively, electric and magnetic dipole moment densities. $\mathbf{P}$ is called the (electric) *polarization*, and **M** the *magnetization* or magnetic polarization. The remaining sections in § 6.1 are concerned with explaining the ideas behind this interpretation, and developing some of its consequences.

6.1.2 *Dipole moments*

Consider the electrostatic field of a neutral molecule treated as a highly localized stationary charge distribution of zero net charge. At all points whose distances from the distribution are large compared with its linear dimensions the field to which it gives rise is approximately the field of an infinitesimal dipole. The vector moment of this dipole is determined from the charge distribution in the following way.

In the general expression (2.10) for the potential of a charge distribution, use the expansion

$$\frac{1}{R} = \frac{1}{r}\left[1 + \frac{\mathbf{r}'.\hat{\mathbf{r}}}{r} + O\left(\frac{r'^2}{r^2}\right)\right], \tag{6.9}$$

where $\mathbf{r} = r\hat{\mathbf{r}}$ locates the field point and $\mathbf{r}'$ the volume element $\mathrm{d}\tau$ as in figure 5.6. This gives

$$\phi = \frac{1}{4\pi\epsilon_0}\left[\frac{q}{r} + \frac{\mathbf{p}.\hat{\mathbf{r}}}{r^2} + \ldots\right], \tag{6.10}$$

where

$$q = \int \rho\,\mathrm{d}\tau \tag{6.11}$$

is the net charge, and

$$\mathbf{p} = \int \mathbf{r}'\rho\,\mathrm{d}\tau \tag{6.12}$$

is called the electric dipole moment of the charge distribution. If q is zero the potential is given approximately by the second term in (6.10), provided only that r is much greater than the largest value of r'; and this term is identical in form with the potential (2.27) of a dipole of moment **p**.

The integral (6.12) is reminiscent of that encountered when locating the centre of gravity of a mass distribution: if the net charge is not zero, **p** depends on the choice of origin, and there is one for which $\mathbf{p} = 0$. But if the net charge is zero then obviously **p** has a unique value independent of the choice of origin.

The expansion (6.9) could be continued explicitly, giving higher order terms in (6.10). That following the dipole term involves the *quadrupole* moment, which is a second rank tensor having nine components. The inclusion of quadrupole and higher order multipole moments is a refinement not needed here.

Consider next the magnetostatic field of a molecule treated as a highly localized steady current distribution; a similar analysis expresses the field in terms of the magnetic dipole moment **m**. That

$$\mathbf{m} = \frac{1}{2}\int \mathbf{r}' \wedge \mathbf{J}\,\mathrm{d}\tau \tag{6.13}$$

has in fact already been established by direct appeal to Ampère's dipole law (see (2.57)). However it is instructive to derive the result by using (6.9) in the vector potential (2.63). This yields

$$\mathbf{A} = \frac{\mu_0}{4\pi}\left[\frac{1}{r}\int \mathbf{J}\,\mathrm{d}\tau + \frac{1}{r^2}\int (\mathbf{r}'.\hat{\mathbf{r}})\,\mathbf{J}\,\mathrm{d}\tau + \dots\right]. \tag{6.14}$$

Now the vector potential of a magnetic dipole is given by (2.67), and it therefore remains to be shown that

$$\int \mathbf{J}\,\mathrm{d}\tau = 0, \tag{6.15}$$

and that

$$\int (\mathbf{r}'.\hat{\mathbf{r}})\,\mathbf{J}\,\mathrm{d}\tau = \tfrac{1}{2}\hat{\mathbf{r}}\wedge\int \mathbf{J}\wedge\mathbf{r}'\,\mathrm{d}\tau. \tag{6.16}$$

To prove (6.15) it need only be noted that, since div $\mathbf{J}$ is zero and grad $x = (1, 0, 0)$, evidently

$$\mathrm{div}\,(x\mathbf{J}) = J_x. \tag{6.17}$$

Hence the volume integral of J_x (and likewise that of J_y and J_z) is seen to be zero by the familiar argument that it can be transformed to a surface integral which vanishes because $\mathbf{J}$ is presumed spatially bounded (cf. problem 1.14).

A similar technique also establishes (6.16). Note first that

$$\mathrm{div}\,(x^2\mathbf{J}) = 2xJ_x, \quad \mathrm{div}\,(yz\mathbf{J}) = yJ_z + zJ_y, \tag{6.18}$$

implying

$$\int x'J_x\,\mathrm{d}\tau = 0, \quad \int y'J_z\,\mathrm{d}\tau = -\int z'J_y\,\mathrm{d}\tau. \tag{6.19}$$

From (6.19), and similar results for other components, it is evident that

$$\int \mathbf{r}'(\mathbf{J}.\mathbf{k})\,\mathrm{d}\tau = -\int \mathbf{J}(\mathbf{r}'.\mathbf{k})\,\mathrm{d}\tau, \tag{6.20}$$

where $\mathbf{k}$ is independent of $\mathbf{r}'$ but is otherwise arbitrary. Then, since

$$\hat{\mathbf{r}}\wedge(\mathbf{J}\wedge\mathbf{r}') = \mathbf{J}(\hat{\mathbf{r}}.\mathbf{r}') - \mathbf{r}'(\hat{\mathbf{r}}.\mathbf{J}), \tag{6.21}$$

(6.16) follows at once from (6.20) by taking $\mathbf{k}$ to be $\hat{\mathbf{r}}$.

Note that, by virtue of (6.15), $\mathbf{m}$ is independent of the choice of origin.

If an aggregate of neutral molecules gives rise to measurable electrostatic or magnetostatic fields it must yield a significant space averaged charge or current density. However, the way to evaluate these average densities is by no means obvious; for if the volume integrals of the actual charge and current densities are zero for each individual molecule then the net charge and the net current within any surface enclosing a large number of molecules are themselves also zero, unless the surface happens to cut through one or more molecules. Admittedly, for the charge density for example, it might be questioned whether the difficulty could not be circumvented by averaging separately the positive (proton) charge and the negative (electron) charge, and then adding. But to the degree of approximation to which the separate problems are straightforward the enormous individual fields would mutually cancel. The field sought is the comparatively small difference field, whose calculation involves a higher order approximation, which, from this point of view, is the origin of the difficulty.

The clue to progress, that the results of averaging are likely to be expressible in terms of the dipole moments of the molecules, is to be found in the analysis just presented. The details are worked out in § 6.1.3. First, though, it is instructive to introduce the concept of idealized dipole moment *densities*, and to see how they are related to charge and current densities.

Imagine initial charge densities ρ_+ and ρ_-, where ρ_- is everywhere negative, and $\rho_+ = -\rho_-$ so that the combined density is everywhere zero. Then suppose that the negative charge suffers some general small displacement $\mathbf{l}$: that is to say, the quantity of charge that crosses any infinitesimal plane surface element $\mathrm{d}\mathbf{S}$, due to the displacement, is $\rho_-\mathbf{l}.\mathrm{d}\mathbf{S}$, where $\mathbf{l}$ may be position and time dependent. Now proceed to the limit $l \to 0, \rho_+(=-\rho_-) \to \infty$, with $\rho_+ l$ remaining finite. There is thus created an electric dipole moment density

$$\mathbf{P} = \rho_-\mathbf{l}. \tag{6.22}$$

As the notational link with (6.4) and (6.5) indicates, a dipole moment density $\mathbf{P}$ as defined is equivalent to a current density

$$\dot{\mathbf{P}}, \tag{6.23}$$

and a charge density

$$-\operatorname{div}\mathbf{P}. \tag{6.24}$$

The value of the current density is simply determined by the rate $\dot{\mathbf{P}}.\mathrm{d}\mathbf{S}$ at which charge crosses a surface element $\mathrm{d}\mathbf{S}$; and the charge density, being initially zero, then follows from the charge conservation relation.

A further point should be noted. If the charge density ρ_+ is zero outside some surface S, but not zero, except perhaps at isolated points, inside S, then the displacement of ρ_- leaves a layer of uncompensated charge on S. On each element $\mathrm{d}\mathbf{S}$ of S the layer contains charge of amount $\rho_-\mathbf{l}.\mathrm{d}\mathbf{S}$, and therefore represents, in the limit, a surface charge density

$$\mathbf{P}.\mathbf{n}, \tag{6.25}$$

being the component of $\mathbf{P}$ at the surface in the direction of the outward normal. This surface density, on a surface outside which $\mathbf{P}$ is zero, arises from the discontinuity in $\mathbf{P}$ across the surface. It is, of course, additional to the volume density (6.24), though it could be derived from the latter by envisaging a continuous passage of $\mathbf{P}$ from a finite value to zero over an arbitrarily short distance.

In the time independent case (6.23) vanishes, and there is simply an equivalent charge distribution. If, further, $\mathbf{P}$ is strictly uniform throughout some bounded region, and zero elsewhere, this equivalent distribution is represented solely by the surface density (6.25) on the boundary of the region.

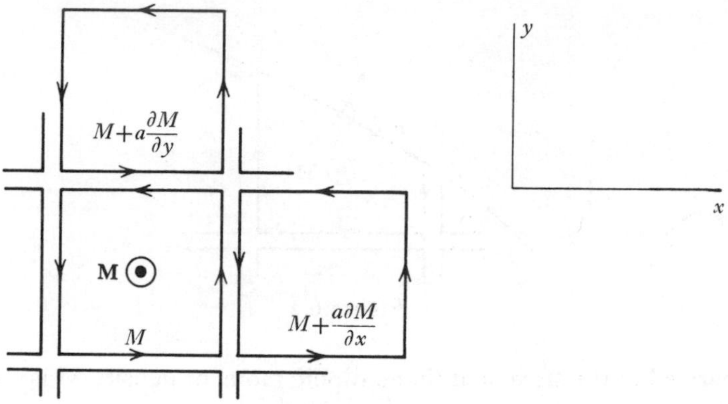

Figure 6.1

An idealized magnetic dipole moment density can be constructed as follows.

Consider two mutually orthogonal sets of parallel planes which form a matrix of narrow cylinders, each with square cross-section of side a. Let there flow round the surface of each cylinder, transversely to its length, a current of surface density M, which is allowed to vary both along each cylinder and from cylinder to cylinder, and may also be time dependent. Then in the limit $a \to 0$ the construct represents a magnetic dipole moment density $\mathbf{M}$ which is parallel to the cylinders and has magnitude M.

The surface currents of two adjacent cylinders flow in opposition in the interface common to the cylinders, and hence leave a residual current in that interface only if the value of M changes from one cylinder to the other. For example, if the cylinders lie in the z direction the net surface current density in the common interface parallel to the yz plane is

$$\left(0, -a\frac{\partial \mathbf{M}}{\partial x}, 0\right), \tag{6.26}$$

since M can be treated as a continuous variable as the limit $a \to 0$ is approached (see figure 6.1).

Now let the current be envisaged not as a flow encircling each cylinder but rather as a net flow throughout each plane of the orthogonal sets of planes forming the matrix. Viewed in this way the current is seen to have volume density

$$\left(\frac{\partial M}{\partial y}, \quad -\frac{\partial M}{\partial x}, 0\right). \tag{6.27}$$

The special form of (6.27) applies to the case $\mathbf{M} = (0, 0, M)$, but can be generalized at once by representing arbitrary $\mathbf{M}$ as the superposition of vectors parallel to the coordinate axes. The general result can evidently be

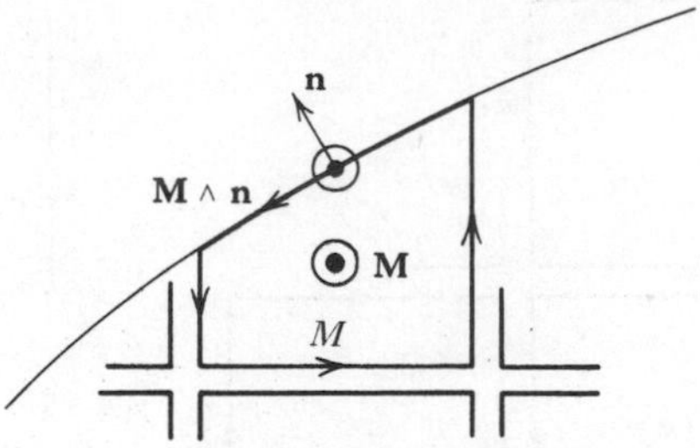

Figure 6.2

summarized in the statement that a dipole moment density **M** is equivalent to a current density

$$\operatorname{curl} \mathbf{M}. \tag{6.28}$$

If the dipole moment distribution terminates at some closed surface S, being zero everywhere outside S, the current matrix representation must include an uncompensated flow of current in S of surface density M. As indicated in figure 6.2, this can be expressed vectorially as a surface current density

$$\mathbf{M} \wedge \mathbf{n}, \tag{6.29}$$

where $\boldsymbol{n}$ is the unit vector in the direction of the outward normal to S. The surface density (6.29) is additional to the volume density (6.28), though it could be derived from the latter by envisaging a continuous passage of **M** to zero over an arbitrarily short distance.

If **M** is strictly uniform throughout some bounded region, and zero elsewhere, the equivalent current distribution is represented solely by the surface current density (6.29) on the boundary of the region.

To summarize: it has been shown first that every space averaged charge and current density associated with the bound charge in a medium can be written in the form $-\operatorname{div}\mathbf{P}$ and $\dot{\mathbf{P}} + \operatorname{curl}\mathbf{M}$, respectively; and secondly, that if a medium were representable by electric and magnetic dipole moment densities **P** and **M**, then it would contribute charge and current densities of the stated form. Now (conductivity apart) the molecular structure of media suggests that on a macroscopic scale they should be approximately representable in this way, with **P** and **M** being the sum, per unit volume over a macroscopic volume element, of the individual molecular dipole moments **p** and **m** respectively. That this is indeed the case is shown in the next section.

6.1.3 *Space averaging*

For the purposes of this section it is convenient to reserve ρ and **J** for the actual microscopic charge and current densities, with $\bar{\rho}$ and $\bar{\mathbf{J}}$ for the corresponding macroscopic quantities obtained by suitable space averaging.

The space average of, say, ρ at a point X is broadly conceived to mean the net charge per unit volume in a macroscopic volume element containing X. However, as has already been emphasized, the concept as it stands is imprecise in the case where the charge is grouped into neutral molecules, because whether or not there is any net charge in a volume element depends critically on the detailed shapc of the boundary of the element.

A way of overcoming the difficulty is to define averages by means of a suitably specified 'weighting function' W, so that, for example,

$$\bar{\rho}(\mathbf{r}, t) = \int W(\mathbf{R})\,\rho(\mathbf{r}', t)\,d\tau, \tag{6.30}$$

where as usual $\mathbf{r}'$ is the position vector of the volume element $d\tau$, $\mathbf{R} = \mathbf{r} - \mathbf{r}'$, and the integration is over all space.

The function W need not be specified explicitly. It must of course satisfy

$$\int W(\mathbf{R})\,d\tau = 1. \tag{6.31}$$

Also, for some length l which is macroscopically small but much greater than the molecular spacing, $W(\mathbf{R})$ must be sufficiently close to zero when $R > l$ for the region so defined to contribute negligibly to the integral (6.30); whereas for $R < l$ W must be of the order of $(4\pi l^3/3)^{-1}$.

These requirements on W would, of course, be met by taking it to be

$$W_0(\mathbf{R}) = \begin{cases} (4\pi l^3/3)^{-1} & \text{for} \quad R < l, \\ 0 & \text{for} \quad R > l. \end{cases} \tag{6.32}$$

For this choice of weighting function the average charge density at X becomes the net charge in the sphere of radius l with centre at X divided by the volume of the sphere; and, in essence, this is just the definition of averaging whose pitfalls have been stressed. They can be avoided by imposing a further restriction on W, namely that its scale of variation, for all values of its argument, be much greater than the scale of a molecule.

Stated more precisely, it is required that

$$W(\mathbf{r} - \mathbf{r}') \simeq W(\mathbf{r}) - \mathbf{r}' \cdot \operatorname{grad} W(\mathbf{r}) \tag{6.33}$$

afford a good approximation for all values of r' less than or comparable to the linear dimensions of a molecule. The ability to replace W in this way by the early terms in its Taylor expansion is the key to simplifying the analysis; it would not be possible with $W = W_0$, because of the jump in W_0 across $R = l$.

With the adoption of any weighting function that is smooth in the sense described the microscopic charge solely belonging to a single electron or molecule yields a correspondingly smooth average charge density. For example, a true 'point' electron at the origin, of charge $-e$, gives

$$\bar{\rho}(\mathbf{r}) = -eW(\mathbf{r}). \tag{6.34}$$

For a single molecule at the origin $\rho(\mathbf{r}')$ is zero when r' exceeds the maximum molecular dimension, and $W(\mathbf{R})$ in (6.30) can therefore be approximated by (6.33). This gives the approximation

$$\bar{\rho}(\mathbf{r}) = qW(\mathbf{r}) - \mathbf{p}.\text{grad}\, W(\mathbf{r}), \tag{6.35}$$

where q is the net charge and $\mathbf{p}$ the dipole moment of the molecule (cf. (6.11), (6.12)).

For an electron at $\mathbf{r}_e$ or a molecule at $\mathbf{r}_m$, rather than at the origin, it is only necessary to replace the argument of W in (6.34) by $\mathbf{r}-\mathbf{r}_e$ and in (6.35) by $\mathbf{r}-\mathbf{r}_m$. Since $\mathbf{p}.\text{grad}\, W = \text{div}\,(\mathbf{p}W)$ a summation over all free electrons and all molecules therefore gives

$$\bar{\rho} = \rho_f - \text{div}\,\mathbf{P}, \tag{6.36}$$

where

$$\rho_f = -e\Sigma W(\mathbf{r}-\mathbf{r}_e) + \Sigma q W(\mathbf{r}-\mathbf{r}_m) \tag{6.37}$$

$$\mathbf{P} = \Sigma \mathbf{p} W(\mathbf{r}-\mathbf{r}_m). \tag{6.38}$$

The corresponding expression for the average current density can be found in a similar way, although the analysis is rather more elaborate.

For a single point electron at $\mathbf{r}_e$, with velocity $d\mathbf{r}_e/dt = \mathbf{v}_e$, the average current density is

$$\bar{\mathbf{J}}(\mathbf{r}) = -e\mathbf{v}_e W(\mathbf{r}-\mathbf{r}_e). \tag{6.39}$$

For a single molecule at $\mathbf{r}_m$, with velocity $d\mathbf{r}_m/dt = \mathbf{v}_m$,

$$\bar{\mathbf{J}}(\mathbf{r}) = \int (W - \mathbf{r}'.\text{grad}\, W)\rho(\mathbf{r}')\,[\mathbf{v}_m + \mathbf{v}(\mathbf{r}')]\, d\tau \tag{6.40}$$

approximately, where the argument $\mathbf{r}-\mathbf{r}_m$ of W is left understood, $\mathbf{r}'$ now denotes the radius vector from $\mathbf{r}_m$ to $d\tau$, and $\mathbf{v}$ is the velocity of the charge relative to the molecule. Evidently

$$\bar{\mathbf{J}} = \mathbf{J}_1 + \mathbf{J}_2 + \mathbf{J}_3, \tag{6.41}$$

where

$$\mathbf{J}_1 = qW\mathbf{v}_m \tag{6.42}$$

$$\mathbf{J}_2 = -(\mathbf{p}.\text{grad}\, W)\mathbf{v}_m + W\int \rho\mathbf{v}\, d\tau, \tag{6.43}$$

$$\mathbf{J}_3 = -\int \rho\mathbf{v}(\mathbf{r}'.\text{grad}\, W)\, d\tau, \tag{6.44}$$

the argument $\mathbf{r}'$ of ρ and $\mathbf{v}$ in the integrands also being left understood from now on.

On summing over all free electrons and all molecules, $\mathbf{J}_1$, together with (6.39), contributes to the average current density the amount

$$\mathbf{J}_f = -e\Sigma\mathbf{v}_e W(\mathbf{r}-\mathbf{r}_e) + \Sigma q\mathbf{v}_m W(\mathbf{r}-\mathbf{r}_m). \tag{6.45}$$

Thus $\mathbf{J}_2$ and $\mathbf{J}_3$ might be expected to contribute $d(\mathbf{p}W)/dt$ and $\text{curl}\,(\mathbf{m}W)$, where $\mathbf{p}$ is given by (6.12), and $\mathbf{m}$ by (6.13) with $\mathbf{J} = \rho\mathbf{v}$.

To see whether this is so, note that

$$\begin{aligned}\operatorname{curl}(\mathbf{m}W) &= -\mathbf{m}\wedge\operatorname{grad}W\\ &= -\frac{1}{2}\int\rho(\mathbf{r}'\wedge\mathbf{v})\wedge\operatorname{grad}W\,d\tau\\ &= \frac{1}{2}\int\rho[\mathbf{r}'(\mathbf{v}.\operatorname{grad}W)-\mathbf{v}(\mathbf{r}'.\operatorname{grad}W)]\,d\tau,\end{aligned}\tag{6.46}$$

which may be compared with (6.44). Now the integrals in (6.46) are of precisely the form of those in (6.20), with $\mathbf{J} = \rho\mathbf{v}$ and $\mathbf{k} = \operatorname{grad} W$. Admittedly the equality corresponding to (6.20) is not general, for it holds only under the *steady* current condition $\operatorname{div}\mathbf{J} = 0$. But by including the additional terms $x^2 \operatorname{div}\mathbf{J}$ and $yz \operatorname{div}\mathbf{J}$ in (6.18) it appears that (6.20) is correct to the present approximation, since the neglected terms are integrals whose integrands are quadratic in x', y', z' and such integrals have already been implicitly ignored by using the approximation (6.33) in which the Taylor series is truncated after the linear term. The individual integrals in (6.46) are therefore in effect equal, and so

$$\mathbf{J}_3 = \operatorname{curl}(\mathbf{m}W).\tag{6.47}$$

Finally, since the argument of W is $\mathbf{r}-\mathbf{r}_m$ and $d\mathbf{r}_m/dt = \mathbf{v}_m$,

$$\frac{d}{dt}(\mathbf{p}W) = \frac{d\mathbf{p}}{dt}W - \mathbf{p}(\mathbf{v}_m.\operatorname{grad}W).\tag{6.48}$$

Moreover

$$\begin{aligned}\frac{d\mathbf{p}}{dt} &= \int\mathbf{r}'\dot{\rho}\,d\tau\\ &= -\int\mathbf{r}'\operatorname{div}'(\rho\mathbf{v})\,d\tau\\ &= \int\rho\mathbf{v}\,d\tau,\end{aligned}\tag{6.49}$$

where the last line follows by writing $x'\operatorname{div}'(\rho\mathbf{v}) = \operatorname{div}'(x'\rho\mathbf{v}) - \rho v_x$, with two similars, and discarding the integrals of the divergences. Thus, from (6.43), (6.48) and (6.49)

$$\mathbf{J}_2 = -\mathbf{v}_m(\mathbf{p}.\operatorname{grad}W) + \mathbf{p}(\mathbf{v}_m.\operatorname{grad}W) + \frac{d}{dt}(\mathbf{p}W);$$

that is,

$$\mathbf{J}_2 = \frac{d}{dt}(\mathbf{p}W) + \operatorname{curl}(\mathbf{p}\wedge\mathbf{v}_m W).\tag{6.50}$$

Summation over all free electrons and all molecules therefore yields the average current density in the form

$$\bar{\mathbf{J}}(\mathbf{r}) = \mathbf{J}_f + \dot{\mathbf{P}} + \operatorname{curl}\mathbf{M},\tag{6.51}$$

where $\mathbf{J}_f$ is given by (6.45), $\mathbf{P}$ by (6.38), and

$$\mathbf{M} = \Sigma(\mathbf{m} + \mathbf{p}\wedge\mathbf{v}_m)\,W(\mathbf{r}-\mathbf{r}_m).\tag{6.52}$$

The approximations (6.38) and (6.52) to **P** and **M** are as anticipated, except for the second term in (6.52). On reflection though it is not surprising that the current of electric dipoles in motion appears as a magnetic dipole contribution.

Further stages of approximation can be obtained by taking more terms of the Taylor expansion in (6.33). The quadratic term corresponds to the effect of electric quadrupole moments, and the refinement is rarely likely to be significant in practice.

The main point is that in most circumstances **P** and **M** are, in principle at least, determinable from a knowledge of the molecular dipole moments **p** and **m** and their distribution. The behaviour of the molecules is in turn conditioned by the electromagnetic field, so that the medium is characterized by relations between **P** and **M** and the field vectors, which, by virtue of (6.6) and (6.7), can be stated as relations between **E**, **B**, **H** and **D**. Such matters are pursued later. The immediate concern, however, is to supplement the earlier treatment of electrostatics and magnetostatics by discussing some formal theoretical aspects in which the vectors **P** and **M** play a central role.

6.1.4 *Polarization in electrostatics*

The governing equations of the macroscopic electrostatic field due solely to dielectric polarization are

$$\operatorname{curl}\mathbf{E} = 0, \quad \epsilon_0 \operatorname{div}\mathbf{E} = -\operatorname{div}\mathbf{P}. \tag{6.53}$$

It has been shown that a polarized dielectric is represented as the seat of volume charge density $-\operatorname{div}\mathbf{P}$ and surface charge density $\mathbf{P}.\mathbf{n}$; or, equivalently, of dipole moment density **P**. The potential may therefore be written either

$$\phi = -\frac{1}{4\pi\epsilon_0}\int\frac{\operatorname{div}'\mathbf{P}}{R}\,d\tau + \frac{1}{4\pi\epsilon_0}\int\frac{\mathbf{P}.d\mathbf{S}}{R}; \tag{6.54}$$

or, from (2.27),

$$\phi = \frac{1}{4\pi\epsilon_0}\int\frac{\mathbf{P}.\mathbf{R}}{R^3}\,d\tau. \tag{6.55}$$

The volume integrals are taken over the region occupied by the dielectric and the surface integral over the surface of the dielectric.

The equality of (6.54) and (6.55) can be confirmed by writing the integrand in the latter as

$$\mathbf{P}.\operatorname{grad}'(1/R) = \operatorname{div}'(\mathbf{P}/R) - (\operatorname{div}'\mathbf{P})/R,$$

of which the integral of the first term on the right hand side transforms to the surface integral in (6.54).

The associated electric field is given by $\mathbf{E} = -\operatorname{grad}\phi$. It can also be written as a direct expression of Coulomb's law

$$\mathbf{E} = -\frac{1}{4\pi\epsilon_0}\int\frac{\mathbf{R}}{R^3}\operatorname{div}'\mathbf{P}\,d\tau + \frac{1}{4\pi\epsilon_0}\int\frac{\mathbf{R}}{R^3}\mathbf{P}.d\mathbf{S}, \tag{6.56}$$

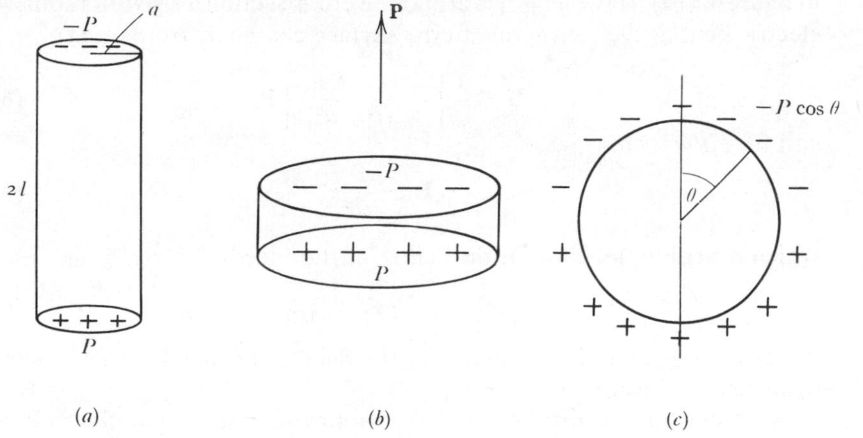

Figure 6.3

which corresponds to taking the gradient operation under the integral signs in (6.54). Note, however, that using the dipole field (2.31) and integrating over the dielectric to get

$$\mathbf{E} = \frac{1}{4\pi\epsilon_0}\int \frac{1}{R^3}\left[3\frac{\mathbf{P}.\mathbf{R}}{R}\mathbf{R}-\mathbf{P}\right] d\tau, \tag{6.57}$$

which corresponds to taking the gradient operation under the integral sign in (6.55), is only valid for field points exterior to the dielectric: at interior points the integral in (6.57) diverges.

That (6.57) is meaningless at interior points is not surprising in view of the limit process involved in the concept of dipole moment density; but the matter is worth a closer examination with reference to the physics.

Suppose that a small volume element of the dielectric is excised. Then the right hand side of either (6.56) or (6.57) gives the vacuum electric field at any point inside the cavity thus created, the volume integrals being taken over the remainder of the dielectric and the surface integral over its boundary including the cavity boundary. However, the value of the field at a fixed interior point depends on the shape of the cavity, even if the maximum chord of the cavity is assumed to tend to zero and the polarization **P** in the remainder of the dielectric is assumed to be unaffected by the excision of the cavity. Three special cases of particular interest exemplify this statement.

Note first that the sole effect of the excision of an arbitrarily small cavity is essentially to introduce surface charge density $-\mathbf{P}.\mathbf{n}$ on the cavity boundary, **n** being the unit vector along the outward normal to the boundary (and hence pointing into the dielectric). The contribution to the field of the volume density $-\operatorname{div}\mathbf{P}$ of the excised matter is negligible.

Now consider a 'needle' cavity that is a narrow cylinder parallel to **P**, as in figure 6.3 (*a*). If the length is $2l$ and the cross-section a circle of radius a the electric field at the centre due to the surface charge is, from (3.22),

$$\mathbf{E} = \frac{1}{\epsilon_0}\left[1 - \frac{l}{(l^2+a^2)^{\frac{1}{2}}}\right]\mathbf{P}, \tag{6.58}$$

and for $a/l \ll 1$ this gives

$$\epsilon_0 \mathbf{E} \simeq \frac{a^2}{2l^2}\mathbf{P} \tag{6.59}$$

which is negligible. Thus at the centre of the 'needle' cavity

$$\mathbf{E}_{\text{cavity}} = \mathbf{E}_{\text{original}}, \tag{6.60}$$

where the right hand side stands for the field that existed at the same point prior to the excision of the cavity.

Next, consider a 'disc' cavity that is a broad cylinder parallel to **P**, as in figure 6.3 (*b*). The corresponding result is evidently

$$\epsilon_0 \mathbf{E}_{\text{cavity}} = \epsilon_0 \mathbf{E}_{\text{original}} + \mathbf{P}. \tag{6.61}$$

Finally, consider a spherical cavity of radius a, as in figure 6.3 (*c*). In this case the surface charge density is $-P\cos\theta$. The corresponding potential is

$$\phi = \begin{cases} -\dfrac{P}{3\epsilon_0}\dfrac{a^3}{r^2}\cos\theta, & r > a, \\[2ex] -\dfrac{P}{3\epsilon_0}r\cos\theta, & r < a, \end{cases} \tag{6.62}$$

since this solution of Laplace's equation is continuous across $r = a$ and has jump $P\cos\theta/\epsilon_0$ in its normal derivative. Thus at all points inside the cavity

$$\epsilon_0 \mathbf{E}_{\text{cavity}} = \epsilon_0 \mathbf{E}_{\text{original}} + \tfrac{1}{3}\mathbf{P}. \tag{6.63}$$

The excision of a small cavity can be adopted as the basis of an 'operational' definition of the field inside a dielectric by measuring the force on a test charge placed inside the cavity. The physical excision of a cavity does in general affect the polarization **P** in its vicinity, but in the common case where **P** is proportional to **E** the results (6.60) and (6.61) for 'needle' and 'disc' cavities remain valid. For the 'needle' cavity the force is a measure of **E** in the dielectric, whereas for the 'disc' cavity it is a measure of $\mathbf{D}/\epsilon_0$ in the dielectric, as (6.6) indicates.

When **P** is proportional to **E** the boundary value problem posed by the physical excision of a cavity can be treated exactly, in a straightforward manner, when the cavity is ellipsoidal. In the limits of an infinitely thin prolate spheroid and an infinitely flat oblate spheroid, with the axis of symmetry along **E** in each case, the results conform to (6.60) and (6.61) respectively. This is not surprising because the continuity of tangential **E** and normal **D** is not significantly violated by the assumption that **P** is unaffected by the excision of the cavity. For a spherical cavity, on the other hand, there is a relation analogous to (6.63) but differing from it.

The spherical cavity boundary value problem with **P** proportional to **E** has in effect already been worked out, in the case of an infinite homogeneous dielectric, when the electric field in the absence of the cavity is uniform. For the potential is obtained from the potential (3.136), which pertains to a homogeneous dielectric sphere placed in a uniform field $\mathbf{E}_0$, merely on substituting $1/\kappa$ for κ; after the substitution κ represents the dielectric constant of the dielectric from which the cavity is excised. Thus

$$\phi = \begin{cases} -E_0 r\cos\theta - \dfrac{\kappa-1}{2\kappa+1}a^3E_0\dfrac{\cos\theta}{r^2}, & r > a, \\ -\dfrac{3\kappa}{2\kappa+1}E_0 r\cos\theta, & r < a. \end{cases} \tag{6.64}$$

Hence the electric field everywhere inside the cavity is

$$\mathbf{E}_{\text{cavity}} = \frac{3\kappa}{2\kappa+1}\mathbf{E}_0 = \mathbf{E}_0 + \frac{\kappa-1}{2\kappa+1}\mathbf{E}_0. \tag{6.65}$$

Now
$$\mathbf{P} = \mathbf{D} - \epsilon_0\mathbf{E} = \epsilon_0(\kappa-1)\mathbf{E}, \tag{6.66}$$

so that for comparison with (6.63) the relation (6.65) can be put in the form

$$\epsilon_0\mathbf{E}_{\text{cavity}} = \epsilon_0\mathbf{E}_0 + \frac{1}{2\kappa+1}\mathbf{P}, \tag{6.67}$$

which is close to the same result only if $\kappa \simeq 1$.

In any situation where the polarization of a dielectric is uniform the associated charge density is solely the surface density $\mathbf{P}.\mathbf{n}$. For example, when a homogeneous dielectric sphere is placed in a uniform field the potential is (3.136), and so **P** is uniform, being

$$\mathbf{P} = \epsilon_0(\kappa-1)\mathbf{E} = \epsilon_0\frac{3(\kappa-1)}{\kappa+2}\mathbf{E}_0. \tag{6.68}$$

It is then easily verified that the contribution to the potential due to the polarization of the dielectric (that is, (3.136) less $-E_0 r\cos\theta$) is indeed the negative of (6.62).

It is also worth remarking that a simple direct derivation of the field of charge density $P\cos\theta$ on the surface of a sphere is obtained from the observation that the charge density, being equivalent to dipole moment density **P**, can be conceived as arising, in the limit $l \to 0$, from two spherical charge distributions of volume density $\pm P/l$, respectively, whose centres have separation $\mathbf{l} = l\hat{\mathbf{l}}$. For $r > a$ the field is palpably that of a dipole of moment

$$\tfrac{4}{3}\pi a^3 P\hat{\mathbf{l}}, \tag{6.69}$$

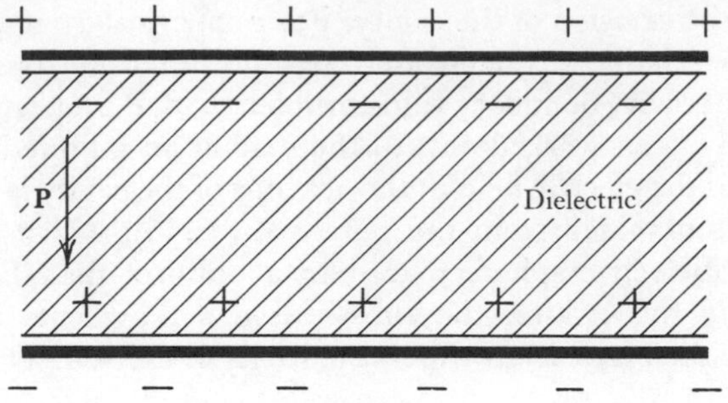

Figure 6.4

whereas for $r < a$ it is seen, by the application of (2.19) to each sphere, to be

$$\mathbf{E} = -\frac{P}{3\epsilon_0}\hat{\mathbf{1}}. \tag{6.70}$$

These results are identical with the field whose potential is the negative of (6.62).

A particularly simple case where **P** is uniform (apart from edge effects) is in a parallel plate capacitor with a homogeneous dielectric filling. The field due to the free charge on the metal plates of the capacitor polarizes the dielectric uniformly. The effect of the dielectric is therefore solely that of uniform surface charge on the faces of the dielectric. This charge partially neutralizes that on the metal plates and therefore depresses the electric field (see figure 6.4).

6.1.5 *Magnetization in magnetostatics*

The governing equations of the macroscopic magnetostatic field due solely to magnetized matter are

$$\operatorname{div}\mathbf{B} = 0, \quad \operatorname{curl}\mathbf{B} = \mu_0 \operatorname{curl}\mathbf{M}. \tag{6.71}$$

A magnetized body is represented as the seat of volume current density $\operatorname{curl}\mathbf{M}$ and surface current density $\mathbf{M}\wedge\mathbf{n}$; or, equivalently, of dipole moment density **M**. The vector potential may therefore be written either

$$\mathbf{A} = \frac{\mu_0}{4\pi}\int\frac{\operatorname{curl}'\mathbf{M}}{R}\,d\tau + \frac{\mu_0}{4\pi}\int\frac{\mathbf{M}\wedge d\mathbf{S}}{R}; \tag{6.72}$$

or, from (2.67),

$$\mathbf{A} = \frac{\mu_0}{4\pi}\int\frac{\mathbf{M}\wedge\mathbf{R}}{R^3}\,d\tau. \tag{6.73}$$

The equality of (6.72) and (6.73) can be confirmed by writing the integrand in the latter as

$$\mathbf{M}\wedge\operatorname{grad}'(1/R) = -\operatorname{curl}'(\mathbf{M}/R) + (\operatorname{curl}'\mathbf{M})/R,$$

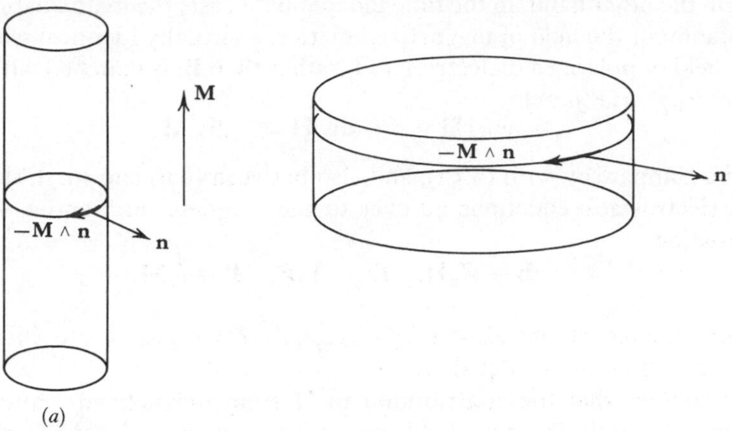

Figure 6.5

of which the integral of the first term on the right hand side transforms to the surface integral in (6.72).

The associated magnetic field is given by $\mathbf{B} = \operatorname{curl} \mathbf{A}$. The curl operation can be taken under the integral signs in (6.72) for all locations of the field point, but in (6.73) only for points exterior to the magnetized body.

An 'operational' definition of the field inside magnetized matter can be given in terms of cavity fields. Suppose a small circular cylindrical cavity is excised parallel to $\mathbf{M}$, and that $\mathbf{M}$ in the remainder of the material is unaffected. Then the sole effect of the excision is to introduce a surface current density $-\mathbf{M} \wedge \mathbf{n}$ on the cavity boundary, $\mathbf{n}$ being the unit vector along the outward normal to the boundary. The cavity therefore acts as a solenoid, and it is clear that, for a 'needle' cavity (figure 6.5(*a*))

$$\mathbf{B}_{\text{cavity}} = \mathbf{B}_{\text{original}} - \mu_0 \mathbf{M}, \tag{6.74}$$

whereas for a 'disc' cavity

$$\mathbf{B}_{\text{cavity}} = \mathbf{B}_{\text{original}}. \tag{6.75}$$

For an ideal permanent magnet a comparison of (6.7) with (4.170) shows that $\mu_0 \mathbf{M}$ is the local remanence $\mathbf{B}_\mathrm{R}$, so that $\mathbf{M}$ is predetermined and the cavity analysis just given is directly applicable. Thus in the 'disc' cavity the force on a moving charge (or the couple on a current loop) is a measure of $\mathbf{B}$ in the magnetized matter; whereas for the 'needle' cavity it is a measure of $\mu_0 \mathbf{H}$, as (6.7) indicates. These results are also applicable to a substance for which $\mathbf{M}$ is proportional to $\mathbf{B}$, for reasons similar to those set out in the two paragraphs following (6.63).

A comparison of what has so far been said in this section with the corresponding part of §6.1.4 shows a broad analogy between the two, with the difference that $\mathbf{E}$ is a conservative field ($\operatorname{curl} \mathbf{E} = 0$) whose source is charge, whereas $\mathbf{B}$ is a solenoidal field ($\operatorname{div} \mathbf{B} = 0$) whose source is current.

On the other hand, in the time independent case, the mathematical representation of the field of magnetized matter is virtually identical with that of the field of polarized dielectrics if **H**, rather than **B**, is matched with **E**. For, from (6.7) and (6.71),

$$\operatorname{curl}\mathbf{H} = 0, \quad \operatorname{div}\mathbf{H} = -\operatorname{div}\mathbf{M}; \tag{6.76}$$

and a comparison with (6.53), and also between (6.6) and (6.7), shows that the electrostatic equations go over to the magnetostatic under the transformation

$$\mathbf{E} \to Z_0\mathbf{H}, \quad \mathbf{D} \to Y_0\mathbf{B}, \quad \mathbf{P} \to \frac{1}{c}\mathbf{M}, \tag{6.77}$$

where the constants $Z_0 = 1/Y_0 = (\mu_0/\epsilon_0)^{\frac{1}{2}}$, $c = (\epsilon_0\mu_0)^{-\frac{1}{2}}$ are included to preserve dimensional equality.

It follows that the contribution to **H** from magnetized matter can be treated theoretically as the field arising from a volume density $-\operatorname{div}\mathbf{M}$ and a surface density $\mathbf{M}.\mathbf{n}$ of hypothetical 'magnetic charge'. There is thus a scalar potential representation

$$\mathbf{H} = -\operatorname{grad}\phi, \tag{6.78}$$

with
$$\phi = -\frac{1}{4\pi}\int\frac{\operatorname{div}'\mathbf{M}}{R}\,d\tau + \frac{1}{4\pi}\int\frac{\mathbf{M}.d\mathbf{S}}{R}. \tag{6.79}$$

Equivalently
$$\phi = \frac{1}{4\pi}\int\frac{\mathbf{M}.\mathbf{R}}{R^3}\,d\tau, \tag{6.80}$$

which form leads back to the interpretation of **M** as magnetic dipole moment density, and should be compared with (6.73). For field points external to the magnetized matter the curl of (6.73) is $-\mu_0$ times the gradient of (6.80), but differs from it by **M** at interior points; the explanation being that the equivalence to a dipole of current flowing steadily within an arbitrarily small region fails at points interior to the region.

It can be seen directly that

$$\operatorname{curl}\mathbf{A} + \mu_0\operatorname{grad}\phi = \mu_0\mathbf{M}; \tag{6.81}$$

for obvious alternative forms to (6.73) and (6.80) are

$$\mathbf{A} = \frac{\mu_0}{4\pi}\operatorname{curl}\int\frac{\mathbf{M}}{R}\,d\tau, \tag{6.82}$$

$$\phi = -\frac{1}{4\pi}\operatorname{div}\int\frac{\mathbf{M}}{R}\,d\tau, \tag{6.83}$$

so that, from (A. 5), the left hand side of (6.81) is

$$-\frac{\mu_0}{4\pi}\nabla^2\int\frac{\mathbf{M}}{R}\,d\tau, \tag{6.84}$$

which is recognized to be $\mu_0\mathbf{M}$ from the known solution to Poisson's equation.

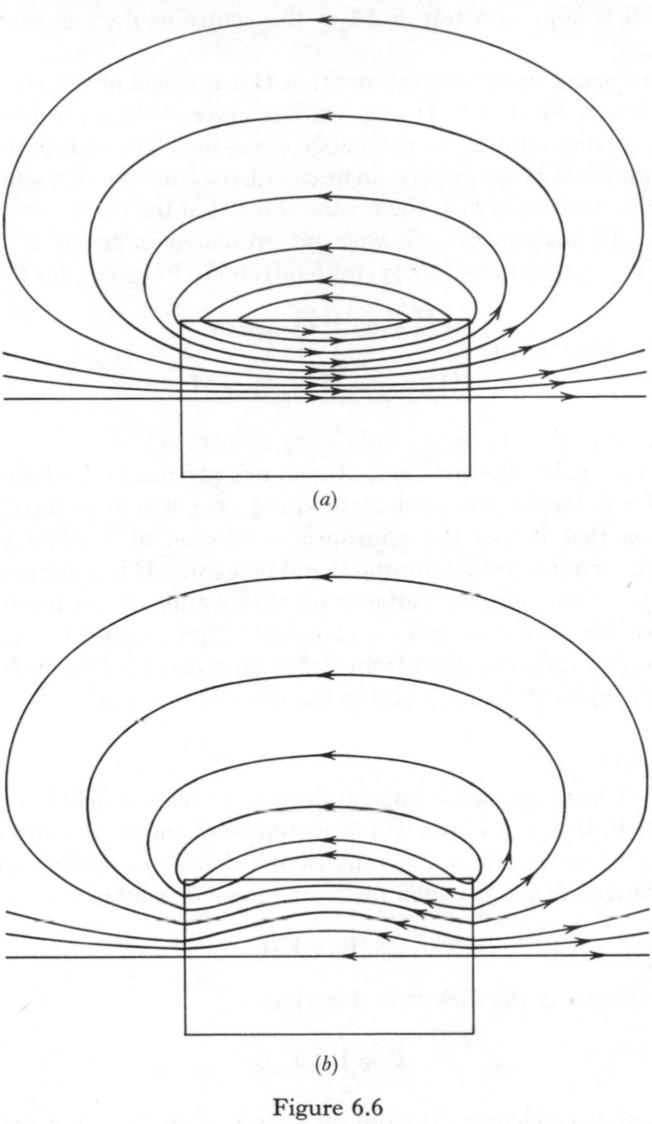

(a)

(b)

Figure 6.6

As an explicit illustration of the alternative ways of treating problems involving magnetized matter consider the uniform cylindrical permanent magnet discussed in the final paragraph of §4.5.3. Through (4.171), with $\mathbf{B}_R = \mu_0 \mathbf{M}$, the earlier discussion alluded to the fact that $\mathbf{B}$ was the field of a surface current density $\mathbf{M} \wedge \mathbf{n}$. The $\mathbf{B}$ lines are therefore as shown in figure 6.6(*a*) (cf. figure 4.10), there being a discontinuity in tangential $\mathbf{B}$ across the curved surface of the magnet. The magnitude of $\mathbf{B}$ on the axis of the magnet can be calculated from (4.87). In particular, for a narrow cylinder, **B** is approximately $\mu_0 \mathbf{M}$ at the centre of the axis and $\frac{1}{2}\mu_0 \mathbf{M}$ at the ends.

The alternative description shows that **H** is the field of surface 'magnetic charge' density $\mathbf{M}.\mathbf{n}$. The **H** lines are therefore as shown in figure 6.6(*b*), there being a discontinuity in normal **H** across the plane ends of the magnet. The magnitude of **H** on the axis can be calculated from the analogue of (3.22), being approximately zero at the centre and $\frac{1}{2}\mathbf{M}$ at the ends.

The cavity analysis can likewise be presented in terms of 'magnetic charge'. The application of (6.77) to (6.60) and (6.61) gives, for the 'needle' cavity,

$$\mathbf{H}_{\text{cavity}} = \mathbf{H}_{\text{original}}, \tag{6.85}$$

and for the 'disc' cavity

$$\mathbf{H}_{\text{cavity}} = \mathbf{H}_{\text{original}} + \mathbf{M}, \tag{6.86}$$

which are equivalent to (6.74) and (6.75), respectively.

Consider, finally, the problem of a uniformly magnetized sphere. The solution for the scalar potential, stated in (4.172), was there justified by the observation that it was the appropriate solution of Laplace's equation meeting the conditions that normal **B** and tangential **H** be continuous across the surface of the sphere. Variants on this argument are implicit in the discussion of the uniformly polarized dielectric sphere in § 6.1.4. Thus (apart from sign) (4.172) is obtained from (6.62) on replacing P/ϵ_0 by B_R/μ_0; and likewise (6.69) yields (4.174), and (6.70) gives **H** in $r < a$.

6.1.6 *Forces*

A lump of dielectric placed in a non-uniform electrostatic field $\mathbf{E}_0$ experiences a net force **F**. If $\mathbf{E}_0$ is an invariable vacuum field, and $\mathbf{E}_\mathrm{b}$ denotes the actual additional field due to the actual microscopic density ρ_b of the bound charge in the polarized dielectric, then the actual force density is

$$\rho_\mathrm{b}(\mathbf{E}_0 + \mathbf{E}_\mathrm{b}). \tag{6.87}$$

By integrating over the dielectric this gives

$$\mathbf{F} = \int \rho_\mathrm{b} \mathbf{E}_0 \, d\tau \tag{6.88}$$

since a stationary charge distribution does not produce any net force on itself.

Now in (6.88) ρ_b may be replaced by its space average, as defined in § 6.1.3, on the assumption that the scale of variation of $\mathbf{E}_0$ greatly exceeds the range over which the weighting function W is appreciable. This assumption is unlikely to be false for the given field in which the dielectric is placed, though it is not, of course, true of $\mathbf{E}_\mathrm{b}$. Thus

$$\mathbf{F} = -\int \mathbf{E}_0 \, \mathrm{div}' \, \mathbf{P} \, d\tau + \int \mathbf{E}_0 (\mathbf{P} . d\mathbf{S}); \tag{6.89}$$

or equivalently, as (3.43) indicates,

$$\mathbf{F} = \int (\mathbf{P}.\mathrm{grad}')\mathbf{E}_0 \mathrm{d}\tau. \tag{6.90}$$

The form (6.90) puts in evidence the fact that the net force is zero if $\mathbf{E}_0$ is uniform over the dielectric.

A lump of permeable or magnetized matter placed in a non-uniform magnetostatic field likewise experiences a net force. Its direct expression in terms of the average of the bound current density is

$$\mathbf{F} = -\int \mathbf{B}_0 \wedge \mathrm{curl}'\,\mathbf{M}\,\mathrm{d}\tau - \int \mathbf{B}_0 \wedge (\mathbf{M} \wedge \mathrm{d}\mathbf{S}); \tag{6.91}$$

but a formally simpler expression is

$$\mathbf{F} = \int (\mathbf{M}.\mathrm{grad}')\mathbf{B}_0 \mathrm{d}\tau, \tag{6.92}$$

which is what (6.90) becomes under the transformation (6.77).

The dielectric and magnetic bodies also, in general, experience net couples $\mathbf{G}$; and expressions for $\mathbf{G}$ analogous to those for $\mathbf{F}$ can be written down. In the case when the given fields $\mathbf{E}_0$ and $\mathbf{B}_0$ are uniform, at least over the region occupied by the bodies, the simplest forms are, for the dielectric

$$\mathbf{G} = \int \mathbf{P}\,\mathrm{d}\tau \wedge \mathbf{E}_0, \tag{6.93}$$

and for the magnet

$$\mathbf{G} = \int \mathbf{M}\,\mathrm{d}\tau \wedge \mathbf{B}_0. \tag{6.94}$$

The formulae so far given require for their evaluation a knowledge of $\mathbf{P}$ and $\mathbf{M}$. This may be quite hard to come by, since, except for ideal permanent magnets (or their rare dielectric counterpart, the so-called *electret*) it involves the solution of a boundary value problem. There is also the integration to do. An alternative approach, which may ease the calculation, or, more importantly, permit effective use of approximations, is to evaluate the energy, from which the force and couple can be derived.

As as simple example of the use of energy consider the force of attraction between the plates of a parallel plate capacitor with a homogeneous dielectric filling. The mechanical work done in increasing the plate separation must equal the increase in energy, provided the plates are insulated so that the charges on them remain constant. Now the energy per unit area of plate is $\frac{1}{2}\sigma^2 x/\epsilon$, where $\pm\sigma$ are the surface charge densities, x is the plate separation, and ϵ the permittivity of the dielectric. Hence

$$F\,\delta x = \frac{\sigma^2}{2\epsilon}\,\delta x; \tag{6.95}$$

and since the electric field is σ/ϵ this gives

$$F = \tfrac{1}{2}\sigma E. \tag{6.96}$$

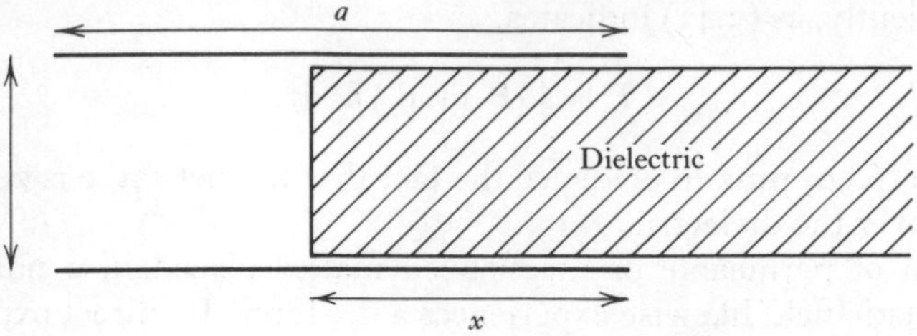

Figure 6.7

The result (6.96) may seem obvious; but notice something at first sight paradoxical. If the plates were not quite in contact with the dielectric, leaving tiny air gaps, (6.96) is unquestionably correct; but E, being the magnitude of the electric field at the inner surface of the plates, has then, of course, the value σ/ϵ_0, not σ/ϵ. How can the presence of tiny air gaps produce a finite change in the force?

The answer is that the dielectric itself is under the tension of electric forces, and may therefore communicate a mechanical pressure to the plates. This effect is included in (6.95), whose derivation assumed that the region between the plates remained filled with dielectric during the increase in plate separation, as would happen, for example, if the plates were immersed in a gaseous dielectric. In the air gap hypothesis it is assumed that the dielectric is retained by some agency independent of the capacitor plates; for example, by its internal structure were it a solid. The energy increment for an increment δx in plate separation is then $\frac{1}{2}\sigma^2\delta x/\epsilon_0$ per unit area of plate.

It may be mentioned in passing that the physical effect on a dielectric of the stress of electric forces necessarily involves thermodynamic considerations, and is outside the scope of this book.

As a more impressive demonstration of the use of energy consider a parallel plate capacitor partly filled with homogeneous dielectric, as illustrated in figure 6.7. The problem is to find the net force tending to pull the dielectric further into the region between the plates (cf. problem 3.15).

Let the plates have separation d and length a, of which length x is occupied by dielectric, and treat the problem as two-dimensional. To the first approximation the electric field is uniform between the plates and zero elsewhere (see §3.4.4), but the direct calculation of force from this approximation would yield nothing. At first sight it looks as though it is necessary to go to the trouble of evaluating the correction field associated with edge effects. However, the change in energy corresponding to a small change δx in x can be found from the approximate field. For if the voltage across the plates is kept constant, and neither end of the dielectric is close to either end of the plates, the displacement of the dielectric alters the field E significantly only in the region between x and $x+\delta x$. The corresponding energy increment, is, therefore,

$$\tfrac{1}{2}d(\epsilon-\epsilon_0)\,E^2\,\delta x \qquad (6.97)$$

per unit width. Since the voltage is kept constant the mechanical work necessary to effect the displacement is the negative of (6.97), as explained in the final paragraph of § 3.2.5. Hence the force per unit width tending to pull the dielectric further into the capacitor is

$$\tfrac{1}{2}d(\epsilon-\epsilon_0)\,E^2. \tag{6.98}$$

There is yet another important method of calculating electrostatic forces. Consider a spatially bounded, but otherwise quite arbitrary, static charge density ρ in vacuum. The total force on it is

$$\mathbf{F} = \int \rho\mathbf{E}\,\mathrm{d}\tau = \epsilon_0\int \mathbf{E}\,\mathrm{div}\,\mathbf{E}\,\mathrm{d}\tau. \tag{6.99}$$

This can be transformed to a surface integral by writing the x component as

$$F_x = \epsilon_0\int \mathrm{div}\,(E_x\mathbf{E})\,\mathrm{d}\tau - \epsilon_0\int \mathbf{E}\,.\,\mathrm{grad}\,E_x\,\mathrm{d}\tau. \tag{6.100}$$

The first integral on the right hand side of (6.100) transforms immediately, and clearly contributes

$$\epsilon_0\int \mathbf{E}(\mathbf{E}\,.\,\mathrm{d}\mathbf{S})$$

to $\mathbf{F}$. For the second, note that $\mathrm{grad}\,E_x = \partial\mathbf{E}/\partial x$, since $\mathrm{curl}\,\mathbf{E} = 0$; its contribution to $\mathbf{F}$ is therefore

$$-\epsilon_0\int \mathrm{grad}\,(\tfrac{1}{2}\mathbf{E}^2)\,\mathrm{d}\tau = -\epsilon_0\int \tfrac{1}{2}\mathbf{E}^2\,\mathrm{d}\mathbf{S}.$$

Thus

$$\mathbf{F} = \epsilon_0\int [\mathbf{E}(\mathbf{E}\,.\,\mathbf{n}) - \tfrac{1}{2}\mathbf{E}^2\mathbf{n}]\,\mathrm{d}S, \tag{6.101}$$

where the integration is over any surface that encloses all the charge, and $\mathbf{n}$ is the unit vector along the outward normal.

By taking the charge density ρ to be that associated with a polarized dielectric (6.101) gives the net force on the dielectric in terms of the field over any surface lying entirely outside the dielectric.

To illustrate (6.101) consider the force per unit area on a surface carrying charge density σ. Use the notation of § 3.1.3, where a different treatment is given, and refer to figure 3.6. If the surface in (6.101) is taken to be Σ, and $\mathbf{n}$ is in the z direction, it is seen that

$$\mathbf{F} = \epsilon_0[\mathbf{E}_2 E_{2z} - \mathbf{E}_1 E_{1z} - \tfrac{1}{2}(\mathbf{E}_2{}^2 - \mathbf{E}_1{}^2)\,\mathbf{n}]\,\mathrm{d}S.$$

But

$$E_{2x} = E_{1x}, \quad E_{2y} = E_{1y} \quad \text{and} \quad E_{2z} - E_{1z} = \sigma/\epsilon_0.$$

Hence

$$\mathbf{F} = [E_x\sigma,\ E_y\sigma,\ \tfrac{1}{2}(E_{2z}+E_{1z})\sigma]\,\mathrm{d}S,$$

which confirms that the force per unit area is indeed (3.17).

The magnetic counterpart to these results is obtained from the net force

$$\mathbf{F} = \int \mathbf{J}\wedge\mathbf{B}\,\mathrm{d}\tau \tag{6.102}$$

on a steady current distribution. A transformation of (6.102) to a surface integral leads to the same result as the application of (6.77) to (6.101), namely

$$\mathbf{F} = \mu_0 \int [\mathbf{H}(\mathbf{H}.\mathbf{n}) - \tfrac{1}{2}\mathbf{H}^2\mathbf{n}]\,\mathrm{d}S. \tag{6.103}$$

The combination of (6.101) and (6.103) gives net force

$$\mathbf{F} = \int \mathbf{T}.\mathrm{d}\mathbf{S}, \tag{6.104}$$

where $\mathbf{T}$ is defined by the statement that for any vector $\mathbf{k}$

$$\mathbf{T}.\mathbf{k} = \epsilon_0[\mathbf{E}(\mathbf{E}.\mathbf{k}) - \tfrac{1}{2}\mathbf{E}^2\mathbf{k}] + \mu_0[\mathbf{H}(\mathbf{H}.\mathbf{k}) - \tfrac{1}{2}\mathbf{H}^2\mathbf{k}]. \tag{6.105}$$

In fact $\mathbf{T}$ is a second rank tensor, and because (6.104) is analogous to the way in which the force due to stresses is represented in elasticity and hydrodynamics it is known as the *electromagnetic* (or *Maxwell*) *stress tensor.*

6.2 Free electron gas

6.2.1 *Charged particle dynamics*

In the remainder of this chapter simple microscopic particle models of different media are considered classically, and some indication is given of the extent to which their theory explains the electromagnetic characteristics of actual media.

The discussion naturally involves the motion of charge particles under the influence of electric and magnetic fields, and a few remarks are made on this topic by way of introduction.

As indicated in § 1.1.2 the Newtonian law of motion for a point charge e of mass m is

$$m\dot{\mathbf{v}} = e(\mathbf{E} + \mathbf{v} \wedge \mathbf{B}), \tag{6.106}$$

where $\mathbf{v}$ is the velocity and $\dot{\mathbf{v}}$ the acceleration of the charge, and $\mathbf{E}$, $\mathbf{B}$ may be taken to be the electromagnetic field excluding that due to the point charge itself. Equation (6.106) is basic to a vast range of natural phenomena and laboratory devices. In application it may often be necessary to take the relativistic rather than the Newtonian form of the equation, and occasionally necessary to allow for the self field of the particle, but neither of these corrections are required here.

The energy equation corresponding to (6.106), obtained on taking its scalar product with $\mathbf{v}$, is

$$\frac{\mathrm{d}}{\mathrm{d}t}(\tfrac{1}{2}m\mathbf{v}^2) = e\mathbf{E}.\mathbf{v}. \tag{6.107}$$

The magnetic force, being at right angles to $\mathbf{v}$, does no work. If $\mathbf{E} = -\mathrm{grad}\,\phi$, which can only happen when $\mathbf{B}$ is time independent, (6.107) has the first integral

$$\tfrac{1}{2}m\mathbf{v}^2 + e\phi = \text{constant}. \tag{6.108}$$

If $\mathbf{E}$ is zero the speed v is constant. Consequently $\dot{\mathbf{v}}$ is along the principal normal to the trajectory and of magnitude v^2/ρ_p, where ρ_p is the principal radius of curvature. Hence (6.106) gives

$$\rho_p = \frac{mv}{eB\sin\theta}, \tag{6.109}$$

where θ is the angle between $\mathbf{B}$ and $\mathbf{v}$, known as the *pitch angle.* In the special case in which the direction of $\mathbf{B}$ is fixed, parallel to the z axis say, the motion is some constant velocity in that direction, superposed on motion in the x, y plane with constant speed $v_\perp$ and radius of curvature

$$\rho = \frac{mv_\perp}{eB} \tag{6.110}$$

that is inversely proportional to B.

Consider now the case in which both $\mathbf{E}$ and $\mathbf{B}$ are constant and uniform.

If the field is purely electric the particle simply has uniform acceleration $e\mathbf{E}/m$ (though when the speed gets comparable to c a relativistic treatment is essential).

If the field is purely magnetic the motion is some constant velocity along $\mathbf{B}$, superposed on uniform circular motion at right angles to $\mathbf{B}$ with radius given by (6.110). The angular velocity of the circular motion is

$$\Omega = eB/m, \tag{6.111}$$

known as the *gyro* frequency, and a negatively charged particle circulates in the sense of a right-handed screw about $\mathbf{B}$. The gyrofrequency of an electron in a magnetic field of 10^{-4} weber m^{-2} (1 gauss) is about 1.8×10^7 sec^{-1}.

If both $\mathbf{E}$ and $\mathbf{B}$ are present the motion along $\mathbf{B}$ is an acceleration $eE_\parallel/m$, and the motion perpendicular to $\mathbf{B}$ is governed by the equation

$$m\dot{\mathbf{v}}_\perp = e(\mathbf{E}_\perp + \mathbf{v}_\perp \wedge \mathbf{B}), \tag{6.112}$$

where the suffices $\parallel$ and $\perp$ denote components parallel and perpendicular to $\mathbf{B}$. A particular solution for $\mathbf{v}_\perp$ in (6.112) is obviously the constant velocity

$$\mathbf{v}_0 = \mathbf{E} \wedge \mathbf{B}/B^2. \tag{6.113}$$

To find the general solution write

$$\mathbf{v}_\perp = \mathbf{v}_0 + \mathbf{u}. \tag{6.114}$$

Substitution of (6.114) in (6.112) gives

$$m\dot{\mathbf{u}} = e\mathbf{u} \wedge \mathbf{B}, \tag{6.115}$$

so that $\mathbf{u}$ corresponds to motion in the case when $\mathbf{E}$ is absent, that is, uniform circular motion. The general motion perpendicular to $\mathbf{B}$ is therefore uniform circular motion at the gyro frequency superposed on the 'drift' velocity $\mathbf{v}_0$: the path is a cycloid, having the form shown in figure 6.8 (a) if v_0 exceeds the speed of circular motion and in figure 6.8 (b) otherwise.

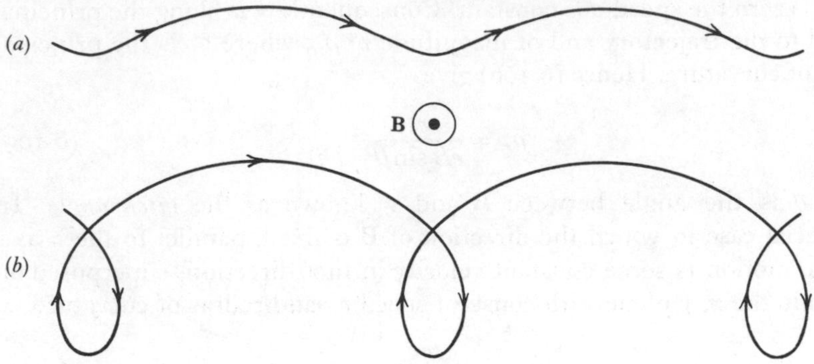

Figure 6.8

Suppose a charged particle were acted on by the field of an electromagnetic wave. For such a wave the order of magnitude of **E** is likely to be c times that of **B**. Consequently the ratio of magnetic to electric force in (6.106) is of order v/c, and the former is therefore usually negligible. When this is so the equation for the position vector **r** of the particle is

$$m\frac{\mathrm{d}^2\mathbf{r}}{\mathrm{d}t^2} = e\mathbf{E}(\mathbf{r}, t). \qquad (6.116)$$

Now suppose that for a monochromatic wave of angular frequency ω there is a motion in which the particle makes only limited excursion from its mean position, at the origin say. Then the *linear* equation

$$m\frac{\mathrm{d}^2\mathbf{r}}{\mathrm{d}t^2} = e\mathbf{E}(0, t) \qquad (6.117)$$

is a valid approximation to (6.116), provided the maximum excursion is much less than the scale of the field structure, which would generally be represented by the wavelength over 2π, namely c/ω.

Equation (6.117) predicts simple harmonic notion for each cartesian component of the displacement of the particle. In the complex representation, with time factor $\exp(\mathrm{i}\omega t)$ understood,

$$-m\omega^2\mathbf{r} = e\mathbf{E}(0), \qquad (6.118)$$

and correspondingly $$\mathrm{i}m\omega\mathbf{v} = e\mathbf{E}(0). \qquad (6.119)$$

In terms of these approximations the inequalities $v/c \ll 1$ and $r \ll c/\omega$ required to validate them are equivalent. The inequality is likely to hold, since $e/(mc) \simeq 600$ coulomb sec $\mathrm{m}^{-1}\,\mathrm{kg}^{-1}$ for an electron, and then affords *a posteriori* justification for the solution.

6.2.2 *Conductivity*

One of the simplest particle models is a gas of free electrons, with a background of positive charge to preserve overall neutrality. The classical treatment of this model has some relevance to conductors and much to ionized gases. The former are discussed first.

To model a metallic conductor it is supposed that there is a fixed lattice of positive ions through which each conduction electron has, in the absence of an applied field, a free motion randomly punctuated by collisions. The velocities are randomized by the collisions: the average speed characterises the temperature, but the average velocity, and therefore the average current, is zero (except for a small statistical fluctuation, recognized in circuits as 'noise').

Now let a steady electric field $\mathbf{E}$ be applied to the conductor. At any instant of time the velocity of a conduction electron is then

$$\mathbf{v} = \mathbf{v}_c - \frac{e}{m}\mathbf{E}\tau, \tag{6.120}$$

where $-e$ is the charge of the electron, m its mass, τ is the elapsed time since its last collision and $\mathbf{v}_c$ its velocity just after that collision. Moreover the average current density (6.45) is

$$\mathbf{J} = -Ne\bar{\mathbf{v}}, \tag{6.121}$$

where N is the number of conduction electrons per unit volume, and $\bar{\mathbf{v}}$ the velocity averaged over them. On the reasonable assumption that the average of $\mathbf{v}_c$ is zero it follows that

$$\mathbf{J} = \frac{Ne^2}{m\nu}\mathbf{E}, \tag{6.122}$$

where ν, being the reciprocal of the average of τ, is the *collision frequency*.

This result reproduces the constitutive relation $\mathbf{J} = \sigma\mathbf{E}$, which is known to be widely satisfied by many substances covering an immense range of values of the conductivity σ (see §4.1.1). The general idea of the frictional effect of collisions is correct, but it must be stressed that the conduction electrons in a metal are too tightly packed for a classical treatment not to lead to quite false conclusions in some respects. The model value

$$\sigma = Ne^2/(m\nu) \tag{6.123}$$

does not therefore bear too close a scrutiny. However, some idea of the order of magnitudes involved is instructive.

If $N \simeq 10^{29}\,\mathrm{m}^{-3}$ (the number of copper atoms per cubic metre) it is striking that under ordinary conditions the value of the average velocity $\bar{\mathbf{v}}$ is orders of magnitude less than the thermal speed of the electrons, as remarked in § 1.3.3. Moreover $e^2/m \simeq 3\times 10^{-8}$ coulomb^2 kg^{-1}, so for copper, for which $\sigma \simeq 6\times 10^7$ mho m^{-1}, (6.123) gives $\nu \simeq 0.5\times 10^{14}$ sec^{-1}. With this figure in mind it may be noted that the simple theory is equally applicable to cases in which the applied electric field is time dependent but not too rapidly varying. For evidently (6.120) remains valid to a good approximation provided the variation of $\mathbf{E}$ is characterized only by frequencies much less than ν. With this proviso, then, (6.122) is obtained as before.

Collisions keep the electrons to a steady average velocity even though they are permanently under the influence of a constant electric field. The analogy with a body moving under a given constant force in a medium producing a frictional force proportional to the velocity is brought out by writing the relation between $\mathbf{E}$ and $\bar{\mathbf{v}}$ in the form

$$e\mathbf{E}+m\nu\bar{\mathbf{v}} = 0. \tag{6.124}$$

This can be regarded as an 'average' equation of motion, expressing the balance between the electric and frictional forces in the steady state, which is reached in a time of order $1/\nu$. Such an 'average' equation of motion is widely used in more general contexts, having the advantage of comparative simplicity at the cost of losing some of the refinements of a statistical treatment. It is now used to explain the *Hall current*: this arises when a conductor is subjected to an applied magnetic field in addition to an electric field, being a component of current flow in the direction perpendicular to both the electric and the magnetic field.

If the applied magnetic field is $\mathbf{B}$, equation (6.124) is replaced by

$$e\mathbf{E}+e\mathbf{v}\wedge\mathbf{B}+m\nu\mathbf{v} = 0, \tag{6.125}$$

where the bar over $\mathbf{v}$ is dropped without confusion. This equation can be written

$$\frac{e}{m}\mathbf{E}+\Omega\mathbf{v}\wedge\mathbf{b}+\nu\mathbf{v} = 0, \tag{6.126}$$

where $\mathbf{b}$ is the unit vector along $\mathbf{B}$ and Ω is the gyro frequency (6.111).

To solve (6.126) for $\mathbf{v}$ in terms of $\mathbf{E}$ take the vector product with $\mathbf{b}$ to get

$$\frac{e}{m}\mathbf{E}\wedge\mathbf{b}-\Omega\mathbf{v}+\Omega(\mathbf{v}\,.\,\mathbf{b})\,\mathbf{b}+\nu\mathbf{v}\wedge\mathbf{b} = 0.$$

In the third term replace $\mathbf{v}\,.\,\mathbf{b}$ by $-(e/m\nu)\,\mathbf{E}\,.\,\mathbf{b}$, obtained by taking the scalar product of (6.126) with $\mathbf{b}$; and in the fourth term substitute for $\mathbf{v}\wedge\mathbf{b}$, again from (6.126). Thus

$$\frac{m}{e}(\nu^2+\Omega^2)\,\mathbf{v} = -\nu\mathbf{E}+\Omega\mathbf{E}\wedge\mathbf{b}-\frac{\Omega^2}{\nu}(\mathbf{E}\,.\,\mathbf{b})\,\mathbf{b}. \tag{6.127}$$

The current density $-Ne\mathbf{v}$ is evidently affected by the magnetic field, apart from the special case when $\mathbf{B}$ is parallel to $\mathbf{E}$. Numerical values previously quoted indicate that the inequality $\Omega \ll \nu$ is likely to hold, so that effectively

$$\mathbf{J} = \sigma\mathbf{E} + \sigma_{\mathrm{H}}\mathbf{b} \wedge \mathbf{E}, \tag{6.128}$$

where σ is the direct conductivity (6.123) and

$$\sigma_{\mathrm{H}} = \Omega\sigma/\nu \tag{6.129}$$

is the Hall conductivity. Thus there is a comparatively small component of current perpendicular to both the electric and magnetic field.

It may happen, naturally or by design, that the Hall current is interrupted at the boundary of the conducting medium. It then builds up a charge density on the boundary which in turn gives rise to an additional electric field. A steady state is quickly reached in which the additional field, $\mathbf{E}_{\mathrm{H}}$ say, exerts on the moving charges constituting the current a force that just cancels the force of the magnetic field.

The appearance of $\mathbf{E}_{\mathrm{H}}$ is known as the *Hall effect*, and is the basis of various measurements, which, in particular, give further information about the mechanism of conductivity, as is now explained.

Suppose that current I flows along a rectangular strip of material of width d and thickness h. The application of a uniform magnetic field $\mathbf{B}$ normal to the plane of the strip results in a momentary flow of Hall current across the strip. The transverse current builds up charges of opposite sign along the two edges of the strip, and continues until the field $\mathbf{E}_{\mathrm{H}}$ of these charges is just sufficient to counter it. In the steady state

$$\mathbf{E}_{\mathrm{H}} = -\mathbf{v} \wedge \mathbf{B}$$

at each point in the strip, where $\mathbf{v}$ is the drift velocity of the conduction particles, assuming them to be of one species. A measurement of $\mathbf{E}_{\mathrm{H}}$ and $\mathbf{B}$ thus yields $\mathbf{v}$ and *a fortiori* the sign of the conduction charges, since this determines whether the drift velocity and the current are in the same or opposite directions.

Moreover, if q is the charge of each conduction particle and N is their number per unit volume, then $I = Nqvdh$, and the voltage $E_{\mathrm{H}}d$ across the strip can be written

$$V = R\frac{B}{h}I, \tag{6.130}$$

where

$$R = \frac{1}{Nq} \tag{6.131}$$

is called the *Hall coefficient*. Thus the transverse voltage can be increased by reducing h; and its measurement provides an experimental value of Nq. With copper, for example, it is found that $R = 5.5 \times 10^{-11}\,\mathrm{m^3\,coulomb^{-1}}$, from which, identifying q as the charge of an electron, it is deduced that $N = 1.14 \times 10^{29}\,\mathrm{m^{-3}}$, implying 1.34 conduction electrons per atom.

The Hall coefficient is comparatively large in semi-conductors by virtue of the lower value of N. In germanium, for example, it is about 10^{-3}, with $N \simeq 0.6 \times 10^{21}\,m^{-3}$.

6.2.3 *Plasmas*

A gas that is partly or fully ionized is commonly called a *plasma*. There is some similarity between a plasma and the free electrons in a conductor, particularly since the positive ions in a plasma, though free like the electrons, may usually be taken to be (statistically) stationary by virtue of their much greater mass. However the free electron concentration in plasmas is likely to be very much less than that in metallic conductors; it might be $10^{18}\,m^{-3}$ in a laboratory plasma, and reaches not more than about $10^{12}\,m^{-3}$ in the ionosphere, for example. As a consequence the validity of classical theory is rarely in doubt. Moreover collisions can be comparatively unimportant.

In §6.2.2 a simple theory of conductivity was presented, applicable at frequencies less than about $10^{13}\,sec^{-1}$. As analogous theory of plasmas can be given: for frequencies much less than the electron collision frequency the previous treatment holds, but the main focus of interest is on higher frequencies. In fact the essence of plasma theory is typified by the case in which the collision frequency is negligible. Consideration of this case leads to results quite different from those so far obtained, as is now shown.

In the absence of collisions each free electron executes uninterruptedly its motion under the influence of the electromagnetic field. If the field has angular frequency ω an electron which would be at rest at the origin in the absence of the field has simple harmonic motion given by (6.118), (6.119), with e replaced by $-e$. A concentration of N free electrons per unit volume correspondingly contributes the current density

$$\mathbf{J} = -Ne\mathbf{v} = -\mathrm{i}\frac{Ne^2}{m\omega}\mathbf{E}. \tag{6.132}$$

Before discussing this result it should be stressed that it is based on the rather sweeping assumption that the thermal motion of the electrons may be disregarded. A crude justification would be that the random velocities produce no effect on average in the absence of the field, so that only the perturbation motion due to the field need be considered. A proper statistical treatment is vastly more elaborate, and gives a refined result to which (6.132) is nevertheless a good approximation for many practical purposes. Specifically, (6.132) is recovered in the limit as the temperature of the plasma tends to zero, and the present theory is often referred to as that for a 'cold' plasma.

Another point to stress is that the treatment is based on a linear approximation. This was made explicit in the derivation of (6.119). It is also implicit in the presumption in (6.132) that N represents the unperturbed concentration of free electrons, that is, their number density in the absence of a field. The perturbation concentration, if any, is neglected in the expression for **J** since it is there multiplied by **v** to give a term quadratic in perturbation quantities.

Return now to develop the consequences of (6.132). This constitutive relation should be contrasted with that for a conductor, (6.122). The factor i in (6.132) implies that **J** and **E** are in phase quadrature, which means that the plasma behaves not as a conductor but rather as a dielectric. This is at once evident by substituting from (6.132) into the appropriate Maxwell equation

$$\operatorname{curl} \mathbf{H} = \mathrm{i}\omega\epsilon_0 \mathbf{E} + \mathbf{J} \tag{6.133}$$

to get

$$\operatorname{curl} \mathbf{H} = \mathrm{i}\omega\epsilon_0 \kappa \mathbf{E}, \tag{6.134}$$

where

$$\kappa = 1 - p^2/\omega^2, \tag{6.135}$$

$$p^2 = Ne^2/(\epsilon_0 m). \tag{6.136}$$

The quantity p is known as the *plasma frequency*. Inserting the values of ϵ_0, and of e and m for the electron,

$$p \simeq 56 N^{\frac{1}{2}} \text{ sec}^{-1}. \tag{6.137}$$

The additional contribution to the current density (6.132), and hence to p^2, from the motion of the positive ions is obtained solely by replacing m by the ion mass (assuming singly ionized molecules) and is therefore negligible.

For $\omega > p$ it is apparent that the plasma behaves as a dielectric with dielectric constant κ given by (6.135). The treatment of plane wave propagation in § 5.3.1 is wholly applicable. The refractive index (5.153) is

$$n = \kappa^{\frac{1}{2}} = (1 - p^2/\omega^2)^{\frac{1}{2}}, \tag{6.138}$$

and the speed of phase propagation is

$$c/n. \tag{6.139}$$

The dispersion relation (6.135) between ω and the wave number $k = n\omega/c$ is conveniently written

$$c^2k^2 = \omega^2 - p^2, \tag{6.140}$$

which is notably identical in form with (5.129) for a rectangular waveguide mode.

Since n is less than unity the phase speed exceeds c. However, the group speed (5.135) is

$$v_{\mathrm{g}} = cn, \tag{6.141}$$

as was calculated for the waveguide mode. It is shown in § 6.2.4 that (6.141) is also the speed of energy propagation.

For $\omega < p$ the dielectric constant becomes negative and the corresponding wave number purely imaginary, as for an evanescent waveguide mode. The result is an evanescent wave which oscillates everywhere in phase and decays exponentially in the direction normal to the equiamplitude planes.

Since monochromatic waves cannot be propagated in a plasma whose electron concentration is so great that $p > \omega$ the plasma is said to be *overdense* to waves of such frequencies. That the ionosphere between certain heights is overdense to medium and *a fortiori* low frequency radio waves is the essential reason for the success of long range broadcasting; the waves cannot penetrate the barrier and their energy is confined to the vicinity of the earth. The point is illustrated by the model problem in which a plane wave in vacuum is normally incident on the plane face of a semi-infinite homogeneous plasma. The reflection coefficient for the electric field is

$$\frac{1-n}{1+n} \tag{6.142}$$

(cf. (5.174) with $\alpha = \alpha' = \frac{1}{2}\pi$ and $\mu' = \mu = \mu_0$), and if n is purely imaginary this has modulus unity, indicating *total* reflection with a specific phase change. A proper investigation of the ionosphere problem involves a gradual change of electron concentration, and hence refractive index, with height, and also consideration of obliquely incident waves.

The case $\omega = p$ is special, with $\kappa = 0$, so that $\operatorname{curl} \mathbf{H} = 0$: physically, the electron current cancels the displacement current. A possible solution is one in which there is no magnetic field at all, and the electric field satisfies

$$\operatorname{curl} \mathbf{E} = 0, \tag{6.143}$$

$$\operatorname{div} \mathbf{E} = \rho/\epsilon_0, \tag{6.144}$$

ρ being the net charge density created by variations in the electron concentration. These equations are formally those of electrostatics: the charge density has the form

$$\rho = \rho_0(\mathbf{r})\, e^{ipt}, \tag{6.145}$$

where $\rho_0(\mathbf{r})$ is arbitrary; and the corresponding spatial dependence of the electric field may be calculated from potential theory. Such disturbances are called *plasma oscillations*. The simple form derived here applies only to a 'cold' plasma. If thermal velocities are taken into account it is found that there can be waves of 'electrostatic' type ($\mathbf{H} = 0$) with frequencies in the neighbourhood of the plasma frequency.

The association of plasma oscillations with a charge density is in marked contrast to the plane wave solutions previously considered, in which $\operatorname{div} \mathbf{E} = 0$ so that no charge density is created. Note in fact that for an arbitrary time variation the relation

$$\dot{\mathbf{J}} = \frac{Ne^2}{m}\mathbf{E}, \tag{6.146}$$

corresponding to (6.132), together with

$$\operatorname{div} \mathbf{E} = \rho/\epsilon_0 \tag{6.147}$$

and

$$\operatorname{div} \mathbf{J} + \dot{\rho} = 0, \tag{6.148}$$

yields

$$\ddot{\rho}+p^2\rho = 0, \tag{6.149}$$

with solution (6.145). Hence any charge density *must* oscillate at the plasma frequency.

In a partially ionized plasma it may well be necessary to take account of the collisions of the electrons with the neutral molecules. To a first approximation this can be done by including the frictional force $-m\nu\mathbf{v}$ in the equation of motion, as discussed in § 6.2.2, ν being the collision frequency. Equation (6.132) is then replaced by

$$\mathbf{J} = -\mathrm{i}\frac{Ne^2}{m(\omega-\mathrm{i}\nu)}\mathbf{E}, \tag{6.150}$$

and (6.135) by

$$\kappa = 1-\frac{p^2}{\omega(\omega-\mathrm{i}\nu)} = 1-\frac{p^2}{\omega^2+\nu^2}-\mathrm{i}\frac{\nu p^2}{\omega(\omega^2+\nu^2)}, \tag{6.151}$$

where κ is now the *complex* dielectric constant. The effect of collisions is therefore to introduce conductivity.

$$\sigma = \frac{\epsilon_0 \nu p^2}{\omega^2+\nu^2}; \tag{6.152}$$

this, of course, gives rise to attenuation of a wave propagated through the medium (cf. §5.3.1), and absorption measurements can be used to estimate the value of the collision frequency.

6.2.4 *Energy density and the velocity of energy propagation*

The simple collisionless plasma of § 6.2.3 affords an example of a medium that can be characterized by a frequency dependent permittivity $\epsilon(\omega)$, and it is instructive to use the model to illustrate the concept of energy density in such a medium.

Consider, for $\omega > p$, a linearly polarized plane wave field associated with the relations (6.138)–(6.141). Fundamentally, the field is the vacuum field generated by the current due to the motion of the electrons, as (6.133) states. There is therefore field energy whose time averaged density is given by

$$\tfrac{1}{4}(\epsilon_0\mathbf{E}.\mathbf{E}^*+\mu_0\mathbf{H}.\mathbf{H}^*); \tag{6.153}$$

and there is also kinetric energy of the electrons, the time averaged density of which is

$$\tfrac{1}{4}Nm\mathbf{v}.\mathbf{v}^*. \tag{6.154}$$

If the magnitude of $\mathbf{E}$ is E_0, that of $\mathbf{H}$ is $Y_0\kappa^{\frac{1}{2}}E_0$ and that of $\mathbf{v}$ is $eE_0/(m\omega)$. The combined time averaged energy density is therefore

$$\mathscr{E} = \tfrac{1}{4}[\epsilon_0(1+\kappa)+Ne^2/(m\omega^2)]\,E_0{}^2 = \tfrac{1}{2}\epsilon_0 E_0{}^2. \tag{6.155}$$

The power flux, on the other hand, comes essentially from the field alone, any contribution from the electrons being at best third order in $\mathbf{v}$. The time averaged power flux density is therefore

$$\mathbf{W} = \tfrac{1}{2}\mathrm{Re}\,\mathbf{E}\wedge\mathbf{H}^* = \tfrac{1}{2}Y_0\kappa^{\frac{1}{2}}E_0{}^2\hat{\mathbf{k}}, \tag{6.156}$$

where $\hat{\mathbf{k}}$ is the unit vector in the direction of propagation.

The associated velocity of energy propagation is

$$\mathbf{W}/\mathscr{E} = \kappa^{\frac{1}{2}}c\hat{\mathbf{k}}. \tag{6.157}$$

It has the same value as the group velocity (cf. (6.141)).

It is clear that the time averaged energy density (6.155) is *not* the same as (6.153) with $\epsilon(\omega)$ replacing ϵ_0. It can, however, be expressed in an analogous way in terms of $\epsilon(\omega)$ and the field vectors alone, without involving the kinetic energy of the electrons explicitly. This formula, applicable to any medium characterized by real permittivity $\epsilon(\omega)$, is now derived.

For an electromagnetic field in a medium the rate of change of energy density is (5.31). Since now $\mathbf{B} = \mu_0\mathbf{H}$ the magnetic part is the time derivative of $\frac{1}{2}\mu_0\mathbf{H}^2$, which represents the magnetic energy density as it would for a vacuum field. For the electric part, however, no such general time integral exists, since, for arbitrary time variation, $\mathbf{D}$ at time t is not expressible in terms of $\mathbf{E}$ at time t. Moreover for variation at a single frequency, which is the present concern, there is no reference level by which to fix the constant of integration.

The latter difficulty can, however, be circumvented by associating with ω an arbitrarily small negative imaginary part $-\mathrm{i}\gamma$, so that the temporal behaviour of the disturbance is specified by

$$\mathrm{e}^{\mathrm{i}\omega t}\,\mathrm{e}^{\gamma t}. \tag{6.158}$$

The disturbance is then essentially monochromatic, with magnitude that builds up slowly from zero at $t = -\infty$ to some finite contemporaneous value. The electric energy density is then given unambiguously at each point by

$$\int_{-\infty}^{t} \mathbf{E}(\tau)\,.\,\dot{\mathbf{D}}(\tau)\,\mathrm{d}\tau. \tag{6.159}$$

Now if, at some point,

$$\mathbf{E} = \mathrm{Re}\,\mathbf{E}_0\,\mathrm{e}^{\mathrm{i}\omega t}, \tag{6.160}$$

then

$$\mathbf{D} = \mathrm{Re}\,\mathbf{E}_0\,\epsilon(\omega)\,\mathrm{e}^{\mathrm{i}\omega t}; \tag{6.161}$$

and the replacement of ω by $\omega-\mathrm{i}\gamma$ gives

$$\mathbf{E} = \mathrm{Re}\,\mathbf{E}_0\,\mathrm{e}^{\mathrm{i}\omega t}\mathrm{e}^{\gamma t}, \tag{6.162}$$

$$\mathbf{D} = \mathrm{Re}\,\mathbf{E}_0\left(\epsilon - \mathrm{i}\gamma\,\frac{\partial\epsilon}{\partial\omega}\right)\mathrm{e}^{\mathrm{i}\omega t}\,\mathrm{e}^{\gamma t} + O(\gamma^2), \tag{6.163}$$

where ϵ and $\partial\epsilon/\partial\omega$ in (6.163) are evaluated at ω. From (6.163),

$$\dot{\mathbf{D}} = \mathrm{Re}\,\mathbf{E}_0\left[\mathrm{i}\omega\epsilon+\gamma\frac{\partial(\omega\epsilon)}{\partial\omega}\right]\mathrm{e}^{\mathrm{i}\omega t}\,\mathrm{e}^{\gamma t}+O(\gamma^2), \tag{6.164}$$

and a little elementary algebra therefore gives

$$\mathbf{E}.\dot{\mathbf{D}} = \tfrac{1}{2}\mathbf{E}_0.\mathbf{E}_0{}^*\gamma\frac{\partial(\omega\epsilon)}{\partial\omega}\mathrm{e}^{2\gamma t}+\mathrm{Re}\,\mathbf{E}_0{}^2\left[\mathrm{i}\omega\epsilon+\gamma\frac{\partial(\omega\epsilon)}{\partial\omega}\right]\mathrm{e}^{2\mathrm{i}\omega t}\,\mathrm{e}^{2\gamma t}. \tag{6.165}$$

When (6.165) is integrated from $-\infty$ up to t the contribution from the latter part of the expression has a factor $\exp(2\mathrm{i}\omega t)$. Time averaging this contribution over a period $2\pi/\omega$, and then letting $\gamma\to 0$, gives zero. Also the integral from $-\infty$ to t of $2\gamma\exp(2\gamma t)$ is $\exp(2\gamma t)$, so the corresponding limit of the time average of (6.165) is

$$\frac{1}{4}\frac{\partial(\omega\epsilon)}{\partial\omega}\mathbf{E}_0.\mathbf{E}_0{}^*. \tag{6.166}$$

In this way the time averaged energy density of a monochromatic disturbance of angular frequency ω is obtained in the form

$$\mathscr{E} = \frac{1}{4}\left[\frac{\partial(\omega\epsilon)}{\partial\omega}\mathbf{E}.\mathbf{E}^*+\mu_0\mathbf{H}.\mathbf{H}^*\right], \tag{6.167}$$

in which $\mathbf{E}$ and $\mathbf{H}$ are now, of course, the complex representations of the field vectors. The factor multiplying $\mathbf{E}.\mathbf{E}^*$ is

$$\epsilon+\omega\,\partial\epsilon/\partial\omega, \tag{6.168}$$

the second term of which vanishes if $\omega = 0$, or if ϵ is effectively independent ω at the frequency concerned.

For the simple model plasma just considered.

$$\epsilon = \epsilon_0\kappa = \epsilon_0(1-p^2/\omega^2), \tag{6.169}$$

so that

$$\frac{1}{4}\frac{\partial(\omega\epsilon)}{\partial\omega}\mathbf{E}.\mathbf{E}^* = \tfrac{1}{4}\epsilon_0(1+p^2/\omega^2)\mathbf{E}.\mathbf{E}^*. \tag{6.170}$$

This confirms the implicit inclusion of the kinetic energy of the electrons in (6.166), for the second term on the right hand side of (6.170) is indeed (6.154) since $\mathbf{v} = -e\mathbf{E}/(\mathrm{i}\omega m)$.

6.2.5 *Child's law*

Much of the development of the theory of plasmas depends on linearizing the governing equations, in the sense that a disturbance from an equilibrium state is treated as a perturbation, and quadratic and higher order terms in perturbation quantities are suppressed. It is extremely difficult to make progress on a wide front in non-linear theory, but particular solutions have been obtained. An important example is afforded by the charge and current distribution in a thermionic valve. This involves a 'space-charge' field, rather than a field in a 'medium', but as regards the basic physics the problem appears appropriately in the present context.

To take the simplest analysis, consider a plane diode; that is, an evacuated space enclosing a parallel plate capacitor formed by the anode and cathode. When heated, the cathode, at zero potential, liberates electrons which are attracted towards the anode whose potential V is positive. These electrons create a negative charge density which affects the electric field distribution and hence the electron flow from cathode to anode that constitutes the current I. The main task is to find, in the steady state, the dependence of I on V.

In the approximation of a plane geometry, in which distance x is measured from the cathode, Poisson's equation (2.35) for the electric potential $\phi(x)$ is

$$\frac{d^2\phi}{dx^2} = -\rho/\epsilon_0, \qquad (6.171)$$

where $\rho(x)$ is the charge density due to the electrons. If $v(x)$ is the electron velocity in the direction of increasing x, and A is the area of each electrode, then

$$I = -A\rho v. \qquad (6.172)$$

The minus sign in (6.172) is included to make I positive; ρ is negative, v positive, and the current flows from anode to cathode. The electron dynamics are conveniently expressed by the energy equation

$$\tfrac{1}{2}mv^2 = e\phi, \qquad (6.173)$$

it being assumed that the electrons are liberated from the cathode with zero velocity.

With suitable boundary conditions the three equations (6.171), (6.172) and (6.173) are sufficient to determine the unknown functions ρ, v and ϕ in terms of the constants of the problem, one of which is I. Elimination of ρ and v gives

$$\frac{d^2\phi}{dx^2} = \gamma^2\phi^{-\frac{1}{2}}, \qquad (6.174)$$

where

$$\gamma^2 = \frac{I}{\epsilon_0 A}\left(\frac{m}{2e}\right)^{\frac{1}{2}}. \qquad (6.175)$$

A first integral of the (non-linear) equation (6.174) is

$$\left(\frac{d\phi}{dx}\right)^2 = 4\gamma^2\phi^{\frac{1}{2}}, \qquad (6.176)$$

where the constant of integration is zero, as written, provided $d\phi/dx$ as well as ϕ is zero at the cathode $x = 0$. This assumption is justified if the voltage V is appreciably less than the 'saturation' voltage beyond which the current is limited by the maximum rate at which it is possible for electrons to leave the cathode, at its particular temperature. At such low voltages the electrons pile up in the vicinity of the cathode until their negative charge density reduces the electric field at the cathode approximately to zero, which represents the steady state.

Since (6.176) can be written

$$\phi^{-\frac{1}{4}}\mathrm{d}\phi = 2\gamma\,\mathrm{d}x, \tag{6.177}$$

a further integration gives

$$\phi^{\frac{3}{4}} = \tfrac{3}{2}\gamma x. \tag{6.178}$$

Now $\phi = V$ at the anode, $x = d$, say; so the square of (6.178), with the restoration of (6.175) for γ^2, gives

$$I = \frac{4\epsilon_0 A}{9d^2}\left(\frac{2e}{m}\right)^{\frac{1}{2}} V^{\frac{3}{2}}. \tag{6.179}$$

That the space-charge limited current varies as the three-halves power of the voltage is known as Child's law.

In the calculation the transverse electric field and the magnetic field of the current are neglected. The force of the latter tends to 'pinch' the current, but is down by a factor v^2/c^2 on that of the former, which tends to expand it.

6.3 Dielectrics

6.3.1 *Molecular dipole moments*

The electromagnetic properties of dielectrics, be they gases, liquids or solids, arise from the bound charges in the atoms or molecules of which they are composed. The macroscopic electric effect of the bound charges is expressible in terms of the polarization **P**, which is, to an adequate approximation, essentially the vector sum per unit volume of the individual molecular dipole moments, as signified in (6.38).

Molecules are called *polar* if they have a permanent intrinsic dipole moment, *non-polar* otherwise. A molecule must have some asymmetry to be polar; atoms, being symmetrical, are non-polar.

Apart from the rare case of electrets (ferro-electrics) the polarization of a dielectric is non-zero only if the dielectric is under the influence of an external source of electric field. For a dielectric of non-polar molecules each molecule acquires a dipole moment induced by the field, and these moments, being similarly directed over macroscopically small volumes, can give a measurable **P**. Since, as previously mentioned, the effect arises from the distortion by an electric field of the electronic charge in each molecule it may be called *deformation* polarization.

For a dielectric of polar molecules deformation polarization makes a contribution to **P**, but there is also a contribution arising from the partial alignment along the electric field of the intrinsic dipole moments of the individual molecules. This latter contribution is called *orientation* polarization. The extent of the field alignment is temperature dependent because it is in competition with thermal randomization of the directions of the moments; in the absence of a field the directions are entirely random and there is no average effect.

Apart from their different temperature dependence, the two types of polarization also have a quite different dependence on the frequency of the external field source. At sufficiently high frequencies only deformation polarization is significant, but at sufficiently low frequencies orientation polarization, if present, may well be dominant.

A reasonable estimate of the order of magnitude of molecular dipole moments can be obtained from simple classical models, even though these cannot be stable. Envisage an atom as a negative charge $-e$ spread uniformly throughout a sphere of radius a, together with a positive charge $+e$ located at the centre of the sphere. Under the influence of a constant electric field $\mathbf{E}$ the centre O of the negative charge takes up vector displacement $\boldsymbol{\xi}$ from the positive charge, such that the attractive force between the charges balances the forces $\pm e\mathbf{E}$. The attractive force is easily calculated: the part of the negative charge that contributes is that in a sphere of radius ξ, namely $-e(\xi/a)^3$, and this may be presumed located at O (see figure 6.9); the attractive force is therefore $e^2\xi/(4\pi\epsilon_0 a^3)$. Hence

$$\frac{e^2}{4\pi\epsilon_0 a^3}\boldsymbol{\xi} = -e\mathbf{E}, \tag{6.180}$$

and the induced dipole moment is

$$\mathbf{p} = -e\boldsymbol{\xi} = 4\pi\epsilon_0 a^3\mathbf{E}. \tag{6.181}$$

In general the relation between the electric field acting on a molecule and the induced dipole moment is written

$$\mathbf{p} = \epsilon_0\alpha\mathbf{E}, \tag{6.182}$$

and α is known as the *polarizability* of the molecule. The simple model gives

$$\alpha = 4\pi a^3. \tag{6.183}$$

Actual polarizabilities vary among atoms over an order of magnitude on either side of $10^{-29}\,\text{m}^3$, so there is broad agreement if a is taken to be $10^{-10}\,\text{m}$, which, being 1 Å, is of the order of an atomic radius.

It should also be noted that these figures confirm what has been assumed in the model, that the displacement ξ is less than the radius a. In fact it is much less. Even for an electric field as strong as $1.4\times 10^6\,\text{volt m}^{-1}$ the substitution of the charge of an electron, 1.6×10^{-19} coulomb, for e, and of $10^{-10}\,\text{m}$ for a, into (6.180) gives only

$$\xi = 10^{-15}\,\text{m}. \tag{6.184}$$

The corresponding dipole moment has magnitude

$$p = 1.6\times 10^{-34}\,\text{coulomb m}. \tag{6.185}$$

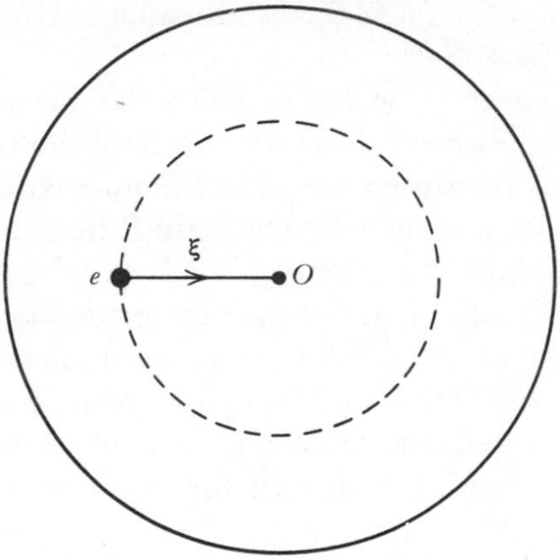

Figure 6.9

Another important feature brought out by the model is that the atom has a natural vibration frequency, ω_0 say, the attractive force associated with displacement $\boldsymbol{\xi}$ being $m\omega_0{}^2\boldsymbol{\xi}$. In terms of this parameter evidently

$$\alpha = \frac{e^2}{\epsilon_0 m\omega_0{}^2}. \tag{6.186}$$

The model predicts

$$\omega_0{}^2 = \frac{e^2}{4\pi\epsilon_0 ma^3}, \tag{6.187}$$

where m, the mass of the negative charge, is much less than that of the positive charge. If e and m have values for the electron, and $a = 10^{-10}$ m, (6.187) yields

$$\omega_0 = 1.6\times 10^{16}\,\text{sec}^{-1}, \tag{6.188}$$

which is the correct order of magnitude for actual atoms. The existence of natural vibration frequencies gives rise to resonances in the response of molecules to a varying field, and thus accounts for so-called 'anomalous' dispersion (see §6.3.3).

Turning now to polar molecules, an order of magnitude estimate for the intrinsic dipole moment would be simply the charge of an electron times, once more, 10^{-10} m. That is,

$$p \sim 1.6\times 10^{-29}\,\text{coulomb m}, \tag{6.189}$$

which is about right. Note that p here exceeds (6.185) by the factor 10^5, despite the large field used to calculate the induced dipole moment. Anything like complete alignment of the polar molecules in a dielectric would produce an enormous effect which is not observed in practice.

6.3.2 *The field acting on a molecule*

One of the difficulties in proceeding from the behaviour of individual molecules to that of dielectrics is in determining the relation between the electric field that is directly responsible for polarizing or orienting each molecule and the space averaged electric field **E** that appears in the macroscopic Maxwell equations. The former is the actual field $\mathbf{E}_m$ acting on the individual molecules, excluding, for each molecule, its 'self' field; and the extent to which $\mathbf{E}_m$ and **E** may differ is now examined.

Consider, at a particular time, a particular molecule, at O, say. Imagine a spherical surface S, with centre at O, that is small on the macroscopic scale but contains many molecules, and write for the molecule at O

$$\mathbf{E}_m = \mathbf{E}_{ext} + \mathbf{E}_{int}, \tag{6.190}$$

where $\mathbf{E}_{ext}$ is the contribution from all sources whatsoever exterior to S, and $\mathbf{E}_{int}$ is the contribution from all molecules interior to S other than the molecule at O. Since a macroscopic analysis is valid for $\mathbf{E}_{ext}$ the calculation described in §6.1.4 that leads to (6.63) gives

$$\mathbf{E}_{ext} = \mathbf{E} + \frac{1}{3\epsilon_0}\mathbf{P}, \tag{6.191}$$

where **E** and **P** are the macroscopic electric and polarization vectors. Thus

$$\mathbf{E}_m = \mathbf{E} + \frac{1}{3\epsilon_0}\mathbf{P} + \mathbf{E}_{int}. \tag{6.192}$$

The simplest medium to consider is a gas dilute enough for both $\mathbf{P}/\epsilon_0$ and $\mathbf{E}_{int}$ to be negligible compared with **E**. That $P \ll \epsilon_0 E$ is synonymous with the fact that the dielectric constant is close to unity. Moreover, a condition sufficient to ensure $E_{int} \ll E$ can be obtained by noting that, were

$$\mathbf{E}_m = \mathbf{E}, \tag{6.193}$$

then, for the time independent case, $\mathbf{E}_{int}$ would be of order $\alpha E/d^3$, where α is the effective molecular polarizability and d the average distance between a molecule and its neighbour. The inequality is therefore verified *a posteriori* provided d is sufficiently great; in deformation polarization, for example, d need only be substantially in excess of the scale of a molecule. In the time harmonic case the dipole field (5.233), (5.234) shows that the result likewise holds up to angular frequencies ω for which the wavelength $2\pi c/\omega$ is greater than d; at higher frequencies the gas cannot legitimately be treated as a continuum.

For other media, in particular liquids and solids, the second term on the right hand side of (6.192) is, of course, significant if the dielectric constant is not close to unity. The importance of the third term, though, can only be assessed by examining particular models. It can be shown that for the special case in which the molecules form a cubic lattice

$$\mathbf{E}_{\text{int}} = 0,$$

giving the result

$$\mathbf{E}_{\text{m}} = \mathbf{E} + \frac{1}{3\epsilon_0}\mathbf{P}. \tag{6.194}$$

This formula may also be applied when the molecules are distributed at random; to dense gases, for example.

The case of gases sufficiently dilute for (6.193) to be valid is now discussed, and some of the consequences of (6.194) are developed in §6.3.4.

6.3.3 *The dielectric constant of dilute gases*

For a gas of non-polar molecules the application of (6.193) is straightforward. Since

$$\mathbf{p} = \epsilon_0 \alpha \mathbf{E}_{\text{m}} \tag{6.195}$$

it follows that

$$\mathbf{P} = \epsilon_0 N \alpha \mathbf{E}. \tag{6.196}$$

Thus **P** is proportional to **E**.

If in general there exists a linear relation of the form

$$\mathbf{P} = \epsilon_0 \chi \mathbf{E}, \tag{6.197}$$

χ is known as the *electric susceptibility*. Then

$$\mathbf{D} = \epsilon_0 \mathbf{E} + \mathbf{P} = \epsilon_0(1+\chi)\mathbf{E},$$

so that the dielectric constant is

$$\kappa = 1 + \chi. \tag{6.198}$$

The present theory gives

$$\chi = N\alpha, \tag{6.199}$$

or in terms of the natural frequency ω_0 in (6.186),

$$\chi = \omega_{\text{p}}^2/\omega_0^2, \tag{6.200}$$

where

$$\omega_{\text{p}}^2 = Ne^2/(\epsilon_0 m). \tag{6.201}$$

The expression for the frequency ω_{p} is formally the same as that for the plasma frequency (6.136), but the symbols N, e and m in the latter refer, of course, to free electrons, whereas here they refer to the number of neutral gas atoms and the electronic charge and mass which each contains.

For a hydrogen atom $\omega_{\text{p}} = 56N^{\frac{1}{2}}$, as in (6.137); and since, at 1 atmosphere and 0 °C, $N = 2.7 \times 10^{25}\,\text{m}^{-3}$, substitution into (6.200) of the experimental value for hydrogen,

$$\chi = 2.7 \times 10^{-4}, \tag{6.202}$$

therefore suggests

$$\omega_0 \simeq 1.8 \times 10^{16}\,\text{sec}^{-1}, \tag{6.203}$$

which is in reasonable agreement with alternative estimates (cf. (6.188)).

The calculation can be extended to the time harmonic case by introducing the inertia term into the equation for the displacement $\boldsymbol{\xi}$ of the electronic charge in the atom. The equation is then

$$\ddot{\boldsymbol{\xi}}+\omega_0^2\boldsymbol{\xi} = -\frac{e}{m}\mathbf{E}, \tag{6.204}$$

so that, for oscillation at angular frequency ω,

$$(\omega_0^2-\omega^2)\boldsymbol{\xi} = -\frac{e}{m}\mathbf{E}, \tag{6.205}$$

and the resulting susceptibility in the form corresponding to (6.200) is

$$\chi = \frac{\omega_p^2}{\omega_0^2-\omega^2}. \tag{6.206}$$

In this over-idealized model resonance at $\omega = \omega_0$ corresponds to an infinity in the dielectric constant. In practice there is some form of damping, and if this is represented in the model by the inclusion of a 'frictional' term $\nu\dot{\boldsymbol{\xi}}$ in the left hand side of (6.204) the complex susceptibility appears as

$$\chi = \frac{\omega_p^2}{\omega_0^2-\omega^2+\mathrm{i}\omega\nu}. \tag{6.207}$$

Separation into real and imaginary parts gives

$$\chi = \chi_r + \mathrm{i}\chi_i, \tag{6.208}$$

where $$\chi_r = \frac{\omega_p^2(\omega_0^2-\omega^2)}{(\omega_0^2-\omega^2)^2+\omega^2\nu^2}, \quad \chi_i = -\frac{\omega_p^2\nu\omega}{(\omega_0^2-\omega^2)^2+\omega^2\nu^2}. \tag{6.209}$$

In considering the frequency dependence of χ it is easy to confirm that, as ω runs from zero to infinity, χ_r passes through stationary values

$$\pm\frac{\omega_p^2}{\nu(2\omega_0\mp\nu)} \tag{6.210}$$

at $$\omega^2 = \omega_0(\omega_0\mp\nu), \tag{6.211}$$

the upper sign specifying a maximum, the lower a minimum. Moreover, $-\chi_i$ has a single stationary value, a maximum, which, for $\nu \ll \omega_0$, is reached close to $\omega = \omega_0$ and is approximately

$$\omega_p^2/(\nu\omega_0). \tag{6.212}$$

The variation of χ_r and $-\chi_i$ with ω is indicated schematically in figure 6.10.

With $|\chi| \ll 1$ the complex refractive index $n = (1+\chi)^{\frac{1}{2}}$ has real and imaginary parts

$$n_r = 1+\tfrac{1}{2}\chi_r, \quad n_i = \tfrac{1}{2}\chi_i, \tag{6.213}$$

approximately. Thus n_r increases with frequency throughout the range from

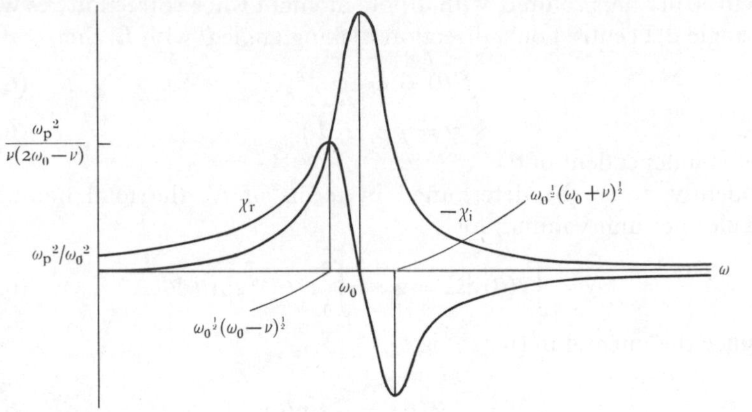

Figure 6.10

zero frequency to a value around that of visible light, in which range the dispersion is traditionally called 'normal'; whereas at higher frequencies, where the dispersion is 'anomalous', n_r first decreases sharply with frequency in the region of predominant absorption, becoming less than unity, and then recovers, ultimately to tend to unity. At X-ray frequencies, and beyond, $\omega \gg \omega_0$ and the electrons are effectively free, with

$$\chi_r \sim -\omega_p^2/\omega^2. \tag{6.214}$$

Actual molecules may have a number of resonant frequencies, each of which contributes its wiggle to the dispersion curve.

Now consider a dilute gas of polar molecules, supposing that each molecule has an intrinsic dipole moment of magnitude p. Under the influence of an electric field there is deformation polarization as already discussed, and also orientation polarization due to some alignment of the dipoles p along the field. The latter is now calculated.

It has already been mentioned that the dipole alignment must be far from complete in practice, since the observed polarization is very much less than the overwhelming amount which full alignment would provide. Thermal agitation tends to randomize the orientations, and a quantitative assessment of its effect in counteracting the field alignment requires the result from statistical mechanics that the relative number of molecules with energy W is proportional to

$$e^{-W/KT}, \tag{6.215}$$

where K is Boltzmann's constant and T is the absolute temperature. The relevant energy here is that of a dipole of moment $\mathbf{p}$ in a field $\mathbf{E}$. This, from (3.41), is

$$W = -\mathbf{p}.\mathbf{E} = -pE\cos\theta, \tag{6.216}$$

where θ is the angle between $\mathbf{p}$ and $\mathbf{E}$. If, therefore, $f(\theta)\,d\Omega$ is the number of molecules, per unit volume, with dipole moment whose direction lies within solid angle $d\Omega$ centred on a direction making angle θ with $\mathbf{E}$, then

$$f(\theta) = n_0 e^{\gamma \cos\theta}, \tag{6.217}$$

where
$$\gamma = pE/(KT), \tag{6.218}$$

and n_0 is independent of θ.

Evidently n_0 can be determined in terms of N, the total number of molecules per unit volume: for

$$N = \int f(\theta)\, d\Omega = 2\pi n_0 \int_0^{\pi} e^{\gamma \cos\theta} \sin\theta\, d\theta; \tag{6.219}$$

and since the integral in (6.219) is

$$\int_{-1}^{1} e^{\gamma x}\, dx = \frac{2}{\gamma} \sinh\gamma \tag{6.220}$$

it follows that
$$n_0 = \frac{N}{4\pi} \frac{\gamma}{\sinh\gamma}. \tag{6.221}$$

The polarization $\mathbf{P}$, which is the average dipole moment per unit volume and is obviously parallel to $\mathbf{E}$, has magnitude

$$\begin{aligned} P &= \int p_0 \cos\theta f(\theta)\, d\Omega \\ &= 2\pi p n_0 \int_{-1}^{1} x e^{\gamma x}\, dx. \end{aligned} \tag{6.222}$$

The integral in (6.222), being the derivative with respect to γ of (6.220), is

$$2\frac{d}{d\gamma}\left(\frac{\sinh\gamma}{\gamma}\right) = 2\frac{\sinh\gamma}{\gamma} L(\gamma), \tag{6.223}$$

where
$$L(\gamma) = \coth\gamma - 1/\gamma \tag{6.224}$$

is called the *Langevin function*. Hence, using (6.221),

$$P = NpL(\gamma). \tag{6.225}$$

The Langevin function has limiting behaviour

$$L(\gamma) = \begin{cases} \frac{1}{3}\gamma + O(\gamma^3) & \text{as} \quad \gamma \to 0, & (6.226) \\ 1 - 1/\gamma + O(e^{-\gamma}) & \text{as} \quad \gamma \to \infty, & (6.227) \end{cases}$$

and is shown in figure 6.11. In practice only the linear approximation (6.226) is relevant to (6.225). For $K = 1.38 \times 10^{-23}$ joule deg^{-1}, and if p is taken to have the value 1.38×10^{-29} coulomb m (cf. (6.189), which is on the high side), then

$$\gamma = 10^{-6} E/T; \tag{6.228}$$

so at room temperature an electric field as strong as 3×10^6 volt m^{-1} only yields $\gamma = 10^{-2}$. Moreover, if the temperature is lowered a change of state occurs before a large value of γ is reached.

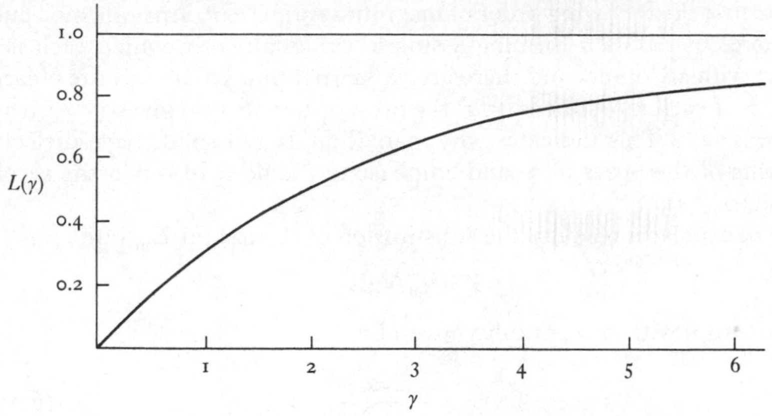

Figure 6.11

Effectively, therefore, for the orientation polarization,

$$\mathbf{P} = \frac{Np^2}{3KT}\mathbf{E}. \tag{6.229}$$

Thus **P** is proportional to **E**, and the corresponding contribution to the susceptibility in the time independent case is

$$\chi = \frac{Np^2}{3\epsilon_0 KT}. \tag{6.230}$$

For a gas at normal temperature and pressure (6.230) gives $\chi = 10^{-3}$ when p is about 2×10^{-30} coulomb m. Such values of p are exceeded in many gases, and the static dielectric constant of such gases is largely determined by the orientation polarization and is markedly higher than that of non-polar gases.

Contrasting features of deformation and orientation polarization that are of particular help in investigating molecular structure through the measurement of dielectric constants are their temperature and frequency dependence. That (6.230) is proportional to $1/T$ can be used to determine p. Also it is the case that as the frequency of the field is increased the rotation of the molecules ceases to respond at a comparatively early stage, so that the dielectric constant can drop far below its static value well before deformation polarization resonance arises.

6.3.4 *The dielectric constant of dense gases, liquids and solids*

When the molecules in a medium are too tightly packed for the simple relation (6.193) to be applicable the problem is inherently much more complicated. Here only the briefest discussion is presented, confined to examining some implications of using (6.194) instead. The latter relation achieves a measure of success where the former fails, but is itself of restricted validity.

Note first the following order of magnitude argument. Imagine molecules as spheres of radius a forming a simple cubic lattice in which each is in contact with six others and there are $1/(8a^3)$ per unit volume. If the polarizability α of each molecule is $4\pi a^3$ the prescription (6.199) gives $\chi = \frac{1}{2}\pi$, and hence $\kappa \simeq 2.6$. This indicates why many liquids and solids have dielectric constants of the order of 3, and emphasizes the need to retain the second term in (6.194).

For non-polar molecules the substitution of (6.194) for $\mathbf{E}_\mathrm{m}$ into

$$\mathbf{P} = \epsilon_0 N\alpha \mathbf{E}_\mathrm{m}$$

yields (6.197) with susceptibility given by

$$\chi = \frac{N\alpha}{1-\frac{1}{3}N\alpha}. \tag{6.231}$$

In the form

$$\frac{\kappa-1}{\kappa+2} = \tfrac{1}{3}N\alpha \tag{6.232}$$

this is known as the Clausius–Mosotti relation.

The analogous result in the time harmonic case is obtained by replacing $\mathbf{E}$ on the right hand side of (6.204) by (6.194), that is by

$$\mathbf{E} - \frac{Ne}{3\epsilon_0}\xi.$$

This simply has the effect of replacing $\omega_0{}^2$ in (6.205) by $\omega_0{}^2 - \frac{1}{3}\omega_\mathrm{p}{}^2$, and the ensuing discussion is only modified in this respect.

Clearly (6.231) is a modification of (6.199), to which it reduces when $N\alpha \ll 3$. When $N\alpha$ is less than but not negligible compared with 3 (6.231) predicts a larger susceptibility than (6.199), and in many cases one in accord with experiment. If α is estimated as α_0 from a measurement of the susceptibility of a dilute gas at atmospheric pressure, measurements at increased pressures, up to 10^3 atmospheres say, show susceptibilities rising to values comparable to unity, being in quantitative agreement with (6.231) for α substantially constant at the value α_0, and *a fortiori* disagreeing with (6.199).

Even more strikingly, similar results have been found for substances compared in gas and liquid phase. For example, an experimental value $\chi = 2.9 \times 10^{-3}$ for CS_2 gas implies $N\alpha = 2.9 \times 10^{-3}$. The assumption that the same value of α holds for the liquid phase, with density 381 times the gas density, therefore gives $N\alpha = 1.105$ for the liquid. The Clausius–Mosotti relation then predicts $\chi = 1.75$ as against a measured value 1.64.

That $N\alpha$ might approach the value 3 suggests the possibility of substances with very high dielectric constants, or even substances that can exist in a polarized state without an external field. Substances capable of spontaneous polarization are called ferro-electric by analogy with ferromagnetism; there are resemblances between the phenomena, but it should be stressed that their origins are dissimilar, for whereas ferro-magnetism is essentially due to the spontaneous alignment of permanent magnetic dipoles (see §6.4.3), ferro-electric substances are non-polar.

Polar liquids present their own problems. If it is supposed that

$$\mathbf{P} = \epsilon_0 \chi' \mathbf{E}_m,$$

where χ' is given by (6.230), then (6.194) implies $\mathbf{P} = \epsilon_0 \chi \mathbf{E}$, where

$$\chi = \frac{\chi'}{1 - \frac{1}{3}\chi'}. \tag{6.233}$$

However, this leads to nonsensical results. For example, χ' is proportional to N, and its measurement for water vapour suggests that it should have the value 13.2 for water. But this makes (6.233) negative, whereas the dielectric constant of water is about 80. Clearly (6.194) is inapplicable.

6.3.5 *The point relation for general time variation*

It has been emphasized that a linear constitutive relation of the form $\mathbf{D} = \epsilon \mathbf{E}$, or equivalently $\mathbf{P} = \epsilon_0 \chi \mathbf{E}$, where $\epsilon = \epsilon_0(1+\chi)$, has comparatively widc applicability only if it is regarded as a relation between the complex representations of the vectors for a time harmonic field of angular frequency ω, and ϵ is recognized to be dependent on ω. Although for practical purposes use of the complex representation leads to the most convenient statement of the constitutive relation, it should be understood that this is synonymous with a linear point relation between the actual vectors, in which $\mathbf{D}(\mathbf{r}, t)$, or $\mathbf{P}(\mathbf{r}, t)$, depend on the values of $\mathbf{E}$ at $\mathbf{r}$ not only at time t but also at all times prior to t. Stated mathematically,

$$\mathbf{P}(\mathbf{r}, t) = \epsilon_0 \int_0^\infty f(\tau)\, \mathbf{E}(t-\tau)\, \mathrm{d}\tau, \tag{6.234}$$

where $f(\tau)$ is the weight to be given to $\mathbf{E}$ at time τ prior to t.

In the time harmonic case (6.234), is, for the complex representation,

$$\mathbf{P} = \epsilon_0 \chi \mathbf{E},$$

where

$$\chi(\omega) = \int_0^\infty f(\tau)\, \mathrm{e}^{-\mathrm{i}\omega\tau}\, \mathrm{d}\tau. \tag{6.235}$$

The form (6.235) is instructive because it leads to some general conclusions about the functional form of $\chi(\omega)$ irrespective of the particular dielectric model. For example, since $f(\tau)$ is of course real, the complex conjugate of χ for real ω is

$$\chi^*(\omega) = \int_0^\infty f(\tau)\, \mathrm{e}^{\mathrm{i}\omega\tau}\, \mathrm{d}\tau = \chi(-\omega). \tag{6.236}$$

The expression (6.207), for example, conforms to this result.

Again, on the reasonable assumption that $f(\tau) \to 0$ as $\tau \to \infty$, the convergence of (6.235) is assured for real ω and the integral tends to zero as $\omega \to \infty$.

Moreover, convergence for real ω *a fortiori* implies convergence for any complex value of ω that has negative imaginary part, so that $\chi(\omega)$ regarded as a function of the complex variable ω has no singularity in the lower half of the complex ω plane, and tends to zero there as $|\omega| \to \infty$. This result depends on the range of integration in (6.235), and hence in (6.234), embracing only positive values of τ; that is to say, on the *causality* requirement that **P** at time t can depend on the values of **E** at times prior to but not subsequent to t.

To take a specific example, suppose that

$$f(\tau) = f_0 \sin(\alpha\tau)\,\mathrm{e}^{-\beta\tau}, \quad \beta > 0. \tag{6.237}$$

By writing $\sin(\alpha\tau)$ in terms of exponentials the integral on the right hand side of (6.235) can be evaluated at once in the form

$$-\tfrac{1}{2}f_0\left(\frac{1}{\omega-\mathrm{i}\beta-\alpha}-\frac{1}{\omega-\mathrm{i}\beta+\alpha}\right), \tag{6.238}$$

which gives

$$\chi = \frac{f_0\alpha}{\beta^2+\alpha^2-\omega^2+2\mathrm{i}\omega\beta}. \tag{6.239}$$

If

$$\beta = \tfrac{1}{2}\nu, \quad \alpha^2 = {\omega_0}^2 - \tfrac{1}{4}\nu^2, \quad f_0 = p^2({\omega_0}^2-\tfrac{1}{4}\nu^2)^{-\frac{1}{2}}, \tag{6.240}$$

then (6.239) is the same as (6.207). Note that, as (6.238) shows explicitly, the singularities of $\chi(\omega)$ are at $\omega = \pm\alpha+\mathrm{i}\beta$; they have a common positive imaginary part, and both therefore lie in the upper half of the complex plane.

6.4 Magnetic media

6.4.1 *Molecular dipole moments*

The primary macroscopic magnetic effect of charges in a medium is expressible in terms of the magnetization **M**. To an adequate approximation **M** is essentially the vector sum per unit volume of the individual molecular magnetic dipole moments as signified in (6.52), the term involving the motion of electric dipoles being ordinarily negligible.

The atoms or molecules of some substances have intrinsic magnetic dipole moments; others do not. The origin of the dipole moment lies both in the orbital motion of electrons round the nucleus and also in the intrinsic 'spin' of each electron, the total moment, possibly zero, being the vector sum of orbital and spin moments. Electron spin is a quantum concept and it is hardly possible to give even the most rudimentary quantitative explanation of magnetic behaviour without reference to the elementary quantum picture of the atom.

To gain some idea of the orders of magnitude involved consider an electron, of charge $-e$ and mass m, in uniform circular motion about a fixed nucleus of charge e. If the angular velocity of the electron is ω and the radius of the circle is a, then classically (but neglecting radiation)

$$ma\omega^2 = \frac{e^2}{4\pi\epsilon_0 a^2}. \tag{6.241}$$

The values of a and ω can be obtained by introducing quantization of the angular momentum. For an order of magnitude estimate take

$$ma^2\omega = \hbar, \tag{6.242}$$

where $\hbar = h/(2\pi)$, h being Planck's constant (1.29). The estimates of a and ω given by (6.241) and (6.242) are

$$a_0 = \frac{4\pi\epsilon_0\hbar^2}{me^2} = 5.3 \times 10^{-11}\,\text{m} \tag{6.243}$$

and

$$\omega_0 = \left(\frac{e^2}{2\epsilon_0 hc}\right)^2 \frac{mc^2}{\hbar} = 4.1 \times 10^{16}\,\text{sec}^{-1}. \tag{6.244}$$

The *Bohr radius* a_0 gives an idea of the size of an atom, and ω_0 of its resonant frequency (cf. the values used in § 6.3).

The magnetic moment of the orbiting electron is given by the product of the circulating current $e\omega/(2\pi)$, which is the rate of passage of charge at any point on the orbit, with the area of the orbit. It is thus

$$p = \tfrac{1}{2}ea^2\omega, \tag{6.245}$$

the symbol p being used here for magnetic dipole moment to avoid confusion with electron mass m. The value of (6.245) if (6.242) is used is

$$p_0 = \frac{e\hbar}{2m} = 0.93 \times 10^{-23}\,\text{amp m}^2, \tag{6.246}$$

known as the *Bohr magneton*. In addition to its relation to the orbital magnetic moment the Bohr magneton is also the value of the spin moment of the electron, and it therefore provides an order of magnitude estimate of the intrinsic moment, if any, of an atom or molecule.

When a molecule is placed in a magnetic field it acquires an induced magnetic moment whether or not it has an intrinsic moment. The induced moment is in the opposite direction to the field and is the origin of the diamagnetic effect mentioned in § 4.5.2.

To appreciate how the induced moment arises consider a model in which an electron at $\mathbf{r}(t)$ is in orbital motion, about a nucleus at the origin, under the influence of some binding radial force $F(r)$. The equation of motion of the electron is

$$m\frac{d^2\mathbf{r}}{dt^2} = F(r)\,\hat{\mathbf{r}}, \tag{6.247}$$

where $\hat{\mathbf{r}}$ is the unit vector along $\mathbf{r}$. If now a magnetic field $\mathbf{B}$ is applied the equation is

$$\frac{d^2\mathbf{r}}{dt^2} = F(r)\,\hat{\mathbf{r}}/m - \frac{d\mathbf{r}}{dt} \wedge \mathbf{\Omega}, \tag{6.248}$$

where

$$\mathbf{\Omega} = e\mathbf{B}/m. \tag{6.249}$$

The magnetic force term is reminiscent of the Coriolis force associated with motion relative to rotating axes, and a convenient way of assessing its effect is to refer the motion to axes with the same origin but rotating with a suitably chosen constant angular velocity, $\boldsymbol{\omega}$ say. If $\boldsymbol{\rho} = (\xi, \eta, \zeta)$ is the vector whose components are the coordinates of the electron referred to the rotating axes, then the components along the rotating axes of the velocity of the electron are specified by $\dot{\boldsymbol{\rho}} + \boldsymbol{\omega} \wedge \boldsymbol{\rho}$, where $\dot{\boldsymbol{\rho}} = (\mathrm{d}\xi/\mathrm{d}t, \mathrm{d}\eta/\mathrm{d}t, \mathrm{d}\zeta/\mathrm{d}t)$, and of the acceleration by

$$\ddot{\boldsymbol{\rho}} + 2\boldsymbol{\omega} \wedge \dot{\boldsymbol{\rho}} + \boldsymbol{\omega} \wedge (\boldsymbol{\omega} \wedge \boldsymbol{\rho}). \tag{6.250}$$

The second and third terms in (6.250) yield the Coriolis and centrifugal forces respectively, the equation of motion being

$$\ddot{\boldsymbol{\rho}} + 2\boldsymbol{\omega} \wedge \dot{\boldsymbol{\rho}} + \boldsymbol{\omega} \wedge (\boldsymbol{\omega} \wedge \boldsymbol{\rho}) = F(\rho)\hat{\boldsymbol{\rho}}/m - (\dot{\boldsymbol{\rho}} + \boldsymbol{\omega} \wedge \boldsymbol{\rho}) \wedge \boldsymbol{\Omega}. \tag{6.251}$$

With the choice

$$\boldsymbol{\omega} = \tfrac{1}{2}\boldsymbol{\Omega} \tag{6.252}$$

the equation appears in the comparatively simple form

$$\ddot{\boldsymbol{\rho}} = F(\rho)\hat{\boldsymbol{\rho}}/m + \tfrac{1}{4}\boldsymbol{\Omega} \wedge (\boldsymbol{\Omega} \wedge \boldsymbol{\rho}). \tag{6.253}$$

Moreover the last term is negligible in practice, for the orbital angular frequency is of order ω_0, given by (6.244), and $\Omega \ll \omega_0$ provided B is less than about 10^3 weber m^{-2} (a field of 10^7 gauss). Finally, therefore,

$$m\ddot{\boldsymbol{\rho}} = F(\rho)\hat{\boldsymbol{\rho}}, \tag{6.254}$$

showing by comparison with (6.247) that in the presence of a magnetic field motion relative to a coordinate system rotating with angular velocity $\frac{1}{2}\boldsymbol{\Omega}$ replicates motion in the absence of the field. It may be added that the statement clearly remains true if in addition to the force $F(r)\hat{\mathbf{r}}$ there is a force due to interaction with other electrons that depends only on the relative configuration of the electrons.

That each orbiting electron acquires an extra angular velocity $e\mathbf{B}/(2m)$ means that each contributes an additional magnetic moment directed in opposition to $\mathbf{B}$. A total moment is thus induced in a molecule irrespective of the existence of a moment in the absence of a field.

The order of magnitude of the induced moment can be estimated by again considering an electron in uniform circular motion of radius a. The extra moment produced by the application of field $\mathbf{B}$ normal to the plane of the motion is given by (6.245) with $eB/(2m)$ replacing ω. It is therefore

$$\mathbf{p} = -\frac{e^2a^2}{4m}\mathbf{B}, \tag{6.255}$$

which with the value (6.243) for a gives

$$\mathbf{p} = -1.9 \times 10^{-29}\mathbf{B}. \tag{6.256}$$

Even for a field of 5×10^4 gauss the induced moment is therefore only about 10^{-28} amp m^2, very much less than (6.246).

6.4.2 *Diamagnetism and paramagnetism*

If the selective injection of rudimentary quantum concepts is accepted the discussion of §6.4.1 suggests that, although the physical mechanisms are quite distinct, the response of substances to an applied magnetic field involves many of the same considerations as the response to an applied electric field.

Substances whose atoms or molecules have no intrinsic magnetic moment are analogous to those with non-polar molecules. Their magnetic effect is due to the moments induced by a field, and since the direction of these moments is opposite to that of the field the substance is diamagnetic. Now it is known experimentally that the specific permeability μ/μ_0 even of diamagnetic solids is less than unity only by a quantity of order 10^{-5}. The average field everywhere therefore differs very little from the applied field ($\mu_0 M \ll B$), as for the dilute gases considered in the electric case ($P \ll \epsilon_0 E$), and it would be expected that the field $\mathbf{B}_m$ acting at each molecule to create its moment could be taken to be the average field $\mathbf{B}$ without any appreciable error. Thus if (6.255) were used the moment $N\mathbf{p}$ per unit volume can be written

$$\mathbf{M} = -\frac{Ne^2a^2}{4m}\mathbf{B}. \tag{6.257}$$

Now when $\mathbf{M}$ is proportional to $\mathbf{B}$, as is the case in (6.257), it is also proportional to $\mathbf{H}$; and since the magnetic susceptibility χ happens to be defined by the relation

$$\mathbf{M} = \chi\mathbf{H}, \tag{6.258}$$

the relation $\mathbf{B} = \mu_0(\mathbf{H}+\mathbf{M})$, as in (6.7), gives $\mathbf{B} = \mu\mathbf{H}$, where

$$\mu = \mu_0(1+\chi). \tag{6.259}$$

But for $\chi \lesssim 10^{-5}$ it is pointless in (6.257) to distinguish $\mathbf{B}$ from $\mu_0\mathbf{H}$, and the theory therefore predicts

$$\chi = -\frac{\mu_0 Ne^2a^2}{4m}. \tag{6.260}$$

For a solid, N might be about 3×10^{29} m^{-3}; and since μ_0 is $4\pi\times10^{-7}$ henry m^{-1} the numerical value of $e^2a^2/(4m)$ appearing in (6.256) gives

$$\chi = -0.7\times10^{-5}, \tag{6.261}$$

which is the right order of magnitude. Being proportional to N the susceptibility of a diamagnetic gas is much smaller still. For specific substances a quantitative improvement in the formula (6.260) is achieved by averaging a^2 and summing over all electrons in an atom.

Substances whose atoms or molecules have an intrinsic magnetic moment are analogous to those with polar molecules. In addition to the diamagnetism there is a paramagnetic effect due to the tendency of the molecular moments to align themselves in the direction of the field. The alignment is opposed by thermal agitation, and the calculation of the effect proceeds in a manner entirely analogous to that which led to (6.225). Since the energy of a given dipole of moment $\mathbf{p}$ in a field $\mathbf{B}$ is $-\mathbf{p}.\mathbf{B}$ (cf. (4.142)) the result is

$$M = NpL(\gamma), \tag{6.262}$$

where
$$\gamma = \frac{pB}{KT}, \tag{6.263}$$

and L is the Langevin function (6.224).

If p has the value (6.246) of the Bohr magneton,

$$\gamma = 2.3 \times 10^{-3} B \tag{6.264}$$

at $T = 290$ °K, and at room temperature and above is therefore much less than unity even for magnetic fields as high as 10 weber m^{-2} (10^5 gauss). In this case $L(\gamma)$ is approximately $\frac{1}{3}\gamma$ (see (6.226)) and

$$M = \frac{Np^2}{3KT} B. \tag{6.265}$$

The corresponding paramagnetic susceptibility, though overwhelming the diamagnetic contribution, is still much less than unity, so B in (6.265) can be replaced by $\mu_0 H$ to yield

$$\chi = \frac{\mu_0 N p^2}{3KT}. \tag{6.266}$$

With the value (6.246) for p this gives, for $N = 3 \times 10^{29}$ m^{-3} and $T = 290$ °K,

$$\chi = 2.7 \times 10^{-3}, \tag{6.267}$$

which is the right order of magnitude.

That the susceptibility of paramagnetic substances is inversely proportional to temperature is in agreement with experiment, being known as Curie's law; p may be found from (6.266) by a careful measurement of χ as a function of temperature.

In contrast to the analogous electrostatic result the formula (6.262) can be applicable at such low temperatures that γ is comparable to unity. At $T = 1$ °K, for example, (6.264) is replaced by $\gamma = 0.67B$, and therefore exceeds unity for $B > 1.5$ weber m^{-2}. Of course the question of whether it is legitimate to take the molecular field $\mathbf{B}_\text{m}$ to be the average field $\mathbf{B}$ should be reconsidered. At room temperature the identification might be accepted on the grounds that $\mu_0 M \ll B$. On the other hand if $\gamma = 1$ then $L(\gamma) \simeq 0.3$, and with $p = 0.9 \times 10^{-23}$ amp m^2, $N = 3 \times 10^{29}$ m^{-3}, the formula (6.262) gives $M = 0.8 \times 10^6$ amp m^{-1}. Thus if by analogy with (6.194) $\mathbf{B}_\text{m}$ were representable as $\mathbf{B} + \frac{1}{3}\mu_0 \mathbf{M}$ the magnitude of the second term would be about

0.3 weber m^{-2} compared with $B = 1.5$ weber m^{-2}, which suggests that $\mathbf{B}_m = \mathbf{B}$ could still be a fair approximation.

On the other hand there are plenty of cases in which the distinction between $\mathbf{B}_m$ and $\mathbf{B}$ is of considerable or even overriding significance; the phenomenon of ferromagnetism, in particular, has its origin in the fact that $\mathbf{B}_m$ can be much greater than $\mathbf{B}$, and indeed exceed any readily produced laboratory field.

It is possible to justify use of the formula

$$\mathbf{B}_m = \mu_0(\mathbf{H} + \lambda\mathbf{M}), \tag{6.268}$$

where the value of the dimensionless constant λ is characteristic of the substance. The calculation of λ depends on a quantum theoretical analysis of the interaction between neighbouring atoms. On the basis of (6.268) the result (6.262) is modified only in that (6.263) is replaced by

$$\gamma = \frac{\mu_0 p}{KT}(H + \lambda M). \tag{6.269}$$

This in turn implies that, for the case $\gamma \ll 1$, (6.265) is replaced by

$$M = \frac{\mu_0 N p^2}{3KT}(H + \lambda M), \tag{6.270}$$

which yields (cf. (6.233))

$$\chi = \frac{\chi'}{1 - \lambda\chi'}, \tag{6.271}$$

where χ' now stands for the expression in (6.266).

To show explicitly the dependence of the susceptibility on temperature (6.271) is written

$$\chi = \frac{C}{T - \theta} \tag{6.272}$$

where

$$\theta = \lambda C = \lambda\frac{\mu_0 N p^2}{3K}. \tag{6.273}$$

The temperature dependence exhibited in (6.272) embodies the Curie–Weiss law. The 'Weiss constant' θ may be positive or negative, and the law accounts adequately for the behaviour of many paramagnetic substances in temperature ranges beyond $|\theta|$.

Since (6.270) can be written $\lambda(T/\theta - 1)M = H$ the reduction of T to θ, when θ is positive, suggests that M can remain finite for an arbitrarily small value of H. This observation is a clue to the understanding of ferromagnetism.

6.4.3 *Ferromagnetism*

A substance that remains 'spontaneously' magnetized in the absence of an applied field is the commonplace permanent magnet. The theoretical possibility is hinted at in the Curie–Weiss law, which suggests an infinitely large susceptibility at a characteristic temperature θ. The Curie–Weiss law, however, is only valid when $\gamma \ll 1$, and a much more generally significant result is readily obtained by examining the implication of setting **H** zero when this restriction is dropped.

When $\mathbf{H} = 0$ the relevant relations are

$$\mathbf{B}_\mathrm{m} = \mu_0 \lambda \mathbf{M}, \quad M = NpL(\gamma), \quad \gamma = pB_\mathrm{m}/(KT). \tag{6.274}$$

These are self-consistent provided

$$L(\gamma) = \frac{T}{3\theta}\gamma, \tag{6.275}$$

where θ is given by (6.273).

The condition (6.275) is easily visualized in terms of the intersection of the curve $L(\gamma)$, shown in figure 6.11, with the straight line through the origin of slope $T/(3\theta)$. Since $L(\gamma)$ has slope $\frac{1}{3}$ at the origin it is clear that if $T > \theta$ there is no intersection at a positive value of γ, whereas if $T < \theta$ there is just one such intersection. In other words, spontaneous magnetization is possible at temperatures below a certain temperature characteristic of the substance, known at the Curie temperature.

The ferromagnetic elements iron, nickel and cobalt are found to exhibit spontaneous magnetization below the respective temperatures 1043, 631 and 1394 °K. Above these Curie temperatures they behave as paramagnetic substances obeying a Curie–Weiss law. On the theory given, the Weiss constant θ in (6.272) is, of course, the same as the Curie temperature; however, fitting experimental results to the Curie–Weiss law yields a somewhat different value.

The measured values of the Curie temperatures of the ferromagnetic elements show the enormous strengths of their molecular fields $\mathbf{B}_\mathrm{m}$. The values used in § 6.4.2 indicate that C, defined in (6.273), is of the order of 1 °K, so that λ is of the order of the value of θ in °K. If, therefore, $\mu_0 M$ is 10^4 gauss a measured value of θ of the order 10^3 °K implies a molecular field $\mathbf{B}_\mathrm{m}$ around 10^7 gauss.

Finally it should not go unnoticed that a theory of ferromagnetism on the lines outlined seems to overplay its hand by implying that below the Curie temperature a ferromagnetic substance would always be spontaneously magnetized rather than require magnetizing in the manner described in the discussion on hysteresis in § 4.5.2. The explanation is that spontaneous magnetization is indeed always present in small 'domains', whose dimensions are perhaps a modest fraction of a millimetre, but that the directions of magnetization in the different domains can be so random that a bulk specimen containing many domains exhibits no overall magnetization. The directions can, however, be readily aligned by the application of an external magnetic field. Domain boundaries have been observed directly in the pattern of finely powdered magnetic particles scattered on the surface of a highly polished ferromagnetic crystal.

Problems 6

6.1 Show that, where there is an electrostatic field **E** but no free charge, the bound charge density in a medium of permittivity $\epsilon(\mathbf{r})$ is

$$\rho_b = -\epsilon_0 \frac{\text{grad}\,\epsilon}{\epsilon}.\mathbf{E}.$$

A capacitor whose plates are concentric spheres of radii a and b ($> a$) is filled with a dielectric whose permittivity ϵ is a function of distance r from the centre of the spheres. Show that the volume density of bound charge is

$$\rho_b = -\frac{q\epsilon_0}{4\pi r^2 \epsilon^2}\frac{\partial \epsilon}{\partial r},$$

where q is the free charge on the inner sphere.

Find also the surface density of bound charge, and confirm by integrating ρ_b throughout $a < r < b$ that the net bound charge is zero.

6.2 Establish that the vector potential is continuous across a surface that carries a steady surface current density, and deduce that it is continuous across the surface of a magnetized body.

Find, in suitable spherical polar components, the vector potential of a uniformly magnetized sphere with magnetization **M**.

6.3 Show that, where there is a magnetostatic field **H** but no free current, the bound current density in a medium of permeability $\mu(\mathbf{r})$ is

$$\mathbf{J}_b = -\frac{1}{\mu_0}\mathbf{H} \wedge \text{grad}\,\mu.$$

Uniform surface current density j flows round a long cylindrical solenoid which is parallel to the z axis. The cylinder has rectangular cross-section, and is filled with a medium of permeability $\mu(1 + x/l)$, where μ and l are constants and x is distance from one face of the cylinder. Show that the volume density of bound current in the medium is

$$\mathbf{J}_b = -\frac{j\mu}{l\mu_0}(0, 1, 0).$$

Find also the surface density of bound current, and confirm by integrating $\mathbf{J}_b$ over unit length of the interior of the solenoid that the net integrated bound current is zero.

6.4 Show that no spherically symmetric charge distribution of zero net charge can have a dipole moment.

A charge distribution of density $\rho(\mathbf{r})$ has net charge q. What is the dipole moment of the distribution with density $\rho(\mathbf{r}) - \rho(\mathbf{r} - \mathbf{l})$, where $\mathbf{l}$ is a constant vector?

What is the dipole moment of charges q and $-q$ uniformly distributed throughout spheres of radii a and b, respectively, whose centres are distance d apart?

6.5 Consider the electrostatic field of a parallel plate capacitor at a great distance from the capacitor. If the plates of the capacitor carry given charges $\pm q$, state the dependence of the far field on (*a*) the area of the plates, (*b*) the distance between the plates, (*c*) the distance from the plates, (*d*) the orientation of the plates, (*e*) the permittivity of the dielectric between the plates.

6.6 A permanent magnet in the form of a uniformly magnetized sphere is placed with its centre at O in a field due to other permanent magnets. Explain why the force on the sphere is the force that would be exerted (in the absence of the sphere) on a certain dipole at O.

Two permanent magnets, each consisting of uniformly magnetized material between concentric spheres of radii a and b ($> a$), are placed with their directions of magnetization along the line of centres, which are distance d ($> 2b$) apart. Find the force between the magnets if each has magnetization $\mathbf{M}$.

6.7 A dielectric of permittivity $\epsilon(\mathbf{r})$ is placed in a given electrostatic field $\mathbf{E}_0(\mathbf{r})$. Show that if ϵ is everywhere sufficiently close to ϵ_0 the dielectric is approximately equivalent to a dipole moment density $(\epsilon-\epsilon_0)\,\mathbf{E}_0$.

Examine the case in which the dielectric is a homogeneous slab of susceptibility χ, and $\mathbf{E}_0$ is a uniform field normal to the plane face of the slab. Show that the approximation gives the field in the slab as $(1-\chi)\,\mathbf{E}_0$. Compare this with the exact field, and show that the latter can be recovered, in the form of an infinite series, by iterating the approximation.

6.8 Diamagnetic material of volume 10^{-6} m^3 and susceptibility -10^{-5} is placed in a field of 1 weber m^{-2}. Find the magnitude of the additional field due to the diamagnetic material, at a point distance 1 m from it in the direction of the applied field.

6.9 Show that for a magnetic field solely due to a spatially bounded distribution of permanent magnetization the integral of $\mathbf{H}.\mathbf{B}$ over all space is zero.

6.10 A long cylindrical solenoid of radius a has N turns per unit length and carries a steady current I. Find the magnetic force on a long cylindrical rod of the same radius and uniform permeability μ, if it is held with a length x ($\gg a$) inserted into the solenoid.

6.11 Establish directly the equivalence of (6.91) and (6.92).

6.12 A small piece of paramagnetic matter of susceptibility χ ($\ll 1$) is in a vacuum in a field $\mathbf{B} = \mu_0\mathbf{H}$ created by localized permanent magnets. If the piece of matter has volume V write down the force on it, and show that the work done in withdrawing it to a region where the field is negligible is $\frac{1}{2}\mu_0\chi V\mathbf{H}^2$. Check that the result can be obtained from field energy considerations.

If the operation were repeated with a small permanent magnet of moment $\mathbf{m}$, initially parallel to $\mathbf{B}$, replacing the piece of paramagnetic matter, what value of m would give the same amount of work?

6.13 Use the stress tensor to show that the force per unit area at a point P on a surface carrying steady current density $\mathbf{j}$ in a magnetostatic field is $\frac{1}{2}\mathbf{j}\wedge(\mathbf{B}_1+\mathbf{B}_2)$, where the suffixes 1 and 2 signify the fields at P on respective sides of the surface.

6.14 Use the stress tensor to show that no distribution of stationary charge and steady current exerts a net force on itself.

6.15 Prove that if an actual charge distribution has zero net charge, and dipole moment $\mathbf{p}$, then so also has the average distribution (6.30).

6.16 Find from (6.30) the average charge density $\bar{\rho}$ of an infinitesimal electric dipole of moment $\mathbf{p}$ in the cases (*a*) $W(\mathbf{R})$ given by (6.32), (*b*) $W(\mathbf{R})$ proportional to $\exp(-R^2/l^2)$. At what points does $\bar{\rho}$ have its maximum and minimum values?

$$\left[\int_{-\infty}^{\infty} \mathrm{e}^{-x^2/l^2}\,\mathrm{d}x = \pi^{\frac{1}{2}} l.\right]$$

6.17 Consider the situation described towards the end of §6.2.2 in which current I flowing along a rectangular strip in a magnetic field gives rise to a Hall voltage across the strip. Suppose now that the edges of the strip are joined through a resistance R so that a fraction of the Hall current is allowed to flow. Find the amount by which the voltage applied to maintain current I along the strip must be increased, and check that the further power demanded is equal to that dissipated in R.

6.18 Consider the analysis of current flow in a conducting medium that led to (6.127). Write down (without approximation) the conductivities σ_1 and σ_2 in the relation

$$\mathbf{J} = (\sigma_1, 0, \sigma_2)\, E$$

for the case $\mathbf{E} = (E, 0, 0)$, $\mathbf{B} = (0, B, 0)$.

When the conductor is bounded by planes z = constant, surface charge is established on the planes. This charge gives rise to a z component of electric field, which in the steady state inhibits current flow in the z direction. Show that the corresponding conductivity in the relation $\mathbf{J} = (\sigma_3, 0, 0)\, E$ satisfies $\sigma_3 = \sigma_1 + \sigma_2^2/\sigma_1$.

6.19 Suppose that the motion of a (possibly fluid) conductor is specified by velocity $\mathbf{u}$, which in general is a function of position and time. Show that the replacement of $\mathbf{v}$ by $\mathbf{v}-\mathbf{u}$ where appropriate in the analysis leading to (6.128) establishes that that result has only to be modified by replacing $\mathbf{E}$ by $\mathbf{E}+\mathbf{u} \wedge \mathbf{B}$. What conclusion can be drawn for the case $\sigma \to \infty$?

6.20 An ionized gas occupies the slab $0 < z < a$, and in it there flows a uniform current density $(J, 0, 0)$. Calculate the magnetic field at all points, assuming the fields in vacuum on either side of the slab are equal and opposite. Evaluate $\mathbf{J} \wedge \mathbf{B}$, and find the pressure distribution in the gas which maintains hydrostatic equilibrium.

6.21 Uniform electron and positive ion distributions are confined to a slab of finite thickness with unbounded, parallel plane faces, the whole being neutral. The slab of electrons is now displaced bodily through a small distance x perpendicular to the faces of the slab, the ions remaining fixed. Find the restoring force on each electron.

Find the restoring force when the slab is replaced by an infinitely long circular cylinder, the displacement being perpendicular to its axis.

6.22 How are the resonance frequencies of a cavity in the form of a rectangular box with perfectly conducting walls affected when the box contains uniform plasma?

6.23 Obtain for the refractive index of a (neutral) plasma an expression that takes account of free positive ions as well as electrons. Is the modification to (6.138) significant?

6.24 The plane wave $\mathbf{H} = (0, 0, 1) \exp[-ik_0(x \cos\alpha + y \sin\alpha)]$ is incident from the vacuum half-space $y < 0$ on a homogeneous collisionless electron plasma occupying $y > 0$. Show that the reflection coefficient is

$$\rho = \frac{n \sin\alpha - \sin\beta}{n \sin\alpha + \sin\beta},$$

where $\cos\beta = (1/n)\cos\alpha$, $n = (1 - p^2/\omega^2)^{\frac{1}{2}}$, and p is the plasma frequency.

Consider the value α_B of α for which $\rho = 0$. Show that α_B is real when $\omega > p$ and pure imaginary when $\omega < p/\sqrt{2}$. Indicate the structure of the complete field when $\alpha = \alpha_B$ in each of these cases, noting that the latter involves surface waves that decay in directions away from the interface $y = 0$.

What does $\alpha = \alpha_B$ yield when $p/\sqrt{2} < \omega < p$?

6.25 In a simple theory of ionospheric reflection of radio waves it is supposed that the ionosphere is a collisionless electron plasma in which the number $N(z)$ of electrons per unit volume is a function of height z; and that a normally incident wave of frequency f is reflected at the height $h(f)$ at which the refractive index (6.138) is zero. Find h if

$$N = \begin{cases} 0 & \text{for } z < h_0, \\ N_0(z-h_0)/H & \text{for } z > h_0, \end{cases}$$

where N_0, h_0 and H are constants; and show that in this case the so-called 'equivalent height of reflection' defined by

$$h'(f) \equiv c \int_0^h \frac{dz}{v_g},$$

where v_g is the group velocity (6.141), is given by $h' = 2h - h_0$.

6.26 In the operation of the model of a plane diode analysed in §6.2.5 find as a function of time t the displacement of an electron emitted from the cathode at $t = 0$.

APPENDIX

A.1 *Formulae and theorems in vector calculus*

The components of grad ψ, div $\mathbf{A}$ and curl $\mathbf{A}$ in simple coordinate systems

Rectangular cartesians x, y, z

$$\operatorname{grad} \psi = \left(\frac{\partial\psi}{\partial x}, \frac{\partial\psi}{\partial y}, \frac{\partial\psi}{\partial z}\right), \tag{A.1}$$

$$\operatorname{div} \mathbf{A} = \frac{\partial A_x}{\partial x} + \frac{\partial A_y}{\partial y} + \frac{\partial A_z}{\partial z}, \tag{A.2}$$

$$\operatorname{curl} \mathbf{A} = \left(\frac{\partial A_z}{\partial y} - \frac{\partial A_y}{\partial z}, \frac{\partial A_x}{\partial z} - \frac{\partial A_z}{\partial x}, \frac{\partial A_y}{\partial x} - \frac{\partial A_x}{\partial y}\right), \tag{A.3}$$

$$\nabla^2\Psi = \operatorname{div} \operatorname{grad} \psi = \frac{\partial^2\psi}{\partial x^2} + \frac{\partial^2\psi}{\partial y^2} + \frac{\partial^2\psi}{\partial z^2}, \tag{A.4}$$

$$\nabla^2\mathbf{A} = \operatorname{grad} \operatorname{div} \mathbf{A} - \operatorname{curl} \operatorname{curl} \mathbf{A} = (\nabla^2 A_x, \nabla^2 A_y, \nabla^2 A_z). \tag{A.5}$$

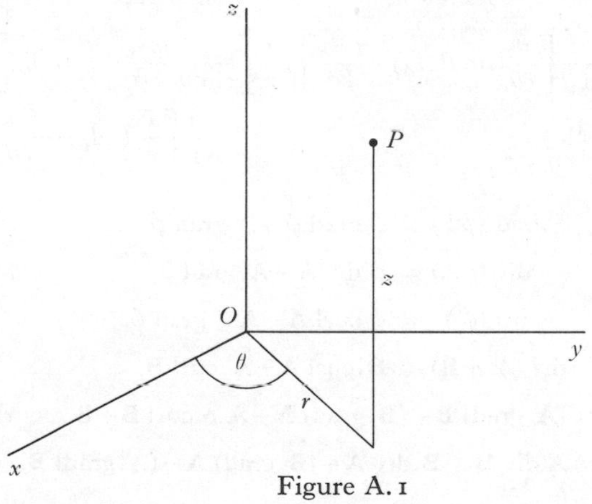

Figure A.1

Cylindrical polars r, θ, z (see figure A.1)

$$\operatorname{grad} \psi = \left(\frac{\partial\psi}{\partial r}, \frac{1}{r}\frac{\partial\psi}{\partial\theta}, \frac{\partial\psi}{\partial z}\right), \tag{A.6}$$

$$\operatorname{div} \mathbf{A} = \frac{1}{r}\frac{\partial}{\partial r}(rA_r) + \frac{1}{r}\frac{\partial A_\theta}{\partial\theta} + \frac{\partial A_z}{\partial z}, \tag{A.7}$$

$$\operatorname{curl} \mathbf{A} = \left[\frac{1}{r}\frac{\partial A_z}{\partial\theta} - \frac{\partial A_\theta}{\partial z}, \frac{\partial A_r}{\partial z} - \frac{\partial A_z}{\partial r}, \frac{1}{r}\frac{\partial}{\partial r}(rA_\theta) - \frac{1}{r}\frac{\partial A_r}{\partial\theta}\right]. \tag{A.8}$$

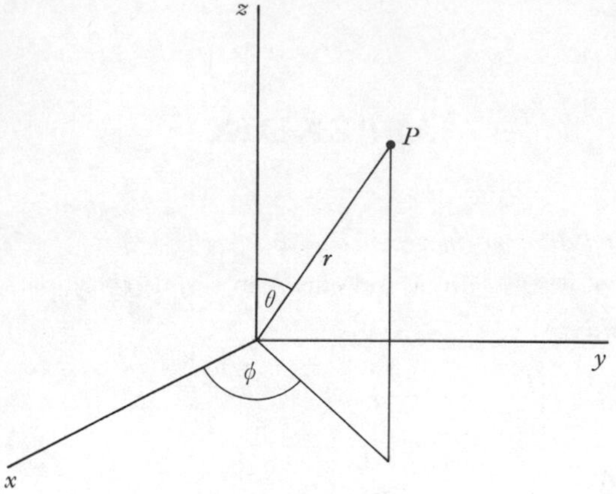

Figure A.2

Spherical polars r, θ, φ (see figure A.2)

$$\operatorname{grad}\psi = \left(\frac{\partial\psi}{\partial r}, \frac{1}{r}\frac{\partial\psi}{\partial\theta}, \frac{1}{r\sin\theta}\frac{\partial\psi}{\partial\phi}\right), \tag{A.9}$$

$$\operatorname{div}\mathbf{A} = \frac{1}{r^2}\frac{\partial}{\partial r}(r^2A_r) + \frac{1}{r\sin\theta}\frac{\partial}{\partial\theta}(\sin\theta A_\theta) + \frac{1}{r\sin\theta}\frac{\partial A_\phi}{\partial\phi}, \tag{A.10}$$

$$\operatorname{curl}A = \left\{\frac{1}{r\sin\theta}\left[\frac{\partial}{\partial\theta}(\sin\theta\, A_\phi) - \frac{\partial A_\theta}{\partial\phi}\right], \frac{1}{r}\left[\frac{1}{\sin\theta}\frac{\partial A_r}{\partial\phi} - \frac{\partial}{\partial r}(rA_\phi)\right], \frac{1}{r}\left[\frac{\partial}{\partial r}(rA_\theta) - \frac{\partial A_r}{\partial\theta}\right]\right\}. \tag{A.11}$$

Some identities are

$$\operatorname{grad}(\phi\psi) = \phi\operatorname{grad}\psi + \psi\operatorname{grad}\phi, \tag{A.12}$$

$$\operatorname{div}(\phi\mathbf{A}) = \phi\operatorname{div}\mathbf{A} + \mathbf{A}.\operatorname{grad}\phi, \tag{A.13}$$

$$\operatorname{curl}(\phi\mathbf{A}) = \phi\operatorname{curl}\mathbf{A} - \mathbf{A}\wedge\operatorname{grad}\phi, \tag{A.14}$$

$$\operatorname{div}(\mathbf{A}\wedge\mathbf{B}) = \mathbf{B}.\operatorname{curl}\mathbf{A} - \mathbf{A}.\operatorname{curl}\mathbf{B}, \tag{A.15}$$

$$\operatorname{grad}(\mathbf{A}.\mathbf{B}) = (\mathbf{A}.\operatorname{grad})\mathbf{B} + (\mathbf{B}.\operatorname{grad})\mathbf{A} + \mathbf{A}\wedge\operatorname{curl}\mathbf{B} + \mathbf{B}\wedge\operatorname{curl}\mathbf{A}, \tag{A.16}$$

$$\operatorname{curl}(\mathbf{A}\wedge\mathbf{B}) = \mathbf{A}\operatorname{div}\mathbf{B} - \mathbf{B}\operatorname{div}\mathbf{A} + (\mathbf{B}.\operatorname{grad})\mathbf{A} - (\mathbf{A}.\operatorname{grad})\mathbf{B}. \tag{A.17}$$

The divergence theorem

$$\int_V \operatorname{div}\mathbf{A}\,d\tau = \int_S \mathbf{A}.d\mathbf{S}, \tag{A.18}$$

where V is the region bounded by the closed surface S.

Related results are

$$\int_V \operatorname{grad}\phi\,d\tau = \int_S \phi\,d\mathbf{S}, \tag{A.19}$$

$$\int_V \operatorname{curl}\mathbf{A}\,d\tau = -\int_S \mathbf{A}\wedge d\mathbf{S}. \tag{A.20}$$

Stokes' theorem

$$\int_S (\text{curl } \mathbf{A}) . \mathrm{d}\mathbf{S} = \oint_C \mathbf{A} . \mathrm{d}\mathbf{s}, \tag{A.21}$$

where S is an open surface bounded by the closed curve C.

A related result is

$$\int_S \mathrm{d}\mathbf{S} \wedge \text{grad } \phi = \oint_C \phi \, \mathrm{d}\mathbf{s}. \tag{A.22}$$

A.2 *A theorem pertaining to the vector potential*

Theorem. If div $\mathbf{B} = 0$ there is a vector $\mathbf{A}$ such that $\mathbf{B} = \text{curl } \mathbf{A}$.

Proof. Take $\mathbf{A} = (A_x, A_y, 0)$. Then

$$\text{curl } \mathbf{A} = \left(-\frac{\partial A_y}{\partial z}, \frac{\partial A_x}{\partial z}, \frac{\partial A_y}{\partial x} - \frac{\partial A_x}{\partial y} \right),$$

and, for a given vector $\mathbf{B}$, the x and y components of curl $\mathbf{A} = \mathbf{B}$ are satisfied by

$$A_y = -\int_{z_0}^{z} B_x(x, y, z')\,\mathrm{d}z' + f(x, y), \tag{A.23}$$

$$A_x = \int_{z_0}^{z} B_y(x, y, z')\, dz' + g(x, y), \tag{A.24}$$

where z_0 is an arbitrary constant, and f and g are functions independent of z but otherwise arbitrary.

Now div $\mathbf{B} = 0$ is

$$\frac{\partial B_x}{\partial x} + \frac{\partial B_y}{\partial y} = -\frac{\partial B_z}{\partial z},$$

and (A.23) and (A.24) therefore give

$$\frac{\partial A_y}{\partial x} - \frac{\partial A_x}{\partial y} = B_z(x, y, z) - B_z(x, y, z_0) + \frac{\partial f}{\partial x} - \frac{\partial g}{\partial y}.$$

Since f and g can evidently be chosen to make the right hand side just $B_z(x, y, z)$ (for example, $g = 0, f = \int_{x_0}^{x} B(x', y, z_0)\,\mathrm{d}x'$ will do), the z component of curl $\mathbf{A} = \mathbf{B}$ can also be satisfied.

A.3 *Mks to Gaussian units*

To convert any formula or equation from rationalized mks units to Gaussian units (unrationalized, with current in electrostatic units), replace

ϵ_0	by	$1/(4\pi)$
ϵ	by	$\epsilon/(4\pi)$
μ_0	by	$4\pi/c^2$
μ	by	$4\pi\mu/c^2$
$\mathbf{B}$	by	$\mathbf{B}/c$
$\mathbf{H}$	by	$c\mathbf{H}/(4\pi)$
$\mathbf{D}$	by	$\mathbf{D}/(4\pi)$
$\mathbf{A}$	by	$\mathbf{A}/c$
$\mathbf{M}$	by	$c\mathbf{M}$

and leave unaltered

$$\rho, \mathbf{J}, \mathbf{E}, \phi, \sigma, \mathbf{P}.$$

Other units in common use are

1 gauss = 10^{-4} weber m^{-2}
1 Å (angstrom) = 10^{-10} m
1 eV (electron volt) = 1.602×10^{-19} joule

A.4 *Some constants*

Electron charge e	1.602×10^{-19} coulomb
Electron mass m	9.108×10^{-31} kg
e/m	1.759×10^{11} coulomb kg^{-1}
e^2/m	2.818×10^{-8} coulomb^2 kg^{-1}
Proton mass m_p	1.672×10^{-27} kg
m_p/m	1836
Vacuum permeability μ_0	$4\pi \times 10^{-7}$ henry m^{-1}
Vacuum speed of light c	2.998×10^8 m sec^{-1}
Vacuum permittivity $\epsilon_0 = 1/(\mu_0 c^2)$	8.854×10^{-12} farad m^{-1}
Vacuum impedance $Z_0 = (\mu_0/\epsilon_0)^{\frac{1}{2}}$	376.7 ohm
Vacuum admittance $Y_0 = 1/Z_0$	2.654×10^{-3} mho
Electron rest energy mc^2	8.186×10^{-14} joule (5.110×10^5 eV)
Classical electron radius $\mu_0 e^2/(4\pi m)$	2.818×10^{-15} m
$e^2/(\epsilon_0 m)$	3183 m^3 sec^{-2}
Planck's constant h	6.625×10^{-34} joule sec
$\hbar = h/(2\pi)$	1.054×10^{-34} joule sec
Fine structure constant $e^2/(2\epsilon_0 hc)$	7.297×10^{-3}
$2\epsilon_0 hc/e^2$	137.0
$e^4/(2\epsilon_0 hc)^2$	5.325×10^{-5}
$mc/\hbar$	2.589×10^{12} m^{-1}
$mc^2/\hbar$	7.764×10^{20} sec^{-1}
Bohr radius $a_0 = 4\pi\epsilon_0\hbar^2/(me^2)$	5.292×10^{-11} m
Bohr magneton $e\hbar/(2m)$	9.273×10^{-24} amp m^2
Boltzmann's constant K	1.380×10^{-23} joule deg^{-1}

INDEX